As we approach the end of the twentieth century, public and social attention is focusing increasingly on the detection and assessment of changes in our environment. This unique volume addresses the potential implications of global warming for fisheries and the societies which depend on them. Using a 'forecasting by analogy' approach, which draws upon experiences from the recent past in coping with regional fluctuations in the abundance or availability of living marine resources, it is shown how we might be able to assess our ability to respond to the consequences of future environmental changes induced by a potential global warming. Leading researchers and thinkers from disciplines as diverse as biology, anthropology, political science, and economics present a series of integrated case studies from around the globe to create a major work in this field.

Climate variability, climate change, and fisheries

Climate variability, climate change, and fisheries

Edited by

MICHAEL H. GLANTZ

National Center for Atmospheric Research, Boulder, Colorado

CAMBRIDGE
UNIVERSITY PRESS

CAMBRIDGE UNIVERSITY PRESS
Cambridge, New York, Melbourne, Madrid, Cape Town, Singapore, São Paulo

Cambridge University Press
The Edinburgh Building, Cambridge CB2 2RU, UK

Published in the United States of America by Cambridge University Press, New York

www.cambridge.org
Information on this title: www.cambridge.org/9780521414401

First published 1992
This digitally printed first paperback version 2005

A catalogue record for this publication is available from the British Library

ISBN-13 978-0-521-41440-1 hardback
ISBN-10 0-521-41440-7 hardback

ISBN-13 978-0-521-01782-4 paperback
ISBN-10 0-521-01782-3 paperback

Contents

1

Introduction

MICHAEL H. GLANTZ

Environmental and Societal Impacts Group
National Center for Atmospheric Research*
Boulder, CO 80307, USA

During the past decade there has been considerable speculation about the possible consequences of a global warming of the atmosphere for terrestrial ecosystems. One of the latest surveys of such impacts was undertaken by the US Environmental Protection Agency (EPA) at the request of the US Congress in its search for policy options with respect to the possible anthropogenically induced climate change (US EPA, 1989). While freshwater ecosystems and two estuarine ecosystems (Apalachicola Bay in Florida and San Francisco Bay in California, USA) were included in this recent EPA survey, marine ecosystems were not. A more recent assessment undertaken by Working Group II of the Intergovernmental Panel on Climate Change (IPCC, 1991) generated some speculation about possible climate change impacts on fish population and on aquatic life.

This volume, *Climate Variability, Climate Change, and Fisheries*, addresses the potential implications for fisheries and societies of the regional impacts of a global warming of the atmosphere. Fisheries case studies were selected for investigation of the responses to changes in their environment. While most of these changes related to biological factors (that is, changes in the abundance of a fish population), some case studies related to abiotic factors, focusing on changes in the availability of fish (that is, a loss of access to commercially exploited fish stocks because of unilateral extensions by nations of their fishing jurisdictions). This study began with the identification of fisheries around the world (see Fig. 1.1) that have undergone changes in availability and abundance, with a preference for fisheries affected by such changes in the past few decades. Some of the cases, however, are

* The National Center for Atmospheric Research is sponsored by the National Science Foundation.

classic ones (e.g., the collapse and reappearance of the Far Eastern sardine). Each chapter provides the general historical background of the fishery, the problems (or prospects) faced as the result of a natural or human-induced change in availability or abundance, and a set of possible lessons to societies that are directly or indirectly dependent on the exploitation of specific living marine resources.

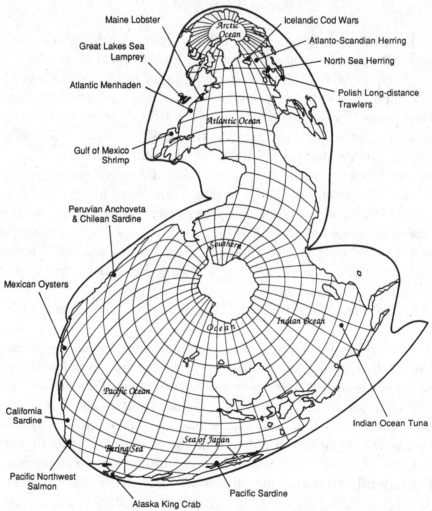

Fig. 1.1 Location of fisheries case studies. Adapted from Athelstan Spilhaus, "Whole Ocean Map," cited in Cousteau, 1981.

The approach taken is referred to as "forecasting by analogy." This is an attempt to forecast society's ability to respond to the consequences of yet-unknown environmental changes that might occur in the future, by looking at societal responses to recent environmental as well as societal (e.g., legal) changes. Some of these changes have been long-term, low grade and cumulative, while others have been short-term and abrupt. This method of "forecasting" regional responses to the regional impacts of global climate change on the abundance or availability of living marine resources has been used in the absence, at this time, of reliable computer-generated regional climate impacts scenarios about the next several decades.

Many studies have already been undertaken on various aspects of the effects of anthropogenic and environmental factors on the viability of specific living marine resources under contemporary climatic conditions (e.g., Troadec, 1990). Clearly, a good base of information is available with which to begin an assessment of the possible regional and local implications of a global atmospheric warming of a few degrees Celsius, as projected by general circulation modeling output. There are also many researchers whose expertise would place them in a good position to address questions about the interrelationship between global changes and fisheries, once they become aware that their research is relevant to global climate change issues.

It is important to note that forecasting by analogy is not an attempt to assess the direct effects of a climate change on the biological aspects of living marine resources. A few such research efforts have already been undertaken (e.g., Bakun, 1990; Bardach & Santerre, 1981; Frye, 1983; Sharp & Csirke, 1983; Shepherd et al., 1984; US Department of Energy (US DOE), 1985; *Fisheries*, 1990). Fish populations are influenced by many elements of their natural environments during all phases of their life cycles. Subtle changes in key environmental variables such as temperature, salinity, wind speed and direction, ocean currents, and strength of upwelling, as well as those affecting predator populations, can sharply alter the abundance, distribution, and availability of fish populations. Human activities can also affect the sustainability of these populations through, for example, the application of a variety of different management schemes or new technologies, each of which could have a different (either beneficial or adverse) consequence for the state of the fishery, years, if not decades, into the future.

Interactions within the marine environment are acknowledged to be extremely complex. The proposed sustained global warming of the atmosphere adds to that complexity. An obvious environmental effect of a global warming would be changes in sea surface temperatures, which, in turn, would have an effect on fish populations during all life stages. However, as a recent DOE report noted, "the production of fish biomass in the oceans is governed by interactions among numerous physical, chemical, and biological processes" (US DOE, 1985, p. 97), not just temperature. Surprises, that is, counter-intuitive responses of marine organisms, should not be ruled out. According to the DOE report (US DOE, 1985, p. 98), "Whatever CO_2-induced climate–fisheries interactions occur on a global scale, there will be local areas or specific fisheries that display the opposite effects." Figures 1.2a and 1.2b depict in a generalized way some of the complexities associated with the direct and indirect effects of climate on the marine environment and on the life stages of fish populations. Thus, the relationship between climate change and fisheries will not be easy to define and most likely will have to depend, at least for the near future, on generalizations derived from case-by-case assessments of past and present experiences. Such assessments can provide first approximations or "guesstimates" about how fisheries *might* (not will) respond to climate-related environmental stresses, until we improve our understanding about how a global climate change will manifest itself in the regional marine environment.

There has been considerable speculation about what a warming of the atmosphere by several degrees Celsius will do to regional climate and to human activities presently attuned to that climate. The basis for that speculation comes mainly from various atmospheric general circulation model (GCM) outputs as a result of sensitivity studies associated with the equivalent of a CO_2 doubling. Speculation about future climate regimes has also been drawn from historical analogues such as the Medieval Optimum (about AD 800–1100) and the Little Ice Age (about 1550 to 1850), and from other paleoclimate analogues including the Altithermal (4,000–8,000 years ago), and epochs tens of thousands as well as millions of years ago when the earth's atmosphere was much warmer than it is at present.

Other approaches to gain a glimpse of the future have also been pursued. For example, composites of the warmest Arctic summers

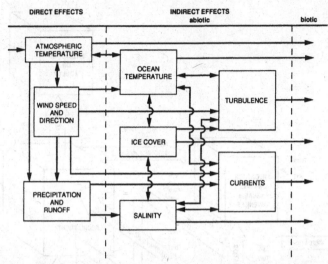

Fig. 1.2a Major climatic pathways affecting the abiotic environment of fishes. Increased atmospheric CO_2 directly affects climate and dissolved CO_2. CO_2 indirectly affects seawater temperature, salinity, ice cover, turbulence, and currents. All of these abiotic effects have biotic consequences (US DOE, 1985).

have provided analogues to global warming based on the view that a global warming will be greatest in the polar regions (e.g., Jäger & Kellogg, 1983). Even the various advanced GCMs yield somewhat divergent pictures of temperature and precipitation changes that will result from a warmer earth, especially when one compares their regional projections in detail (e.g., Schlesinger & Mitchell, 1987). This raises the troubling question about which GCM to use for climate-related impact analyses.

There is also considerable disagreement about how a global average warming might translate into climate changes (i.e., temperature and precipitation) at the regional and local levels. At present the spatial resolution of general circulation models of the atmosphere is too coarse for the generation of regional scenarios that can be useful for reliable and credible social impact assessment. In addition, none of these GCMs as yet has defined an effective oceanic component. This, however, has in no way hindered speculation about regional and local climate changes and their socioeconomic impacts. In the absence of such scenarios, we have relied on the historical record in an attempt to forecast societal responses to climate change by analogy.

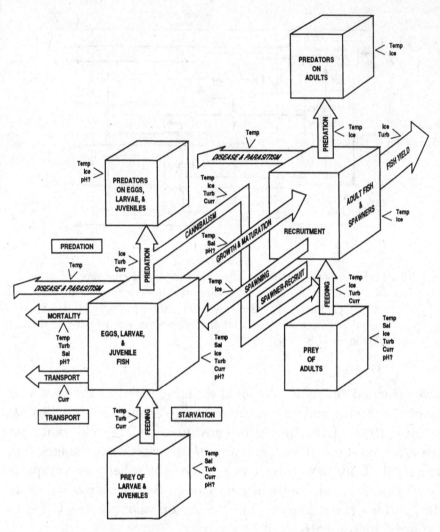

Fig. 1.2b Major biotic processes affecting fish production and the abiotic factors that modify these processes. The four major hypotheses concerning control of fishery abundance are related to the major processes controlling production and mortality of early life history stages: reproductive output, starvation, predatory (including cannibalistic) losses, and transport losses. To represent an actual fishery environment, several such interlocking diagrams would be needed to depict multiple species (US DOE, 1985).

Since regional climatic changes that might be associated with a global warming are not yet well understood, there is a need to produce information that will be of value regardless of the magnitude (or direction) of those changes. In this regard, forecasting by analogy might be viewed as providing a win/win approach (as opposed to win/lose) to researchers as well as policymakers. It underscores the value of improving our understanding about how societies respond to environmental stress. It provides decisionmakers with baseline information about how well societies have responded to the consequences of past environmental changes, even in the absence of an anthropogenically induced warming of the atmosphere. Whether the atmosphere warms, cools or stays as it has been for the past several decades, it is important to improve our understanding of the interactions between human activities and climate variability. The information gathered in these and other forecasting by analogy studies around the globe (e.g., Glantz, 1988; Antal & Glantz, 1988; Magalhães & Neto, 1989; Ninh et al., 1991) can be used to develop ways to mitigate the societal impacts of a variable climate at the regional level.

Analogies have been used to perform a variety of functions, some of which are as follows: (1) For general education: analogies can be used to educate nonspecialists about some aspects of a complex situation by making reference to a different situation about which they already have some information. (2) To educate researchers: more sophisticated analogies can be identified to enable researchers to better understand changes in processes, interrelationships, and sensitivities that might conceivably accompany a global warming. (3) To parameterize complex processes: analogues are used in numerical modeling where there is a need to include important processes related to atmospheric circulation in the model. As a result, there are simple "base" analogies that can be used to generate information about "target" analogies, or at least serve as adequate place holders in the models until those processes become better understood. (4) To forecast future states of systems, such as the atmosphere or society: while an analogy may be used for any one of a variety of purposes, a troublesome use is to forecast a state of the atmosphere or of society several decades into the future. It can, however, be used to make other kinds of projections about the nature of different types of societal responses to cope with a variety of plausible (but not necessarily probable) future regional climatic changes. (5) To generate policy options or responses: plausibility of a physical or societal analogy is not a sufficient condition for use by policymakers, because several plausible but contradictory policies could be formulated based on different analogues drawn from the same pool of objective scientific information. Analogies, however, can be used to identify policy needs in order to eliminate shortcomings in societal responses to environmental change. (6) To fulfill a psychological need: when confronted by unknown situations, analogies can provide us with a feeling of understanding. They provide a first step toward knowing or at least considering the unknown.

Using analogies to gain a glimpse of the future can be advantageous in several ways. Analogies provide a wealth of detail, an ease of communication. Yet, analogies can be developed without a need to provide all details; they can be presented from the perspective of an individual, a sector, a level of government, etc. Even when they are not consistent, they could serve to illuminate different aspects of the future. Also, analogies are conducive to communication, thereby inviting questions and discussions about what can or cannot be told about the future.

To summarize, analogies are an integral part of both physical and social science research with regard to the global warming issue (Glantz, 1991). Analogies are useful heuristic devices that can enhance our understanding. Almost every aspect of the global warming dialogue, from the projection of future production of radiatively active trace gases to the effects of global warming on society, must be explicitly recognized as having been based on analogy. Given the current state of uncertainty surrounding the implications for atmospheric processes, the environment, and societies of an increased loading of the atmosphere with radiatively active trace gases, it is essential that we examine the analogies we use.

There are, however, problems with the use of analogies. First of all, the reason behind making the analogy must be made clear or the analogy will be viewed as either irrelevant, misapplied, or misleading when judged from other perspectives. Secondly, there may be a tendency to "strain" an analogy; one must not read more into it than is there; one must not downplay or ignore important dissimilarities; one must remember that an analogy will not be a perfect replication of what might be expected. Thirdly, sometimes we are forced to make analogies that are inappropriate for cultural or historical reasons. Finally, plausible but mutually inconsistent scenarios can be developed (see, for example, Jamieson, 1988).

Scenarios about future worlds based on human experience have the political and social credibility that computer-generated scenarios lack. Decisionmakers who have been directly involved in problems generated by climatic anomalies of the recent past have already been using that experience as a guide to dealing with current issues. Such experience is being passed on to future decisionmakers, just as the experiences of the 1930s US Great Plains drought or the California sardine or Peruvian anchoveta collapse have been (and continue to be) carried from one generation to the next.

Some atmospheric scientists have argued that the climate of the future will not be like the climate of the past. Therefore, they contend that the past cannot be seriously considered as a useful guide to the future. However, societal responses to regional climate in the near future will most likely be similar to societal responses to the climate-related environmental changes of the recent past. Recent societal responses to variable climatic conditions might provide useful insights into how best to cope with such conditions at least in the near future. Forecasting society's ability to cope with the impacts of climatic variations and change can be achieved through this method. Researchers can identify strengths and weaknesses, successes and failures in the way societies have responded to events that are most likely to recur in the future. Societies can then reduce the weaknesses while capitalizing on the strengths to mitigate those impacts in the future.*

This volume presents a set of case studies from around the world representing a variety of fisheries. Although given some broad guidelines, each contributor to this volume was allowed considerable flexibility in his or her approach to develop the case studies and to identify possible insights into potential societal responses to global warming.

Wooster's chapter on the Alaska king crab discusses the development and collapse of this important fishery. It also identifies management responses to the collapse with the expectation that there are lessons for fisheries managers responding to the impacts of global warming in the Gulf of Alaska/Bering Sea region.

---

* For example, a recent study (Glantz, 1988) using the forecasting-by-analogy approach assessed 10 North American case studies. Five of the climate-related environmental changes considered have occurred since the first workshop was held in June 1987.

The California sardine fishery has become part of American folk-lore as a result of the writings of John Steinbeck. Ueber and Mac-Call describe this classic case of a fishery collapse. The chapter underscores an improvement in the way living marine resources are managed. It also shows how the collapse of the California sardine fishery spawned the rapid development of major fisheries in South Africa and in Peru.

Miller and Fluharty's chapter is centered on the regional implications of the 1982–83 El Niño-Southern Oscillation (ENSO) event and focuses on the difficulties of separating economic pressures on fish populations from environmental ones. Their study also points out how a decline in one area can be accompanied by a sharp increase in fish landings in other adjacent regions.

Condrey and Fuller investigated the history of the Gulf shrimp fishery. They view this fishery as a classic example of an open-access fishery which has been allowed to expand beyond the point of maximum long-term economic benefit. A resource that had been viewed as limitless has in recent decades been threatened by fishing pressures as well as habitat destruction and occasional low streamflow in the Mississippi River.

Although Atlantic menhaden have been uncommon as a food fish, they have several industrial uses. Feingold points out in her chapter that society has had a direct effect on the fortunes of the menhaden fishery as a result, for example, of zoning laws that govern the location of processing plants, of intentional changes in coastal and estuarine habitats, and of increased demands for menhaden-based products.

Everyone associates lobsters with the US State of Maine. In fact, the lobster has been "immortalized" by serving as a graphic design on Maine's license plate. Acheson has reviewed the lobster industry during its decline in the first half of the twentieth century in order to identify possible lessons for changes in lobster availability or abundance that might be associated with global warming.

McGoodwin's chapter on the Mexican oyster fishery evaluates societal responses to adverse changes in the availability of harvestable mollusks along Mexico's south Sinaloan coast. Changes in demographics in this region since the turn of the century have made traditional responses to losses in oyster productivity no longer viable. McGoodwin suggests ways that local fishermen can maintain a degree of flexibility in response to potential environmental changes that might accompany a climate change.

The Great Lakes, considered the "fifth coast" of North America (along with the Atlantic, Pacific, Caribbean and Arctic; for a discussion of this concept see Ashworth, 1987), is the geographic field of research by Regier and Goodier. They investigated the history of the sea lamprey in the Great Lakes as a possible analogue to some unpredictable consequences of global warming. Just as an ecosystem can be caused to undergo serious restructuring with an intrusion of a parasitic species, climate-related environmental changes can also prompt ecosystem restructuring.

Bailey and Steele assessed the North Sea herring, one of the world's most important living marine resources that has supported major fisheries in many northwest European countries for centuries. Their chapter addresses the role of environmental changes as well as the role of perceptions held in management organizations and the fishing industry in this stock's collapse in the mid-1970s.

A Soviet contribution was provided by Krovnin and Rodionov, scientists at the All-Union Research Institute of Marine Fisheries and Oceanography (VNIRO) in the USSR. Their study focused on changes in Atlanto-Scandian herring during the warmer decades of the the 1920s and 1930s. They suggest that a global warming might be favorable for the development of the Atlanto-Scandian herring fishery.

The next two chapters are somewhat different in that they are not based on changes in the physical environment but in the political setting in which fisheries must operate. The first of these by Glantz uses the Anglo-Icelandic conflicts (several of which were referred to as the Cod Wars) as a surrogate for societal responses to changes in the availability of cod. Iceland and UK came into conflict over the exploitation of this valuable resource as a direct result of a series of unilateral extensions by Iceland of its territorial waters between 1952 and 1976.

Russek's chapter assesses the impacts of the creation and implementation of the 200-mile exclusive economic zones (EEZs) by coastal nations worldwide. Poland's long-distance fishing industry was forced to adjust to this precipitous shock or face extinction. This chapter documents how Poland's fleet managed to survive a loss in availability of living marine resources that resulted from international legal decisions.

The history of the Far Eastern sardine fishery extends back at least to the early 1600s. Kawasaki reports on the rise and collapse and rise again of this fishery. The chapter discusses the impacts of these changes in abundance of the Far Eastern sardine population not only in Japan but in Korea and the USSR as well. He notes that coastal communities dependent on the exploitation of this fish population should prepare for the eventuality of yet another decline. He also compares some aspects of this fishery with those of California and Peru.

Caviedes and Fik address the implications of ENSO events for fisheries along the western coast of South America. They conclude that ENSO has a clear and major impact on regional fisheries, specifically the anchoveta in Peruvian waters and the sardine along the Chilean coast. They also highlight the importance of improving ENSO forecasts so that fisheries could be better managed in the face of this recurrent environmental change. Caviedes and Fik suggest a need for regional cooperation in the management of the fisheries of these two countries.

In the final case study, about western Indian Ocean tuna, Sharp discusses the development of the tuna fishery around the Seychelles Plateau. He assesses why this fishery thrives, while similar fisheries in other oceans in recent decades have either been marginally successful or have failed. He then compares the development of the tuna fisheries of the Seychelles and the Maldives.

The concluding section presents a summary of the highlights of each of the case studies and serves as an "executive summary." The information in this section has been drawn from the chapters, as prepared by the contributing authors, with the general findings prepared by Glantz and Feingold.

As a final comment on the forecasting by analogy approach, it is important to note that the purpose of looking back is neither to identify the exact types of climate changes that societies must prepare for nor is it to put emphasis on the most recent aberrations of climate as the most likely forecasts of future climate. The purpose is to determine how flexible (or rigid) societies are or have been in dealing with climate-related environmental changes. We must be aware of past events but we must not get drawn into preparing for them. Societies everywhere have already shown the propensity to prepare for the last climate anomaly by which they were affected. However, such anomalies seldom seem to recur in the same place, with the same intensity, or with the same societal impacts. Decisions today must take into consideration the need to maintain as much flexibility as practicable in the face of future unknowns.

Acknowledgments

I would like to acknowledge the consistent editorial support and coordination activities of Maria Krenz, without which this publication would have remained "in press" for a long time. I would also like to thank Jan Stewart, who has been integrally involved in the production of various drafts of the manuscript. Her technical skills in the TEX formatting language enabled us to produce the final camera-ready copy for publication.

Also, I want to express my sincere appreciation to my research assistant, Lucy Feingold, who was instrumental in organizing the climate and fisheries workshop that launched this research project, and to the contributors to this book for their interest and enthusiasm, as well as their patience and perseverance in the preparation of their manuscripts for publication.

Financial support for this project came from the National Marine Fisheries Service (NOAA) and the Environmental Protection Agency's Climate Change Division. Finally, I would like to thank Sara Trevitt at Cambridge University Press for her editorial support during this project.

References

Antal, E. & Glantz, M.H. (Eds.) (1988). *Identifying and Coping with Extreme Meteorological Events.* Budapest: Hungarian Meteorological Service.

Ashworth, W. (1987). *The Late, Great Lakes: An Environmental History.* Detroit: Wayne State University Press.

Bakun, A. (1990). Global climate change and intensification of coastal ocean upwelling. *Science*, **247**, 198–201.

Bardach, J.E. & Santerre, R.M. (1981). Climate and the fish in the sea. *BioScience*, **31**, 206–15.

Cousteau, J.-Y. (1981). *The Cousteau Almanac: An Inventory of Life on Our Water Planet.* New York: Dolphin Books.

Fisheries (1990). (The entire issue No. 6 is dedicated to the effects of global climate change on fisheries resources.)

Frye, R. (1983). Climatic change and fisheries management. *Natural Resources Journal*, **23**, 77–96.

Glantz, M.H. (Ed.) (1988). *Societal Responses to Regional Climatic Change: Forecasting by Analogy.* Boulder: Westview Press.

Glantz, M.H. (1991). The use of analogies in forecasting ecological and societal responses to global warming. *Environment,* **33**, 10–4 and 27–33.

IPCC (Intergovernmental Panel on Climate Change) (1991). *Climate Change: The IPCC Impacts Assessment.* Report from Working Group II to IPCC. Geneva: World Meteorological Organization/UN Environment Programme.

Jäger, J. & Kellogg, W.W. (1983). Anomalies in temperature and rainfall during warm Arctic seasons. *Climatic Change,* **5**, 39–60.

Jamieson, D. (1988). Grappling for a glimpse of the future. In *Societal Responses to Regional Climatic Change: Forecasting by Analogy,* ed. M.H. Glantz, pp. 73–93. Boulder: Westview Press.

Magalhães, A.R. & Neto, E.B. (1989). Impactos sociais e econômicos de variacões climáticas e respostas governamentais no Brasil. Programa das Nações Unidas para O Meio–Ambiente. Fortaleza: Secretaria de Planejamento e Coordenação do Ceará.

Ninh, N.H., Glantz, M.H. & Hien, H.M. (1991). *Case Studies of Climate-Related Impact Assessment in Vietnam.* UNEP Project Document No. FP/4102-88-4102. Nairobi: United Nations Environment Programme.

Schlesinger, M.E. & Mitchell, J.F.B. (1987). Climate model simulations of the equilibrium climatic response to increased carbon dioxide. *Reviews of Geophysics,* **25**, 760–98.

Sharp, G.D. & Csirke, J. (1983). *Proceedings of the Expert Consultation to Examine Changes in Abundance and Species Composition of Neritic Fish Resources.* Workshop in San Jose, Costa Rica, 18–29 April 1983. Rome: FAO Fisheries Report 291, Vols. 2–3.

Shepherd, J.G., Pope, J.G. & Cousens, R.D. (1984). Variations in fish stocks and hypotheses concerning their links with climate. *Rapports et Proces-Verbaux des Reunions. Conseils International pour l'Exploration de la Mer,* **185**, 255–67.

Troadec, J.-P. (Ed.) (1990). *Man, Marine Fishery and Aquaculture Ecosystems* (in French). Paris: IFREMER.

US DOE (Department of Energy) (1985). *Characterization of Information Requirements for Studies of CO_2 Effects: Water Resources, Agriculture, Fisheries, Forests, and Human Health.* DOE/ER-0236. Washington, DC: Carbon Dioxide Research Division, US DOE.

US EPA (Environmental Protection Agency) (1989). *Policy Options for Stabilizing Global Climate.* Three-Volume Draft Report to Congress. Washington, DC: Office of Policy, Planning, and Evaluation, US EPA.

2

King crab dethroned

WARREN S. WOOSTER

School of Marine Affairs
University of Washington
Seattle, WA 98195, USA

Introduction

The king crab stock in the eastern north Pacific (eastern Bering Sea and Gulf of Alaska; see Fig. 2.1) has varied nearly tenfold in abundance in the last 25 years (Hayes, 1983). Since the late 1960s, the fishery has been the second most valuable Alaskan seafood industry, exceeded in value only by the combined six salmonid species harvested in Alaska (Hanson, 1987).

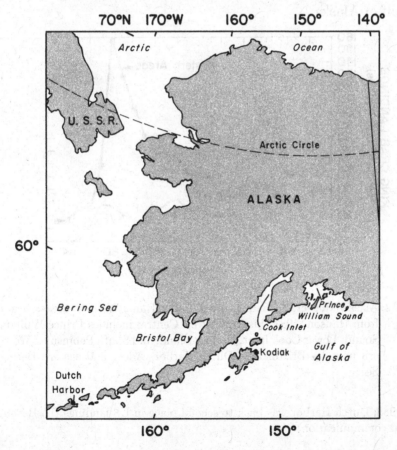

Fig. 2.1 Map of the study area.

The small Alaskan port of Dutch Harbor, a major center for crab processing, was in 1979 the number one US fishing port in dollar volume, handling seafood valued at more than the combined landings of the North American ports of Seattle, Astoria, Ketchikan, Newport, Westport, Charleston, Coos Bay, and Eureka (McLafferty, 1980). However, by 1982 Dutch Harbor was "beginning to look like a ghost town" (Anon., 1983).*

The change took place in 1981, when stock abundance fell precipitously; it has recovered only very slowly since then (Fig. 2.2). Stocks of other king crabs (blue, brown) also shrank as did Tanner crabs. The reasons for the collapse have not been established although various explanations have been offered, including overfishing, predation, disease, and environmental change. Evidence for none of these is very convincing. That the cause was some sort of environmental change is suggested by the widespread nature of the decline including several species in both the Bering Sea and the Gulf of Alaska.

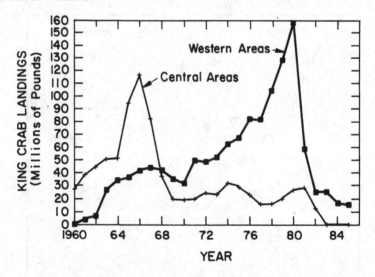

Fig. 2.2 Alaska king crab landings from Central and Western areas. Data from Hanson, 1987 (his Table 2.1). Central includes Prince William Sound, Lower Cook Inlet, Kodiak Island, and South Peninsula. Western includes Bristol Bay, Dutch Harbor, Adak, and eastern Bering Sea.

* By 1988, Dutch Harbor was back to second place in US landings (D. Bevan, personal communication).

Whatever its cause, collapse of the fishery led to economic disaster. The fleet was too large, many vessels were heavily leveraged, and most owners were unable to pay their bills. Unprecedented prices, resulting from low production, threatened loss of all but the luxury markets. Fishermen were faced with foreclosure or diversification – and funds for the latter were scarce (Sabella, 1982). Yet, the eventual solution for the industry was the transfer of effort and investment to other resources.

This is not the first fishery crisis caused by the disappearance of a resource, nor will it be the last. Indeed, such collapses may be more frequent in a future of drastically changed climate. Are there lessons in how the industry responded to this set of events? Could the fishery have been managed more effectively (1) to prevent the collapse, (2) to anticipate the collapse more effectively, or (3) to mitigate the economic and social cost of the collapse? Will there continue to be other resources to absorb the energies of the industry?

Biological and oceanographic background

Three species of king crab are harvested in the eastern North Pacific:
- *Paralithodes camtschatica* (red);
- *Paralithodes platypus* (blue) ;
- *Lithodes aequispina* (brown).

Of these, red king crab is by far the most important and occurs on the shelf in the eastern Bering Sea, Aleutian Islands, edge of the Gulf of Alaska to SE Alaska, and northern British Columbia.*

While adults feed offshore and migrate inshore for spawning, juveniles are found in the littoral zone and shallow water. In the Bering Sea, adults prefer bottom temperatures of 0° to 5.5°C, suggesting a temperature influence on distribution.

Molting and spawning take place in shallow (10–50m) waters in late winter and early spring. Males molt in March–April, females just before spawning in April–May (see Fig. 2.3). Eggs, 50,000 to 400,000 in number, are attached to females and develop for 11 months, normally hatching in April–May (timing can vary by

---

* This summary of king crab biology is based mainly on Hayes, 1983.

more than one month in different years). Five successive larval stages live as plankton in the water column for a total of about six weeks, then settle to the bottom.

KING CRAB LIFE HISTORY
(Paralithodes camtschatica)

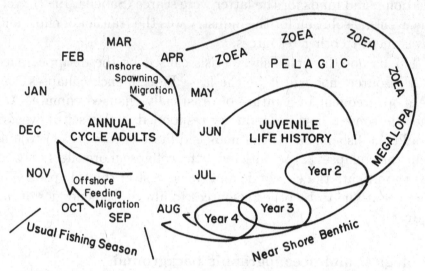

Fig. 2.3 Schematic of king crab life history. (From Hayes, 1983.)

During the first year of growth, juveniles are solitary in relatively shallow water; during the first two to four years they are often in shallow water in dense "pods." King crabs, even adults, tend to segregate by size, sex, and molt condition.

Growth is discontinuous at times of molts. In the absence of a method for direct aging, growth is studied by determining the frequency of molts and the increment of growth per molt. Animals reach about 78 mm in four years and can then be tagged without loss during molt. King crabs enter the fishery at age eight. They grow to about 200 mm in 11 years and there is some molt skipping by males from 145 mm in size.

King crab are bottom-foraging omnivores; there appear to be no significant differences in diets between sexes and sizes of adults. Major food includes starfish, clams, and other mollusks, as well as small crabs, shrimps, other crustaceans, worms, fish, and algae. Predators on king crab include yellowfin sole, Pacific cod, walleye pollock, and halibut (Fukuhara, 1985; Larkin et al., 1990).

In view of this life history, ocean conditions on the inner shelf of the eastern Bering Sea, eastern Aleutians, and Kodiak Island in the winter and spring probably influence, and perhaps determine, the success of recruitment.

The eastern Bering Sea is divided into three domains separated by fronts (Fig. 2.4a,b) which are a function of depth and differ in circulation and vertical mixing (Schumacher & Reed, 1983). Crab eggs, larvae, and juveniles appear to be mostly within the coastal domain where tidal mixing exceeds buoyancy input, where water away from river mouths is mixed vertically, and where the average flow is to the northeast along the Alaska Peninsula and northward along isobaths east of the inner front (Fig. 2.5a,b). Flow is affected by wind events but is principally geostrophic and is driven by interaction of the tides with bathymetry. There is important interannual variability in wind stress, especially in the winter, and in its effects on temperature and ice coverage (up to 80% coverage in March).

In the Gulf of Alaska, the prevailing circulation is westward along the Alaska Peninsula (the Alaska Stream offshore, the Alaska Coastal or Kenai Current inshore) with some flow through Unimak Pass into the Bering Sea (Fig. 2.5b). The coastal circulation is wind driven, coupled with freshwater input along the coast. There is interannual variability in winds and rainfall and runoff with effects on currents, temperature, and salinity. Larval development, hence recruitment success, can be affected by interannual changes in *transport* of larvae to favorable grounds for settling, in *temperature* which determines the duration of larval periods, and in *food supply*, for example as determined by the timing and location of the spring plankton bloom (Larkin et al., 1990).

Early history of the fishery

The commercial harvest of king crab in the eastern Bering Sea began with a Japanese fishery in 1930 (Otto, 1981). Between then and 1939, when the fishery closed with the start of World War II, nearly eight million crabs were taken. Meanwhile, in 1940 the US Congress appropriated funds for Alaska fish surveys, and Lowell Wakefield began to can crab near Kodiak (Blackford, 1979). But the US fishery, which started as a supplement to salmon and halibut, did not really begin until 1947. Until 1965, only Wakefield's deep sea converted trawler (140 ft – 42.7m), with processing on board, was specialized for the fishery. Most of the fleet was much

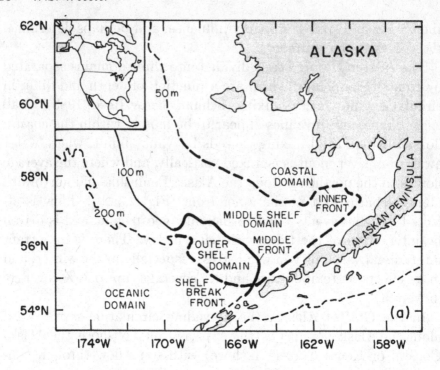

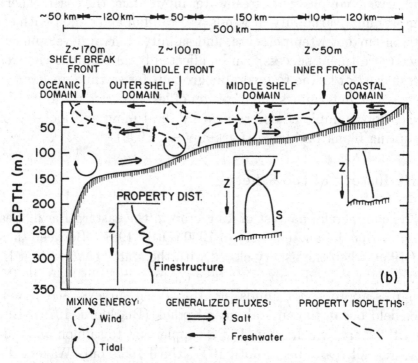

Fig. 2.4 (a) Hydrographic domains and fronts over the southeast Bering Sea shelf, and (b) schematic interpretation of energy balance, fresh and salt water fluxes, and vertical structure. (From Schumacher & Reed, 1983.)

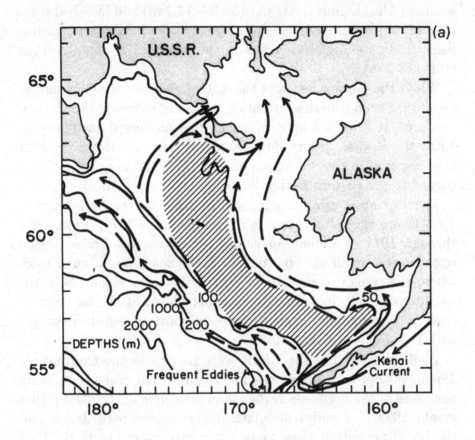

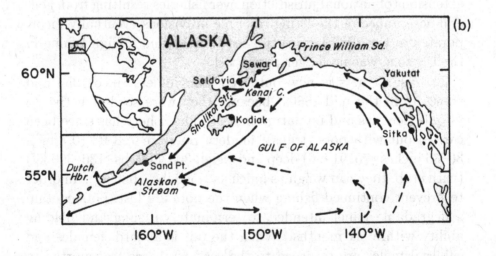

Fig. 2.5 Schematic of long-term mean circulation (a) in the eastern Bering Sea and (b) in the Gulf of Alaska. The Kenai Current is also known as the Alaska Coastal Current. (From Schumacher & Reed, 1983.)

smaller (Alaska-limit boats were 58 ft – 17.7 m) and included some ex-sardine seiners (refugees from the collapsed California sardine fishery). In 1965, of 190 vessels in the fishery, 120 were less than 60 ft (18.3 m).

While the Bering Sea was the site of the first development of significant crab fisheries, attention was soon redirected to the waters around Kodiak where Wakefield had pioneered and, prior to 1969, the Kodiak fishery dominated the harvest. However, after a widespread decline in abundance in 1970, the major fishery returned to the eastern Bering Sea where it has remained.

The Japanese, using tangle-nets, returned to the Bering Sea in 1953 where they dominated crab catches until 1970. From 1959 through 1971, a similar Soviet fishery operated. These foreign operations were affected by the Law of the Sea Convention of 1958, where the coastal state gained jurisdiction over resources of the continental shelf, including "organisms which, at the harvestable stage ... are unable to move except in constant physical contact with the sea bed or the subsoil."

While the US and the USSR were parties to the Convention, Japan was not and, in any case, considered that crabs were living resources of the high seas rather than creatures of the shelf (Miles et al., 1982). A series of bilateral agreements with Japan and the USSR permitted some control over catches and, with the 1977 extensions of national jurisdiction over fisheries resulting from general acceptance of the living resource provisions eventually incorporated in the 1982 Law of the Sea Convention, foreign fishing in the US zone was no longer permitted.

The use of nets is now outlawed (tangle-nets since 1954 and trawls from the mid-1960s) because they were so destructive to illegals (females and immature males). The fishery has since been carried out with pots (traps); modern pots are 7–8 ft^2 (0.7 m^2), 30–36 inches (76–91 cm) deep and weigh 300–800 lbs (136–360 kg) (with crab, they can weigh as much as 1,500 lbs – 680 kg). In order to prevent continued fishing when the pots are lost, pots contain a degradable panel intended to terminate catching and holding ability within six months. When the pot is aboard, females and sublegal males are returned to the sea. Crab are transported to processors in live tanks which exchange water every 20–30 minutes. Water and crab sloshing in these tanks can cause severe stability problems, as do the heavy loads of pots carried on deck.

History of the stock

The king crab population consists of many relatively independent stock units. At least two red and three blue king crab stocks are recognized in the eastern Bering Sea. Statistics often combine several stock units and sometimes more than one species.

While fishery stock assessment is commonly based on catches, this is particularly difficult in the king crab fishery. Factors complicating stock assessment include (1) rapid development of the fishery in the late 1940s and 1950s, so landings are a poor index of abundance in that period; (2) catch is always much larger than landings, and the mortality of returned females and sublegals is poorly known; and (3) there is no method to determine age other than size composition. However, on a relative basis, data on size composition of commercial catches are used to determine the proportion of "recruit" crabs, a rough approximation of year-class strength.

For assessment purposes, regular systematic trawl surveys began in eastern Bering Sea in 1955 and have continued ever since, except for a five-year hiatus in 1962–66. Pot surveys have been made at Kodiak since 1971. The trawl survey pattern was extended from 1958 and covers from the Bristol Bay to the Gulf of Anadyr and Chukchi Sea. From 1971, the station pattern was expanded to include other king and Tanner crab stocks. The surveys allow determination of species composition, sex, carapace size, shell age; in pot surveys, tagged crabs are released and used to estimate population size and estimates of fishery yields for the succeeding one to two years (Hayes, 1983).

The king crab population has varied enormously. In the eastern Bering Sea, for example, legal males have ranged from near 50 million crab at peak in 1978–79 to a low of a few million in 1982–83. Prior to 1969, the Kodiak fishery dominated the harvest. The first boom was in the mid-1960s (Hanson, 1987) but by 1969, the Kodiak fishery was down, never (as yet) to return to earlier levels (Fig. 2.2). There seems to have been a major decline in abundance similar to that which occurred a decade later. Kodiak fishermen switched to the Bering Sea where new stocks were developed. A big growth in harvesting and processing capacity preceded the collapse in early 1980s.

Management of the king crab fishery

Regulations were first promulgated by the Alaska Department of Fish and Game in 1941 and included size limits on male crab, prohibition against landing females or soft-shelled crab, and the requirement for recording landings (Otto, 1986). Management objectives included the following: to protect reproductive potential, to prevent waste, to maintain product quality, and to optimize size at harvest. Fishing was prohibited during molting/mating periods (March–May), allowing a period of growth for newly molted crab. There were prohibitions against trawls and tangle-nets to reduce the handling mortality of females and sublegal males. Management areas recognize distinctions among Alaskan stocks. Regulatory measures established since 1969 have a set of goals and regulations known as "size-sex-season" management.

Big declines in landings in the period 1966–70 resulted in the perception that management measures were inadequate, so additional goals emphasizing biological and economic stability were adopted, with quotas by management area and a system of exclusive area registration (indirect effort control). Pot limits also control effort to some extent and maintain stability as well as protecting local industry by favoring small boats (see below). Management appears to be highly conservative in its prohibition of harvesting females and smaller mature males despite the lack of evidence that recruitment is affected thereby. In part, these restrictions are intended to benefit processors who prefer the larger males; also, through providing for a multiple year-class fishery, they attempt to ensure stable production. King crab regulations are more restrictive than those for other North American crab fisheries and affect allocations as well as conservation (Otto, 1986).

A federal Fishery Management Plan for the Bering Sea and Aleutian Islands was developed in 1984 but has only recently been implemented; it includes various social, economic, and administrative goals (Anon., 1988). A special feature of Alaska king crab management is that, despite the leading role of the federal fishery management system established by the Magnuson Fishery Conservation and Management Act of 1976, the fishery continues to be managed largely by Alaskans for the benefit of Alaskans. The Alaskan fishery is conducted largely by small vessels using shoreside processors in contrast to the Seattle fleet consisting of large vessels with on-board processing; Seattle is the financial and logistical center of the fishery. In view of the successful moves to exclude foreign fishing after 1977, it is ironic that both the Alaskan shoreside processing plants and the Seattle distant-water fleet are increasingly foreign owned.

The crash and its consequences

There was a tremendous growth in Alaskan harvests from 1969 through 1980, especially in Bristol Bay where harvests rose from 8.6 million pounds (3.9 million kg) in 1970 to 130 million pounds (59 million kg) in 1980. Within three years, the fishery collapsed and the Bristol Bay fishery closed. The stock collapsed after 1980 for unknown reasons. In view of the conservative management policies, it was not likely to have resulted from overfishing. However, it has been suggested (Larkin et al., 1990) that the removal of large males and the inadequacy of the remaining small males in performing their conjugal duties weakens the ability of the population to recover after several years of poor recruitment.

Disease or predators (e.g., increases in fin fish such as Pacific cod) have also been proposed as possible causes. A more probable explanation is environmental change favoring poor recruitment and affecting the subarctic ecosystem (both the eastern Bering Sea and the Gulf of Alaska). This is suggested by the observation that declines in red king crab were found in prerecruit males and brood stock females as well as legal males, and in Kodiak and Dutch Harbor as well as the eastern Bering Sea. Similar declines occurred in blue king crab and in Bering Sea Tanner stocks (especially *bairdi*). Whatever the cause, there was a rapid decline in abundance of red king crab and a very slow recovery.

Because of scarcity, nominal wholesale and retail prices tripled between 1980 and 1986. Between 1980 and 1983, ex-vessel revenues to fishermen fell by US$93.2 million, more than 50 percent; processor sales dropped US$178 million (60%), sales from wholesalers dropped by US$304 million (66%) (Hanson et al., 1988). Comparable losses were felt by associated industries such as shipyards and lending institutions. For example, as the US fishery grew, especially in 1970s, the fleet was overcapitalized with too many boats including expensive large crabbers/processors. After the collapse, fishermen had great difficulty in meeting their payments and bankruptcies were common.

Only 32 percent of total shellfish revenues are returned to Alaska, much of the remaining 68 percent being spent in the Seattle area for vessel maintenance/construction, gear and supplies, general consumer goods (Anon., 1983). Most processing and cold storage firms are based in the Seattle area. As noted earlier, generally, small boats and shore processing are characteristic in Alaska, whereas large boats and at-sea processing come from Seattle. An important element in the politics of king crab (and other Alaskan fisheries) is the competition and controversy between Alaskan and Seattle-based components of the industry (Miles, 1989).

Response of the industry to the crash was threefold: (1) shift to replacement species – other king crabs (blue, brown), Tanners (*bairdi, opilio*); (2) shift to other grounds (e.g., St. Matthews); (3) shift to other fisheries, especially groundfish.

(1) *Replacement species*: Blue and brown king crab were the initial targets. The stocks were much smaller than the red king crab, and the blue stock also rapidly declined. Brown king crab is much harder to catch, occurring as deep as 400 m on steep slopes, with lower catch rates and more gear lost, and is less marketable. Attention was also directed to Tanner crab which is generally less desirable. The larger species (*bairdi*) declined; the small species (*opilio*) continues to be abundant but commands a much lower price and cannot really be considered a replacement.

(2) *Other grounds*: Large vessels, especially those with processing capability, are more mobile than small ones. However, fishing on declining stocks in remote areas quickly led to increased cost of production.

(3) *Other fisheries*: Collapse of the king crab fishery coincided with growing US interest in Alaska groundfish stocks, then fished primarily by foreigners. The initial US development was in joint ventures, in which US trawlers, some converted from crabbers at costs averaging US$700,000 per vessel (McNair, 1982), caught groundfish (especially pollock) and transferred cod-ends at sea to foreign processors. In related developments, US efforts were made (e.g., with *Arctic Trawler*) to catch, process (at sea), and market cod fillets. US processing (at sea) was later extended to pollock fillets, and US surimi and minced fish processing and marketing were subsequently developed.

Fishing for pollock is much less lucrative than fishing for crab. "A $3 million crab boat can't afford to fish Shelikov at US$0.04 [per pound], much less the Bering Sea" (Anon., 1983). It is ironical that pollock bought for US$0.04 per pound returns as artificial crab at US$2.25 to US$3 per pound. In 1982, more artificial crab (*kanibo*) was imported from Japan than was real king crab produced in Alaska.

Since the collapse, the US harvesting sector has expanded to harvest the total Alaska groundfish optimum yield, but US processing and marketing are inadequate to handle all of the product, so the industry remains heavily dependent on foreign (Japan) markets (Miles, 1989). The fleet has been transformed to handle a high-volume, low-value product. While it could presumably revert when the king crab population is restored, the cost of reconversion is likely to be high.

Was management effective?

By most criteria, management of the king crab fishery must be seen as a failure. Abundance of the resource was reduced to a very low level, the fishery collapsed, and great financial losses were incurred by many participants in the fishery.

Why was the decline not predicted, and how could it have been averted by a different management policy? The assessment of Otto (1986, pp. 103, 105) is relevant here:

> *From a biological perspective it seems unlikely that further control on fishing effort would have prevented declines in king crab landings.... The management system in place has been successful in preventing growth overfishing and in insuring product quality but has clearly not been able to prevent severe declines in abundance, and hence, stabilize landings.... In retrospect, the set of policies and regulations intended to provide more stable king crab fisheries is flawed, principally because there is an implicit assumption that natural mortality rates on prerecruit and legal crab remain fairly constant from year to year.... I conclude that directed or undirected fishing has not been a major cause of population decline in Bristol Bay red king crab.... Management measures failed to prevent*

recent declines in landings because causes of declines in abundance are not related to fishing, and hence largely beyond control.

In other words, the model used to predict crab abundance and/or the data used with it were inadequate. In particular, the factors of natural mortality (from predation, disease) and recruitment, "causes of declines ... not related to fishing," were essentially unknown. In addition, estimates of incidental mortality from handling large numbers of pots and from returning nonlegals are highly uncertain.

It is conceivable that heavy fishing pressure plus additional high unaccounted-for mortality, plus environmental conditions unfavorable for recruitment conspired to cause the sharp decline. The relative importance of these factors is not known (nor is it likely to be revealed in the near future by present research programs). The answer could have profound implications for management. If, for example, the major cause of fluctuations in abundance is environmentally induced fluctuations in recruitment, much higher catches could be permitted at times when the stock is abundant.

From a nonbiological point of view, management could be faulted not only because of the great costs incurred as a consequence of the collapse but also because the fishery was not managed to maximize sustainable economic return. If the factors controlling stock abundance were understood and maximum acceptable harvest levels could be more adequately specified, it is likely that these levels could be extracted at much lower cost than is now the case. However, even if such a management scheme were designed, there is little evidence of the political will to bring it into existence.

Lessons for the future

Experience with the collapse of the king crab fishery may hold some lessons for a future marked by climate change. The fate of any specific fishery under changed environmental conditions is difficult to predict, but there is a high probability of continued large changes in abundance, on both annual and decadal scales. The king crab case concerns a low-volume, high-value fishery, but there are also historical cases of collapses of high-volume, low-value fisheries (e.g., California sardine, Peru anchovy). Response of the industry to collapse is to diversify, to target other stocks, and to develop new fisheries. Success depends on there being other stocks to turn to and on ingenuity in their utilization. A further problem is to keep transition costs to a tolerable level. This requires great flexibility in the harvesting, processing, and marketing sectors. Present methods for fishery management in the US are clearly ineffective in matching catching capacity to potential resources, a key to minimizing response.

An underlying question is whether there will continue to be alternative stocks as the climate continues to change. The answer is uncertain without intensified research on the causes of fluctuation of marine animal populations. One can speculate as follows:

- There is no convincing evidence that primary production is likely to decline; the total biomass should remain unchanged. The extent to which transfer efficiencies between trophic levels, or the allocation of production among various stocks of commercial significance, will change is unknown, but on average the change is likely to be small.

- Environmental changes tend to favor one stock or group of stocks and disfavor others. Of course, all stocks are not equally desirable nor are they interchangeable, but there is no evidence that the total biomass of presently used commercial species is likely to decline. The species mix, on the other hand, is likely to change from time to time as it has in the past.

- The demand for seafood increases with population growth. Sources for the increase include presently underutilized species (including nonconventional species) and aquaculture. The ecosystem costs of the former (e.g., large scale use of Antarctic krill) are not yet well understood. Aquaculture production will undoubtedly continue to grow, whatever climate changes transpire.

References

Anon. (1983). *Factors and Consequences Associated with Collapse of the King and Tanner Crab and Northern Puget Sound Salmon Fisheries.* Seattle: Natural Resources Consultants.

Anon. (1988). Draft fishery management plan for the commercial king and Tanner crab fisheries in the Bering Sea/Aleutian Islands. Anchorage: North Pacific Fishery Management Council.

Blackford, M.G. (1979). *Pioneering a Modern Small Business: Wakefield Seafoods and the Alaskan Frontier.* Greenwich: JAI Press.

Fukuhara, F.M. (1985). *Biology and Fishery of Southeastern Bering Sea Red King Crab* (Paralithodes camtschatica, *Tilesius*). Northwest and Alaska Fishery Center Proceedings, Report No. 85–11. Seattle: Northwest and Alaska Fishery Center.

Hanson, J.E. (1987). Bioeconomic analysis of the Alaskan king crab industry. Unpublished dissertation. Pullman: Washington State University.

Hanson, J.E., Matulich, S.C. & Mittelhammer, R.C. (1988). Bioeconomic analysis of the 1983 Bering Sea king crab fishery closure. Paper presented at the American Fisheries Society Annual Meeting, Toronto, Canada, 12–15 September 1988.

Hayes, M.L. (1983). Variation in the abundance of crab and shrimp with some hypotheses on its relationship to environmental causes. In *From Year to Year*, ed. W.S. Wooster, pp. 86–101. Seattle: Washington Sea Grant.

Larkin, P.A., Scott, B. & Trites, A.W. (1990). *The Red King Crab Fishery of the Southeastern Bering Sea.* Seattle: Fisheries Management Foundation.

McLafferty, T. (1980). Dutch Harbor. *Pacific Fishing*, **1**, 23–7.

McNair, D. (1982). What the country needs is a good six-cent pollock. *Pacific Fishing*, **3**, 45–53.

Miles, E.L. (1989). *The US/Japan Fisheries Relationship in the Northeast Pacific: From Conflict to Cooperation?* FMF-FRI-002. Seattle: Fisheries Management Foundation and Fisheries Research Institute.

Miles, E.L., Gibbs, S., Fluharty, D., Dawson, C. & Teeter, D. (1982). *The Management of Marine Regions: The North Pacific.* Berkeley: University of California Press.

Otto, R.S. (1981). Eastern Bering Sea crab fisheries. In *The Eastern Bering Sea Shelf: Oceanography and Resources*, Vol. 2, eds. D.W. Hood & J.A. Calder, pp. 1037–66. NOAA Office of Marine Pollution Assessment. Washington, DC: NOAA.

Otto, R.S. (1986). Management and assessment of eastern Bering Sea king crab stocks. *Canadian Special Publication of Fisheries and Aquatic Sciences*, **92**, 83–106.

Sabella, J. (1982). Life after crab in the northeast Pacific. *Pacific Fishing*, **3**, 41–5.

Schumacher, J.D. & Reed, R.K. (1983). Interannual variability in the abiotic environment of the Bering Sea and the Gulf of Alaska. In *From Year to Year*, ed. W.S. Wooster, pp. 111–33. Seattle: Washington Sea Grant.

3

The rise and fall of the California sardine empire

Edward Ueber

Gulf of the Farallones National Marine Sanctuary
San Francisco, CA 94123, USA

and

Alec MacCall

National Marine Fisheries Service
Tiburon, CA 94920, USA

The plane circled slowly, searching. The US Navy pilot and crew had been trained to locate and report the position of the prey under the waves. Once sighted, a message would be sent to the US Navy Air Station ashore which then relayed the sighting to a subchaser or US Coast Guard cutter in the area. The warship would signal 10 to 15 pursuit vessels, inform them of the prey's reported location and the hunt would begin (Scofield, 1920). All the men in the air and on the sea were searching for the bright crescent of light that would be visible during the dark of the moon (Scofield, 1924). The inner edge of the crescent would be green and the outer edge red (Daniel Miller, private communication, 19 September 1989). Once a vessel sighted the crescent of light, the entire attack fleet would employ a number of capture techniques to ensnare the prey.

This was no hunt for an enemy submarine, but the latest twentieth century technology being used in 1919 to assist fishermen off the San Diego area in locating schools of Pacific sardine, *Sardinops sagax*. Sardines were just beginning to be used by the canning industry. Sardine canning started on the US west coast in 1889 at the Golden Gate Packing Company of San Francisco (California). When the San Francisco plant closed in 1893 the equipment was sold to the Southern California Fish Company in San Diego (Thompson, 1926). This company canned sardine in oil, mustard, spices and tomato sauce in two, one, and quarter pound sizes until 1909 (Smith, 1895; Thompson, 1926).

Another cannery started producing canned sardine in 1909 at San Diego but closed in 1913. By 1915 three sardine canneries were in operation, one in San Francisco and two in Monterey. The Monterey plants commenced canning in 1902 (the Booth plant) and 1906 (Monterey Fishing and Packing Company). A San Francisco plant opened sometime between 1900 and 1915 (Schaefer et al., 1951) (Fig. 3.1). The sardine packed at the Booth plant were labeled mackerel until this practice was stopped by the US government in 1910. However, canned sardines from California soon gained a reputation for having flavor and quality equal to the then-preferred French brands.

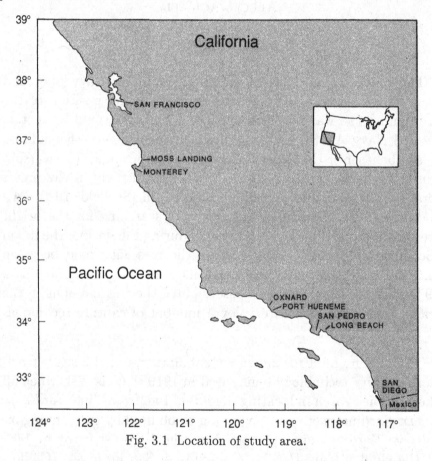

Fig. 3.1 Location of study area.

Sardine canning and reduction had become the largest fishery on the west coast by 1925. This major industry landed 173,000 tons of sardine in California and another thousand in British Columbia (Canada) during that 1925–26 season (Radovich, 1981). The sardine fishery had started out as a supplier of fresh whole fish in

the 1860s and sardines had also been used as bait since the 1880s
(Smith, 1902). The shift to canning from the 1890s to the 1920s
actually created two new industries. The first produced a high
quality and highly valued canned sardine for human consumption;
the second produced protein-rich feed for chickens as well as fertilizer for green plants. The chicken feed and the plant fertilizer
were produced from canning waste, using a process called reduction. The value of this by-product soon caused canners to set up
their own reduction plants at the canneries. By 1920 the increased
demand for sardine meal and fertilizer resulted in some plants using whole fish along with canning waste to produce fish meal, flour,
oil and fertilizer.

The California Department of Fish and Game became concerned
about the direct use of sardine for nonhuman consumption in 1920.
Starting in 1920, and excluding only 1923 and 1924, new laws
were passed to curtail the use of whole fish for reduction in every year through 1941 (Schaefer et al., 1951). The position of
the US Bureau of Fisheries was that "[canned] sardines must sell
at a price that is based on their own cost of production. Production of fish meal and oil can not [sic] continue to dominate
canning" (Beard, 1928). This statement was made because only
plants which canned fish could legally reduce sardines: canned
sardines were being produced and sold at cost or at a loss so that
canneries could obtain enough waste and whole fish for reduction.
Selling at or below cost kept the sales of canned sardines above
what the market would have demanded. The high quality of the
California canned sardine resulted in a product which could be
sold in almost all existing canned sardine markets, thus increasing
the sale of California sardine. The canners also received another
benefit from maintaining the high quality of their canned sardines.
The canning of high quality sardine produces more by-product per
ton of fish landed; because there is an increase in the amount of
offal and unsuitable whole fish, the assured quality is higher. This
meant that more fertilizer and meal could be produced from each
ton of sardine landed.

Although the state and federal governments were in agreement
on the need to reserve the sardine resource for human consumption, the economics of reduction and the legal apparatus mitigated
against this being accomplished. The major loophole in the legal
structure existed when fish were caught and processed outside the

three-mile state jurisdiction. The inability of the state to reserve
the sardine resource for human consumption became clear during
the 1926–27 season, when a Monterey canner towed the concrete
barge *Peralta* outside the state's three-mile jurisdiction and commenced reducing sardines, without even the pretext of canning.
Because of financial, legal, and social problems, this vessel never
successfully obtained sardines. A self-propelled vessel, the SS *Lake
Miraflores*, also tried to obtain fish, but fishermen would not sell
to her off Monterey or Santa Barbara (Fig. 3.2). The same vessel
did obtain some fish off San Pedro, but the operation proved unprofitable. In November 1930 the vessel moved north to the waters
just south of San Francisco; another vessel, the SS *Lansing*, joined
her in 1932. These ventures became profitable and, as a result,
floating reduction plants became common off all the major sardine
ports from San Diego to San Francisco.

Fig. 3.2 SS *Lake Miraflores*, the first reduction ship to operate outside the
        jurisdiction of the State of California, unloading sardines from a purse
        seiner in the early 1930s (Glantz and Thompson, 1981).

Such vessels, along with a few others, operated until 1938, when
oil and meal prices fell and an amendment changed the State of
California's constitution. This new amendment gave the state the
authority to stop offshore reduction plants. Legal proceedings were
not brought to bear on these at-sea reduction plants, because the
vessels had stopped processing by the time the amendment became

law. The reduction ships had landed a total of 778,560 tons of sardine in nine seasons. These nine seasons occurred during some of the best years in the sardine fishery. At-sea purchases of sardines represent an annual average of 16 percent of the sardines landed during the period.

The 1936–37 season saw the entry of Oregon and Washington into the fishery. The landings of sardine in these states, along with those in British Columbia, and California, produced the largest one-season landing of any single species of fish* ever caught on the west coast – 791,334 tons (Table 3.1).†

The 12 seasons from 1934 to 1946 would have to be considered as *pax-sardinia* in the California fishing industry. Landings averaged 599,467 tons a season. World War II prompted good prices for oil, meal, fertilizers, and canned sardine. State fishery biologists had been warning for years that the sardine biomass could not sustain removals over 250,000 tons. However, the industry and federal agencies resisted, thwarting the state biologist's attempt to enact a quota of that size, or indeed any quota at all. "The canneries themselves fought the war by getting the limit taken off fish and catching them all. It was done for patriotic reasons..." (Steinbeck, 1954).

During the next six seasons, from 1946 until 1952, landings averaged 234,068 tons. This amount was about 40 percent of the previous 12-season average. The next 10 seasons, through 1962, recorded average landings of 55,322 tons, or one-tenth the record mid-1930s to mid-1940s seasons. The last six seasons of the fishery produced average landings of 23,985 tons (Table 3.1 and Fig. 3.3). As one fisherman who participated in the last season (1968) said, "In the last year we caught them all in one night" (Louis Mascola, private communication, 6 September 1989). A fishery biologist, when asked to comment on the fishery, said that "It was big while it lasted" (Ralph Silliman, private communication, 18 September 1989).

* There are enough 10-inch sardines in these landings that together, if laid end to end, would reach from the earth to the moon and back (Reinstedt, 1978).

† During the history of this fishery, all landings were reported in short tons (908 kg). Hence the weights in this chapter have not been converted to metric units.

Table 3.1 Sardine catches from the Pacific Coast of North America (from Murphy, 1966).

Season	Pacific Northwest				California						Baja Calif.	Grand Total
	British Columbia	Washington	Oregon	Total	Reduction Ships	Northern California — San Francisco	Monterey	Total	Southern Calif.	Total Calif.		
1916-17							7710	7710	19820	27530		27530
1917-18	80			80		70	23810	23880	48700	72580		72660
1918-19	3640			3640		450	35750	36200	39340	75540		79180
1919-20	3280			3280		1000	43040	44040	22990	67030		70310
1920-21	4400			4400		230	24960	25190	13260	38450		42850
1921-22	990			990		80	16290	16370	20130	36500		37490
1922-23	1020			1020		110	29210	29320	35790	65110		66130
1923-24	970			970		190	45920	46110	37820	83930		84900
1924-25	1370			1370		560	67310	67870	105150	173020		174390
1925-26	15950			15950		560	69010	69570	67700	137270		153220
1926-27	48500			48500		3520	81860	85380	66830	152210		200710
1927-28	68430			68430		16690	98020	114710	72550	187260		255690
1928-29	80510			80510		13520	120290	133810	120670	254480		334990
1929-30	86340			86340		21960	160050	182010	143160	325170		411510
1930-31	75070			75070	10960	25970	109620	146550	38570	185120		260190
1931-32	73600			73600	31040	21607	69078	121725	42920	164645		238245
1932-33	44350			44350	58790	18634	89599	167023	83667	250690		295040
1933-34	4050			4050	67820	36336	152480	256636	126793	383429		387479
1934-35	43000			43000	112040	68477	230854	411371	183683	595034		638054
1935-36	45320	10	26230	71560	150830	76147	184470	411447	149051	560498		632058
1936-37	44450	6560	14200	65210	235610	141099	206706	583415	142709	726124		791334
1937-38	48080	17100	16660	81840	67580	133718	104936	306234	110330	416564		490404
1938-39	51770	26480	17020	95270	43890	201200	180994	426084	149203	575287		670557
1939-40	5520	17760	22330	45610		212453	227874	440327	96939	537266		582876
1940-41	28770	810	3160	32740		118092	165698	283790	176794	460584		493324

California

Season	British Columbia	Washington	Oregon	Total	Reduction Ships	San Francisco	Monterey	Total	Southern Calif.	Total Calif.	Baja Calif.	Grand Total
	Pacific Northwest				Northern California							
1941-42	60050	17100	15850	93000		186589	250287	436876	150497	587373		680373
1942-43	65880	580	1950	68410		115884	184399	300283	204378	504661		573071
1943-44	88740	10440	1820	101000		126512	213616	340128	138001	478129		579129
1944-45	59120	20		59140		136598	237246	373844	181061	554905		614045
1945-46	34300	2310	90	36700		84103	145519	229622	174061	403683		440383
1946-47	3990	6140	3960	14090		2869	31391	34260	199542	233802		247892
1947-48	490	1360	6930	8780		94	17630	17724	103617	121341		130121
1948-49		50	5320	5370		112	47862	47974	135752	183726		189096
1949-50						17442	131769	149211	189714	338925		338925
1950-51						12727	33699	46426	306662	353088		353088
1951-52						82	15897	15979	113125	129104	16184	145288
1952-53							49	49	5662	5711	9162	14873
1953-54							58	58	4434	4492	14306	18798
1954-55							856	856	67609	68465	12440	80905
1955-56							518	518	73943	74461	4207	78660
1956-57							63	63	33580	33643	13655	47298
1957-58							17	17	22255	22272	9924	32196
1958-59							24701	24701	79270	103971	22334	126305
1959-60							16109	16109	21147	37256	21446	58702
1960-61							2340	2340	26538	28878	19899	48777
1961-62							2231	2231	23297	25528	21270	46798
1962-63							1211	1211	2961	4172	14620	18792
1963-64							1015	1015	1927	2942	18384	21326
1964-65							308	308	5795	6103	27120	33223
1965-66							151	151	568	719	22247	22966
1966-67							23	23	321	344	19531	19875
1967-68									71	71	27657	27728

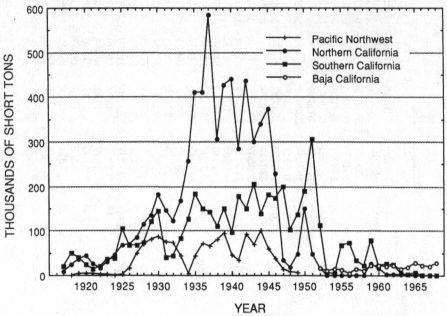

Fig. 3.3 Sardine catches from the Pacific coast of North America. Pacific Northwest includes British Columbia, Washington, and Oregon. Northern California includes reduction ships, San Francisco, and Monterey. (Data from Murphy, 1966.)

The history of the sardine fishery is not just a story of landings, government regulations and industry exploitation; nor is this story being told to affix the blame or determine the cause of the sardine fishery collapse. The causes could have been overfishing (Scofield, 1938; MacCall, 1979), management conflicts (Ahlstrom & Radovich, 1970), climate change (Smith, 1979) or, most likely, a combination of all of these.

Upon following this fishery from the 1860s to its demise in 1968, one not only becomes aware of the vast quantity of sardines harvested and the loss of a very valuable industry, but also of the people who worked in the plants, caught the fish and invested their funds in this colorful dynamic venture. The loss of the sardine industry had ramifications for the west coast fishing community, the State of California, and foreign nations in Central and South America, and Africa. The lessons available on how people, institutions, and society coped with, and learned from, this loss are every bit as important as the lessons learned from the loss of the resource itself.

The California sardine fishery was composed of the aforementioned groups of fishermen, plant workers, and entrepreneurs. These three groups of people are in no way mutually exclusive. Some fishermen and their families were involved in all three activities. Others invested and fished or invested and worked. In the very beginning in San Francisco most of the fishermen were Italian (*San Francisco Chronicle*, 20 July 1885). As the fishery moved to southern California, Portuguese and "Jugo-Slavs ('Austrian')" [sic] fishermen predominated (Skogsberg, 1925). Fishermen from Oregon and Washington in the 1930s were mostly Scandinavian. The period of expansion and large landings covered 21 years from 1925 to 1946. With one exception during that period, these nationality groupings stayed roughly the same throughout the fishing communities. In the late 1920s and 1930s Japanese-American fishermen dominated fishing for sardine out of southern California (Higgins & Holmes, 1921). At the outbreak of World War II, Japanese-American fishermen were removed to concentration camps, never to regain ownership of their vessels or their dominant position in the fishery. Beside the foregoing nationalities, two other nationalities were prominently involved as plant workers: the Mexicans and Chinese. Most of the higher skilled workforce at the plants were related to fishermen or came from the same ethnic background as the fishermen in that port.

The Monterey area, southern California, and the San Francisco area produced the majority of canned sardines and fish meal. The most famous of the sardine canning communities was, and still is, Monterey. Monterey's fame can be linked to its current tourist popularity and also to the writings of John Steinbeck, even though southern California landed as many sardines and San Francisco started earlier. The national and international distribution of the "top of the line" Monterey-canned sardines also contributed to the area's fame and recognition.

During the Great Depression (1929–41), Monterey did not suffer as much as most other areas of the US, because sardine production remained high and even increased in the late 1930s. Although the Monterey canning area became "a poem, a stink, a grating noise" to many Americans (Steinbeck, 1945), it meant bread on the table to those involved in it.

The Legaz family was one of the families that earned its livelihood from fishing. They were of Austrian decent and started their own fishing business in 1912. In that year, with his cousins and brothers, Mr. Legaz bought a small trawler and named it the *Legaz Brothers*. The oldest of the Legaz brothers subsequently bought the *Georgia* in 1917, the *Ansonia* (70 feet – 21 m) in 1927, the *Valencia* (75 feet – 23 m) in 1928, the *Marconia* (80 feet – 25 m) in 1937, and the *Leviathan* (98 feet – 30 m) in 1946. In 1947 he invested in a sardine plant and sold out a few years later, not losing any money. His sons fished with him from age nine or ten, until they could buy their own vessels or skipper one of the other boats in the Legaz fleet. In the 1940s oil was the big moneymaker from sardines, not canning (Louis Legaz, private communication, 1 October 1989). Investing in a plant that reduced sardine was one way for fishing families to get rich.

The Legaz family was one of hundreds of families and thousands of people who fished for sardine for part of a year. A typical multi-fishery pattern consisted of fishing for salmon on the west coast of Alaska from June through September, sardine off California from October through March, and squid fishing near San Pedro and Monterey in April and May and sometimes September. The number of vessels involved in the sardine fishery was substantial. As late as 1956, the October monthly report of the Monterey office of California's Department of Fish and Game states that 150 vessels and eight airplanes were active on one night. During the 1956–57 season, landings were 33,580 tons. Each seiner and lampara vessel had between six and 10 people on board. If you expand this fishermen ratio to the *pax-sardinia* fleet of 500 vessels, then 4,000 fishermen were employed for half a year, or 2,000 fisherman-years of work annually during the 1930s and early 1940s.

The California landings in 1943 were about 500,000 tons, and in 1946 about half that amount. The ex-vessel values were US$10.8 and US$6.9 million, respectively (Pinsky & Ball, 1948). The product value in 1943 exceeded US$29 million and in 1946 is estimated to have been about the same, although the 1946 catch was actually much smaller. If we expand this product value with a conservative consumer price index (CPI) to 1989, the product value would exceed US$210 million (1946 CPI = 52, 1989 CPI = 380, 1967 CPI = 100) (Anon., 1989). Yearly employment was over 9,000 in 1946, probably far greater in 1943, and likely to have been over 25,000 in 1936, based on the relative size of the reported catches.

The wages paid to these 1943 workers was around US$28 million, equivalent to US$200 million in 1989 dollars.

Monterey fishermen continued to can fish until 1957, but reduction ceased by 1950. Most of the fish canned in Monterey during the 1950s were trucked in from southern California. Fishermen received an agreed-upon price regardless of the quantity landed, but had to pay the cost of trucking. The trucking rates and prices were negotiated between the union and the remaining two Monterey buyers in the late 1950s. Fishermen negotiated an ex-vessel price of US$47.50 per ton in 1955, but after paying the cost of trucking, loading, unloading, and ice they received only US$40.00 per ton. In August 1957 the agreed price rose US$5.00 to $52.50 per ton, but fishermen received US$2.50 less, $37.50, because the costs associated with transporting the sardines were increased to US$15.00 per ton. In October 1957, transport costs were raised to US$18.00, lowering the net received ex-vessel price to US$34.50 (see the monthly California Department of Fish and Game reports for the period 1945–65). By 1960 only the bait fishery remained in Monterey, and although the ex-vessel price was sometimes US$200 per ton, sardines were not available. Sportfishermen were paying US$1.00 per sardine in 1968. The fishery continued in southern California until that night in 1968 when the airplanes led the fleet to the last sardine schools.

Capital formation had occurred, from 1889 to 1946, at roughly the same rate as the developing fishery. In 1946, 101 reduction plants were in operation in California. Plant ownership was distributed among a broad spectrum of people. People such as the Legaz family were only one type of owner. Other ownership structures included small investors who had combined funds. These small investors were often groups of fishermen, or cannery workers who, through investment, believed they could give themselves an opportunity to better their position by increasing job security and decreasing middleman costs. Large investors also participated. A number of investors were already wealthy, such as Zellerbach, Fleishhacker and Christopher. Christopher was mayor of San Francisco, and men like Zellerbach and Fleishhacker had friends in high local, state, and federal positions on whom they could call for help, giving these investors more political clout than the small investors. The industry was able to use this clout to either limit laws or block (or at least postpone) effective conservation legislation.

By 1968, 80 years after the fishery had started in San Francisco, it was gone. It had collapsed, crumbled, and disappeared. The collapse had taken 22 seasons. The vessels were gone, the machinery was gone, and the people were no longer sardine workers. Fishermen had for years been witnessing the demise of the resource with fewer and smaller schools of fish almost annually. In the 1950s workers still talked about the sardines coming back. A decent season like 1958–59, when 126,000 tons were landed, would keep people hoping for another five years. Many men finally realized that the sardines would not return; the lucky or smarter fishermen went to other fields, other fisheries or other countries.

The men who had multi-fishery options diversified. Some of the smaller vessels fished market (Dungeness) crab (*Cancer magister*), from November through February, others fished rockfish (*Sebastes spp.*), albacore (*Thunnus alalunga*), and salmon (*Oncorhynchus spp.*). These vessels which changed fisheries would change from eight-man sardine crews to three- or two-man, or even skipper-only crews. The displaced crewmen found employment ashore as painters, gardeners, construction workers, and other similarly skilled positions. Many maintained their relationship with fishing by going to Alaska in the summer, but no longer fished all the year round. Those vessels which could not economically switch from lampara or seine gear fished for squid, anchovy or tuna. In the US none of the displaced fishermen or cannery workers received retraining or assistance from the government to start a new profession.

Men who owned vessels, but could no longer pay the mortgage or upkeep on their vessels, sold them. Smaller vessels, in the 45-foot range, were useful in the above-mentioned alternative fisheries. People who sold and serviced vessels, such as Woodward's in Moss Landing, saw large numbers of vessels being sold, foreclosed, not repaired, and lost at sea. Because of the large number of vessels on the market, even small vessels sold at a loss, but larger and more valuable vessels (75–110 feet – 23–33 m in length) sold for half price and even as low as 10 cents on the dollar (Lillian Woodward, vessel brokerage family, private communication, 19 September 1989). Some of these larger vessels went to Alaska to participate in the expansion of the king crab (*Paralithodes camtschatica*) fishery (see

Wooster, this volume). The low prices of the vessels kept fixed costs down, allowing new owners the luxury of learning to fish and market alternative products while having to cover little more than their operating (i.e., variable) costs.

At the same time as the vessels were being sold and transferred to other west coast fisheries, other vessels and equipment associated with sardine canning and reduction were being sold in international markets to such countries as Peru, Chile, and South Africa. Although much of the equipment pre-dated World War II, a lot of equipment was relatively new and had been used sparingly, since being purchased in the late 1940s when the fishery was in decline (Sal Ferrante, private communication, 20 September 1989).

Peru and Chile were assisted in these purchases by two agencies of the United Nations: the Food and Agriculture Organization (FAO), responsible for increasing the food (particularly protein) supply for Third World nations, and the United Nations Development Program (UNDP), responsible for securing funds for Third World development. Funds supplied by the UNDP were also used to obtain expertise in fishing, canning, reduction, and fishery management. Lampara fishermen from San Francisco and seiners from San Diego went to South America generally for six months to a year to teach fishing methods or as contract skippers. Some men sold their vessels, delivered them in South America and stayed on as skippers of the vessel. As in Alaska, the low price of idle equipment and vessels allowed the sardine and anchoveta (*Engraulis ringens*) fisheries off the west coast of South America to expand rapidly. This expansion occurred even faster, because of the technology transfer attributed to the expertise of the Californians and the low cost of equipment. Sal Ferrante, an experienced canner and reduction plant operator/owner (and other men like him), was hired to establish a new plant in Peru during the 1950s. From 1958 until 1960 he stayed on to run the plant he established. Ferrante was available to do this type of work because, in 1957, he had sold his fertilizer plant in Oxnard, California, to a South African company. The Oxnard plant was almost new and he received a good price for used sardine equipment, a price equal to about 60–70 percent of the cost of building a new facility. Ferrante's plant was dismantled, shipped to South Africa, reassembled and became operational within a matter of months. Except for the location, nothing in the fertilizer plant had been changed, even the name on the door remained "Ferrante Co."

Other businessmen, such as Leo Hart and Craig Johnson, established companies which sold used machinery primarily to South American countries and to Mexico. Machinery was shipped to these locations from San Francisco, Moss Landing, Monterey, Port Hueneme, Long Beach, and San Diego (Fig. 3.1). Fish meal and oil reduction machinery constituted the majority of the equipment sold to South American enterprises. This equipment was generally older and less costly than the equipment shipped to South Africa. Unlike Ferrante's fertilizer plant, most equipment prices rarely exceeded half the cost of new equipment, and some were sold for one-fifth the new value. Locating these surplus machines in the western US was accomplished via a worldwide network of people who had previously worked in the production and management of California's sardine fishery.

Fishery biologists and managers from California became involved in the management of South American fisheries through the United Nations' support of Peru's Instituto del Mar (IMARPE) and the Ministerio de Comercio e Industria. Along with people from California (William Ripply and Frances Clark), Australia (Jeffery Kestevan), England (Phillip Appleyard), and other US citizens (Millian Kravanja and Wilbur Doucet), were local scientists (e.g., Coronel Portillo) attempting to evaluate the sardine and anchoveta resources of the Peru Current. Many of these people had gained knowledge and experience working with California sardine and were attempting to use that experience to manage southern clupeid stocks (William Ripply, private communication, 18 September 1989). Like the industry workers, these scientists had gone south "searching for new raw material" for their scientific skills (Popovici, 1964).

Conclusions and lessons

Global climate change is likely to cause diverse alterations and changes in fisheries around the world. Some fisheries may decline or collapse, while others may increase. The historical collapse of a major fishery like the California sardine fishery provides a number of lessons on how the local society and the national and international fishing industry may be expected to respond to these changes.

Some of those lessons were postulated by Radovich (1981). He addressed the interactions of the politics of fishery management with the biology of the species, and concluded that "the present scarcity of sardines off the coast of California, and their absence off the northwest, is an inescapable climax, given the characteristics and magnitude of the fishery and the behavior and life history of the species" (p. 134). Lessons drawn by Radovich regarding within-fishery dynamics include the following:

- Overfishing can cause fishery collapse rather than a sustained low-level harvest.
- Political process can be controlled through industry's influence to thwart rational management.
- Development-oriented government agencies may contribute to delayed and ineffective management.
- Research can be used to delay solutions as well as to provide solutions.

We would offer a fifth lesson on the internal dynamics of fishery collapse:

- Overfishing is a natural consequence of institutional (government as well as industry) momentum, following the paradigm that "bigger is better," with size being the ultimate measure of "success."

These lessons clearly indicate the path that a new industrial clupeid fishery may be expected to take. More importantly, they indicate that strong management is necessary to counter these destructive tendencies.

New fisheries are eagerly encouraged by many segments of society, industry, and government agencies. This encouragement is often manifested in the form of subsidies:

- Non-market funding of equipment or expertise will cause the fishery to develop more rapidly than would be expected from purely market-driven development (e.g., US Navy and Coast Guard assistance in locating sardines).
- Fishery management can behave like a subsidy in that it encourages investment if its perceived presence engenders optimism, or decreased expectation of risk. In this respect, ineffective management is worse than no management at all, as

fishery development is accelerated but no resource conservation benefit is derived.

Our examination of the international events during and following the collapse of the California sardine fishery provides another set of lessons regarding the development of substitute fisheries. *A substitute fishery will develop more rapidly than would be the case for a newly developed independent fishery.* This rapid development occurs because existing capital, labor, technology, and markets are readily transferred to the substitute fishery.

- Capital, labor, and technology can be obtained at less cost and with less delay than would be required for independent development of an isolated fishery.
- Technology and expertise are available, eliminating the "learning curve."
- Labor available to substitute fisheries (semi-skilled, skilled, and managerial) has provided an opportunity to maintain preferred professional and cultural lifestyles, avoiding the economic risk and cultural hazards of retraining.
- Market development is unnecessary with substitute fisheries, because existing markets are in search of a product. Product prices offered for the substitute product are generally high.

The combination of a highly valued product due to existing unfilled market demand, subsidies, and low start-up cost (due to cheap surplus equipment and labor) provides the economic conditions for rapid industrial development. The lack of normal time delays dangerously accelerates the developmental process.

For these reasons, we expect global warming not only to cause large international relocations of fishing industries, but those relocations will be accomplished by a rapid transfer of industrial structure from collapsing fisheries to emerging or new fisheries. While both the old and new substitute fisheries may be inherently unstable because of climate change, we expect that this rapid transfer of harvesting and processing capacity will exacerbate fishery instability.

The political process of establishing management institutions and the scientific process of developing predictive fishery models are much slower than industrial development of substitute fisheries. Internationally, governments and their fishery management agencies should be prepared to adopt the politically difficult and industrially resisted management policy of deliberately constrained fishery development, and avoid politically popular but destabilizing subsidies. The alternative is likely to be a few years of glory and high profits followed by decades of disillusionment, unemployment and industrial decay.

References

Ahlstrom, E.H. & Radovich, J. (1970). Management of the Pacific sardine. In *A Century of Fisheries in North America*, ed. N.G. Benson, pp. 183–93. Special Publication No. 7 of the American Fisheries Society. Washington, DC: American Fisheries Society.

Anon. (1989). *Consumer Price Index*. Washington, DC: US Bureau of Labor Statistics.

Beard, H.R. (1928). Preparation of fish for canning of sardine. *Report of the United States Commissioner of Fisheries for Fiscal Year 1927*, pp. 67–223. Washington, DC: US Government Printing Office.

Higgins, E. & Holmes, H.B. (1921). Methods of sardine fishing in southern California. *California Fish and Game*, **7**, 219–37.

MacCall, A.D. (1979). Population estimates for the waning years of the Pacific sardine fishery. *CalCOFI Report* No. 20, pp. 72–82. Monterey: California Cooperative Oceanic Fishery Investigations.

Pinsky, P.G. & Ball, W. (1948). *The California Sardine Fishery*. San Francisco: California Congress of Industrial Organizations Council.

Popovici, Z. (1964). Remarks on the Peruvian anchoveta fishery. *Document VIII, Marine Research Committee Minutes of 6 March 1964*. Sacramento: State of California Marine Research Committee.

Radovich, J. (1981). The collapse of the California sardine fishery—What have we learned? In *Resource Management and Environmental Uncertainty: Lessons from Coastal Upwelling Fisheries*, ed. M.H. Glantz & J.D. Thompson, pp. 107–36. New York: John Wiley & Sons.

Reinstedt, R.A. (1978). *Where Have All the Sardines Gone?* Carmel: Ghost Town Publishers.

Schaefer, M.B., Sette, O.E. & Marr, J.C. (1951). Growth of the Pacific Coast pilchard fishery to 1942. *Research Report* No. 29, pp. 1–31. Washington, DC: US Fish & Wildlife Service.

Scofield, N.B. (1920). Commercial fishery notes. *California Fish and Game*, **6**, 29–32.

Scofield, N.B. (1924). The lampara net. *California Fish and Game*, **10**, 66–70.

Scofield, W.L. (1938). Sardine oil and our troubled waters. *California Fish and Game*, **24**, 210–23.

Skogsberg, T. (1925). Preliminary investigation of the purse seine industry of southern California. *California Fish and Game Commission Bulletin*, **9**, 1–95.

Smith, C. (1979). Cited by Radovich (1981) as *San Diego Union*, January, Col. 1, B-2, Col. 4, B-5.

Smith, H.M. (1895). Notes on the reconnaissance of the fisheries of the Pacific coast of the United States in 1894. *US Fish Commission Bulletin*, **16**, 223–8.

Smith, H.M. (1902). The French sardine industry. *US Fish Commission Bulletin*, **21**, 1–26.

Steinbeck, J. (1945). *Cannery Row*. New York: Viking Press.

Steinbeck, J. (1954). *Sweet Thursday*. Dallas: Penguin Books.

Thompson, W.F. (1926). The California sardine and the study of the available supply. *California Fish and Game Commission Bulletin*, **11**, 5–66.

4

El Niño and variability in the northeastern Pacific salmon fishery: implications for coping with climate change

KATHLEEN A. MILLER

Environmental and Societal Impacts Group
National Center for Atmospheric Research
Boulder, CO 80307, USA

and

DAVID L. FLUHARTY

School of Marine Affairs
College of Ocean and Fishery Sciences
University of Washington
Seattle, WA 98195, USA

Introduction

In 1982 and 1983 an intense El Niño in the central and eastern equatorial Pacific Ocean spread warm water far northward along the west coast of North America. This event is believed to have been an important factor contributing to poor salmon harvests along the California, Oregon, and Washington coasts during the 1983 and 1984 seasons and has been largely blamed for the socioeconomic distress experienced by commercial salmon trollers during those seasons. At the time, newspaper headlines that appeared in the US Pacific Northwest followed the lead of distressed commercial harvesters and disappointed sports fishers in proclaiming the El Niño to be a natural disaster with significant impacts on the salmon fishery. To what extent was El Niño responsible for the poor runs of coho and chinook salmon along the US west coast in 1983 and 1984? How large were the actual socioeconomic impacts? To what extent was the reported socioeconomic distress among commercial harvesters a direct result of this event? These questions are complex, and no simple answers can be given. Nevertheless, an examination of the experience of the Pacific Northwest

salmon fishery during this El Niño event can further our understanding of the interactions between climate, biological processes,
and the human activities dependent on those processes.

The purpose of this study is to gain insight into the impacts of
climatic variability on a complex fishery system and, by analogy,
the potential impacts on fisheries of climate change. The fishery
system as defined here encompasses not only the natural history
and biological oceanography of salmon, but societal components as
well. These include scientific research and monitoring, harvesting,
processing, marketing, consumption, and governmental management of the commercial and sport fisheries. Each component of
this system is affected by multiple sources of variability, many of
which are inadequately monitored, documented, and understood
to allow a clear separation of the role of climatic variations from
a host of confounding factors. This chapter is thus, necessarily, a
first cut at describing the major factors affecting this fishery and,
where possible, sorting out their relative importance during the
period surrounding the 1982–83 El Niño event.

Although five species of Pacific salmon (*Oncorhynchus spp.*) are
harvested along the west coast of North America, the discussion
here will focus especially on the chinook (*O. tshawytscha*) and
coho (*O. kisutch*) salmon fisheries off Washington, Oregon, and
California, and on the Fraser River sockeye (*O. nerka*). These
fisheries were apparently most strongly affected by the 1982–83
El Niño event, with reductions in the abundance and size of the
coho and chinook, and an altered migration pattern for the Fraser
River sockeye.

The Washington, Oregon, and California chinook and coho
stocks account for only a small share of the total North American commercial, sport, and subsistence salmon harvest (less than
2% by numbers of fish in 1985). However, they have been locally
important to a large community of commercial salmon harvesters,
sports fishers and Indian harvesters, and salmon are often seen
as an important part of the regional culture. In addition, they
have been the object of extensive biological research regarding the
contribution of oceanographic conditions to variations in their productivity (Nickelson, 1986; Pearcy, 1988; Walters, 1988). Finally,
they are possibly the salmon stocks most susceptible to climate
change as they are closest to the southern range of the genus and

may be more highly stressed than northern stocks by alteration of habitat, hatchery developments, and fishing pressure.

The Fraser River sockeye are an internationally shared resource between Canadian and US (Washington State) harvesters. The altered migration pattern during the 1982–83 El Niño event meant that the majority of these salmon remained in Canadian waters as they returned to spawn, making them unavailable to the US fishery.

This chapter is divided into five sections. The first provides a description of the North American salmon fishery. The second section describes El Niño events in the northeastern Pacific and discusses the mechanisms by which El Niño and other fluctuations in oceanographic conditions can affect the biological productivity of salmon stocks. The third discusses the responses of the scientific research community and of salmon managers to the 1982–83 El Niño event. The fourth section discusses the socioeconomic impacts of this event. The concluding section distills implications from this case study for adaptation to climatic variability and climate change.

The salmon fishery in the northeastern Pacific

Salmon are anadromous fish, spawning in fresh water, spending early life stages in streams and lakes, moving then into coastal estuaries and finally into the open ocean. After a period of one to six years in salt water depending on the species, the mature fish return to spawn and die (Fig. 4.1). Hatchery production cycles are similar to natural runs except that there is artificial spawning and hatching, and the early stages of the salmon's life are spent in a controlled freshwater environment.

In the course of their wide-ranging migration, salmon spawned in North America traverse virtually the whole northeastern Pacific Ocean from approximately 40°N to the Bering Straits and beyond for some species (Fig. 4.2). The distribution picture is complicated by the mixing of salmon stocks of Asian origin (Fredin et al., 1977). For North American stocks, there is a distinct change in species composition of the salmon catch with latitude. For example, the California and Oregon commercial harvest consists almost entirely of coho and chinook, whereas in Washington and British Columbia, chum, pink and sockeye as well as coho and chinook

THE LIFE CYCLE OF THE PACIFIC SALMON

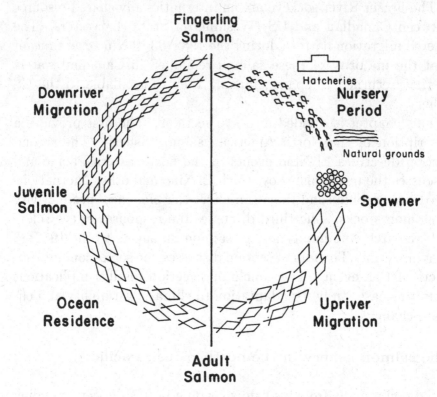

Fig. 4.1 Life cycle of the Pacific salmon.

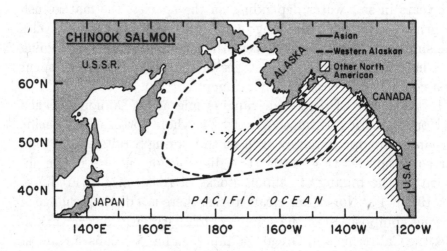

Fig. 4.2 Generalized ocean distribution of chinook salmon from Asia, western Alaska, and other North American areas.

salmon are taken. Large numbers of salmon of all five species are caught off Alaska, with pink, chum and sockeye accounting for over 95 percent of the salmon harvested.*

Harvesters

Salmon are of primary importance to three major groups of North American harvesters: non-Indian commercial harvesters, Indians, and sports fishers. The Indian fishery has deep historical and cultural roots. The salmon fisheries were the mainstay of the northwestern Native American culture and livelihood. When settlers moved into the region they quickly began to exploit the salmon resource, at first for subsistence. By the late nineteenth century, major commercial fisheries were developing along the coast. As the region's population has grown, particularly in the post-World War II era, sport fishing for salmon has become increasingly important.

The commercial fishery, which includes substantial Indian commercial fishing activity, accounts for most of the salmon harvested coastwide. The dollar value of the Pacific salmon fishery in today's terms is measured in the hundreds of millions of dollars and its direct employment is in the order of tens of thousands of persons on a seasonal basis. It has long been one of the most valuable commercial fisheries in the US, with Alaska accounting for most of the harvest (87% in 1987 and 1988) (National Marine Fisheries Service, 1988, 1989).

The commercial salmon fishery in the northeastern Pacific is dominated by two nations – Canada and the US – although Japan has participated through international agreements (Jackson & Royce, 1987; Myers et al., 1987). Incidental catch by groundfish and driftnet fisheries has made other fishing interests, domestic and foreign, indirect parties to North American salmon fisheries (French, 1977).

The commercial harvest of salmon by North Americans takes place in coastal or terminal area fisheries mainly by gillnetting,

---

* Author's computations, data in numbers of fish. Source: INPFC (International North Pacific Fisheries Commission), *Statistical Yearbook*, 1980–85, (published 1983–88).

purse seining, and trolling, with traps, set nets and weirs used in limited areas. As one moves northward along the coast, the commercial harvest of salmon is conducted with an increasing variety of gear types. Trolling is the predominant method of commercial harvest in California and Oregon, while net-type gears are more common in the other jurisdictions where pink, chum and sockeye account for a major proportion of the salmon harvest (e.g., Bathgate, 1984; Schelle & Muse, 1986).

The sport fishery concentrates primarily on coho and chinook, although sport harvests of sockeye and pink appear to be increasing in Alaska, British Columbia, and Washington. Coastwide, sport fishing accounts for a small share of the salmon harvest; even of the harvest of the targeted coho and chinook.

For historical reasons, the competition between these three groups appears to have been most intense in the State of Washington. In that state, the non-Indian commercial fishery competes with a relatively large sport fishery for coho and chinook, while the state's obligation to uphold treaties protecting the Indian salmon fishery was reaffirmed in the US courts by the landmark 1974 Boldt decision and subsequent court cases (*United States* v. *Washington*, 1974).

Public management of the fishery

Public management of the salmon fishery initially arose in an effort to prevent biological overharvesting and to stabilize the runs. It has evolved into a complex monitoring, prediction and allocation system. There are two international commissions (International North Pacific Fisheries Commission and Pacific Salmon Commission), and within the US, two domestic regional fisheries management councils (North Pacific Fisheries Management Council and Pacific Fisheries Management Council). Management of the fishery also involves the federal governments of Canada and the US as well as state, provincial and territorial governments. By its nature, management affects the distribution of the harvest between jurisdictions and between competing groups of harvesters aligned by gear type, tribal affiliation, location and commercial–noncommercial orientation. The management regimes (Young, 1977) for salmon in the northeastern Pacific have regulated fishing through quotas, gear restrictions, license limitations, and seasons

– all of which have varied in their effect and effectiveness (Cooley, 1963; Crutchfield & Pontecorvo, 1969; Netboy, 1980).

Variability in abundance

Large interannual variations in the size of returning salmon runs have been a fact of life throughout the history of the salmon fishery. Commercial harvest statistics provide the only available long-term record of this variability and these statistics are available only for the last 80–100 years (Bell & Pruter, 1958; INPFC, 1979).

Commercial and sport catch statistics are an imperfect measure of variations in stock abundance, for reasons noted by Nickelson & Lichatowich (1984). This presents problems for biologists seeking to understand the causes of variability in the individual stocks contributing to the coastal salmon fisheries. However, since the anadromous nature of salmon makes them quite easy to harvest and since most salmon stocks run a gauntlet of intensely competing gear units as they return to their spawning streams, it seems likely that the variability historically observed in coastwide commercial harvests has had its major roots in variations in stock abundance. At the local level, it is more difficult to ascribe variability in harvests to changes in stock abundance. Due to their migratory nature, salmon from any given area may be harvested far from their native streams. Local harvest rates may also vary due to local weather conditions, variations in market conditions or fishing skills, or due to alterations in management regimes designed to rebuild stocks.

Commercial harvest records are available from 1920 to the present for Washington, British Columbia and Alaska. Shorter time series are available for commercial harvests in Oregon and California and for noncommercial harvests coastwide (INPFC, 1952–85; 1979).

Long-term trends in the total commercial harvest of all salmon species by state and province can be seen in Figs. 4.3a–e. The figures record catch in terms of numbers of fish (thousands) rather than in terms of weight to make use of the longest available time series. The primary value of these figures is to depict the variability inherent in these fisheries, which is a striking feature of all of the time series.

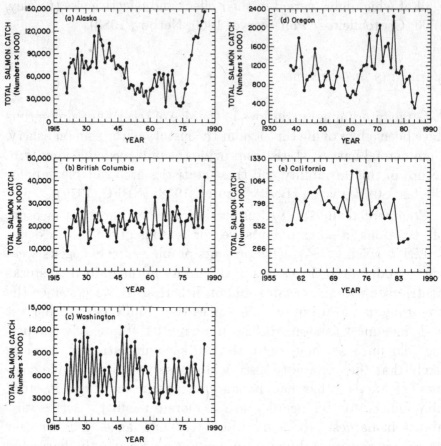

Fig. 4.3 Total commercial salmon harvest, all species, by jurisdiction. (Data source: INPFC, 1952–85; 1979.)

Alaska stands out as being by far the largest salmon producer. After peaking in the mid-1930s, Alaska's production declined to a low in 1959. Alaskan harvests recovered somewhat in the 1960s but dropped to record low levels in the early to mid-1970s. They have made a remarkable recovery to record high levels over the past decade. Some attribute this recovery to anomalously warm temperatures in the Gulf of Alaska and Bering Sea during the period 1976–84 (e.g., Rogers, 1984). Others cite improved management of the fishery (e.g., Royce, 1989). The reduction in the Japanese high-seas mothership fishery, brought about by the implementation in 1977 of the Fishery Conservation and Management Act (Public Law 94-265) and subsequent revisions of the North Pacific Treaty in 1978 and 1986, may also have contributed to recent record harvests in Alaska. However, the available evidence suggests that the recent dramatic growth in Alaskan salmon harvests far exceeds the reduction in Japan's mothership harvest (Harris, 1988; INPFC annual series, 1978–85; National Marine Fisheries Service, 1988; 1989).

Commercial harvests in British Columbia, Washington, Oregon, and California have been quite variable over the period recorded. Sharp changes in harvest levels from one year to the next are not uncommon, and the odd-year runs of pink salmon in Washington result in a clearly sawtoothed pattern for that state's total salmon harvest.

The mid-1970s were extremely productive years for coho and chinook along the southern part of the North American coast (see Figs. 4.4a–d and 4.5a–d). This elevated productivity has been attributed to anomalously cool waters and strong upwelling creating favorable conditions for salmon survival and growth (Nickelson, 1986). However, these record years marked the beginning of a declining trend for coho and chinook harvests. The fact that these harvests were declining immediately prior to the 1982–83 El Niño event makes it all the more difficult to determine the degree to which the low harvests of 1983 and 1984 can be attributed to El Niño.

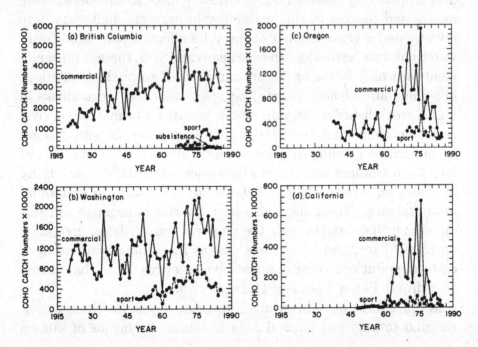

Fig. 4.4 Coho harvests, by user group. (Data source: INPFC, 1952–85; 1979.)

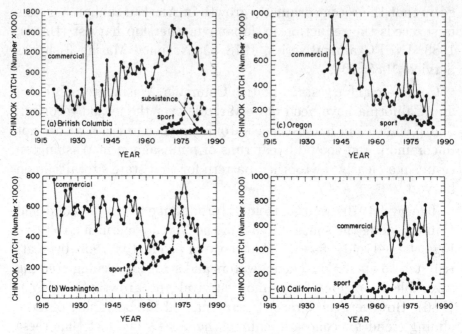

Fig. 4.5 Chinook harvests, by user group. (Data source: INPFC, 1952–85; 1979.)

The causes of the variablity in salmon harvests are complex and only imperfectly understood. In their efforts to understand the nature and sources of this variability in harvests, biologists have investigated a great number of variables affecting salmon in fresh-water habitats, estuaries, and in the ocean. Still, there is no agreement on which factor or factors can be used to predict variations in salmon abundance. Overfishing, particularly of weak stocks in mixed-stock fisheries (Bevan, 1988), variations in marine survival rates, environmental degradation, hatchery programs and changes in international fishing regulations have all been cited as causes of long-term changes in salmon abundance. Streamflow variability in spawning streams is also believed to contribute to variability in abundance. There have also been enormous negative impacts on salmon habitat through the construction of dams, irrigation facilities, dikes, and landfills as well as from industrial and agricultural pollution, forest management practices, and urbanization (Northwest Power Planning Council, 1986).

Major efforts have been made to compensate for these activities and to augment natural runs of salmon by means of salmon

hatchery construction and operation. In addition, there have been
changes in fishing methods, locations and management activities.
The presence of these confounding factors makes the task of identifying the role of climate-related variability in the marine environment almost insurmountable. Nevertheless, some evidence can
be gleaned from El Niño events and from other periods of major
change in the northeastern Pacific salmon fisheries.

Ocean environment, El Niño, and salmon

While a comprehensive discussion of the effects of the marine
environment on salmon is beyond the scope of this chapter, this
section provides a brief summary of the apparent effects of ocean
conditions on salmon both during El Niño events and over longer
time periods. Some plausible explanations of the mechanisms that
might relate changes in ocean conditions to salmon harvests are
discussed.

At sea, salmon are constantly moving (Quinn & Groot, 1984).
Their migratory patterns and behavior are believed to enable
salmon to make use of abundant food resources available on a
site- and time-specific basis in a major ocean ecosystem (Gross et
al., 1988). This would tend to make salmon survival and growth
a potentially good indicator of ocean conditions over a wide area.
Large shifts in ocean temperatures and productivity would likely
be reflected in changes in abundance. However, while considerable progress has been made at unlocking the oceanic "black box"
into which salmon swim as smolt and from which salmon return
as adults, much remains unknown (Walters et al., 1978). A considerable amount is known about the general migration patterns
and behavior of salmon and substantial work has been done to
identify the ocean distribution of each species (see Dodimead et
al., 1963; Favorite et al., 1976; Pearcy, 1984). However, there is
little systematic monitoring of ocean conditions, or of the spatial
distribution of salmon stocks at any given time.

If a marked change in the marine environment, like an El Niño
event, could be adequately monitored, it might provide a sort of
natural experiment. Such an event might yield information about
the effects of climatic variations on salmon and, perhaps, insights
into the potential impacts of climate change (Walters, 1988).

El Niño periods

El Niño is the name given to the occasional invasion of warm surface water from the western equatorial part of the Pacific Basin to the eastern equatorial Pacific, where these events produce a rise in sea level, higher sea surface temperatures and a depressed thermocline (Namias & Cayan, 1981; Cane, 1983; Philander, 1983; Rasmusson & Wallace, 1983; Rasmusson, 1985). El Niño is a recurring, quasi-periodic phenomenon associated with a reversal of barometric pressure between the eastern and western tropical Pacific, known as the Southern Oscillation.

These events are most frequent in the equatorial Pacific but infrequently they affect the west coast of North America as well (see Wooster & Fluharty, 1985). In the northeastern Pacific, evidence of the impacts of strong El Niño events can be seen in elevated sea levels (5–20 cm), higher sea surface temperatures (1–3°C) and warmer subsurface temperatures (to 300 m). Three such events are documented in the northeastern Pacific, north of central California: 1940–41, 1957–58, and 1982–83 (McLain, 1984; Fiedler, 1984; Tabata, 1984a; Cannon et al., 1985).

The chief changes in the ocean environment associated with El Niño are elevated sea surface temperatures, reduced offshore flow, and reduced primary productivity (Barber & Chavez, 1983). The temperature increase would appear to be within the range of tolerance of salmon (Tabata, 1984b) but little is known about the relationship between primary productivity and feeding, growth, and survival of salmon. Similarly, little is known about the effects of reduced offshore flow on adult salmon or on the vulnerability of salmon smolts to predators. However, it is reasonable to expect that lower primary productivity would result in reduced food availability and slower salmon growth. If severe shortage of food existed, salmon might starve, or they might be more susceptible to disease or predation. Finally, salmon in poor condition may not be able to complete the migration pattern or may be less fit reproductively. During the 1982–83 El Niño, a modest research effort was mounted to examine the effects of the event on salmonids. The following effects can be documented with some degree of certainty:

- Coho salmon off the coast of southern Washington, Oregon, and California (the Oregon Production Index (OPI) area)

showed considerably reduced survival compared with predicted survival.

- Coho and chinook salmon in the OPI area were generally lower
  than average in weight during this period.
- There was poor survival of coho salmon smolts in the OPI
  area.
- Sockeye salmon from the Fraser River system altered their
  migratory route around Vancouver Island, with an increased
  proportion returning through Johnstone Strait (Fig. 4.6).

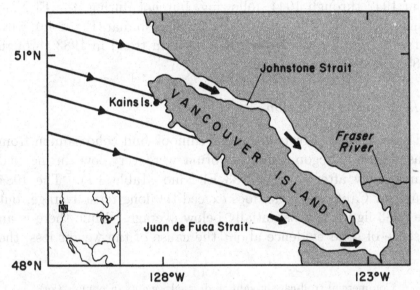

Fig. 4.6 Return migration routes, Fraser River sockeye salmon.

Ideally, the effects observed above could be compared to those
observed during a series of El Niño episodes as a means of establishing a pattern. However, due to the infrequent occurrence of
the impacts of El Niño events in the northeastern Pacific and the
lack of a consistent ocean environment/fisheries monitoring system, only tentative conclusions can be drawn about the former
effects of El Niño events on salmon.

Reduced coho survival

In the OPI area, returns of adult coho salmon are routinely predicted from returns of precocious (i.e., those returning one year
early) "jack" males during the previous year. This index shows a
remarkably good rate of success for the period 1977–86 with only
the El Niño year 1983 falling well below the prediction (Hayes &

Henry, 1985; Pearcy & Schoener, 1987; Pearcy, 1991). Examination of historic catch records confirms that El Niño events and the years just following those events tend to be periods of low coho catches in Washington and Oregon but the low harvests are not single events, they are nested generally in declining catch trends. For example, in Fig. 4.4b it can be seen that the commercial harvest of coho in Washington was relatively low in 1983 and 1984 as well as during the entire period from 1957 through 1964, with a record low in 1960. The Washington coho harvest was also low from 1942 through 1944, following, but not during, the El Niño event of 1940–41. The pattern in Oregon is similar (Fig. 4.4c), with harvests significantly below the long-term trend in 1983 through 1985 and 1958 through 1962.

Reduced size of salmon

Dressed weights of troll-caught chinook and coho salmon from Washington, Oregon, and California were very low during and immediately after the 1982–83 El Niño (Table 4.1). The 1984 figure for California coho does exceed the long-term average, but the 1983 figure is substantially below average. While there is an absence of solid evidence about the cause of the weight loss, the

Table 4.1
Commercial troll-caught salmon dressed weights in pounds (kg)

	Long-term average	1983	1984
Coho			
CA	5.7 (2.59)	4.4 (2.00)	7.4 (3.36)
OR	5.5 (2.50)	3.4 (1.54)	5.1 (2.32)
WA	4.6 (2.09)	4.2 (1.91)	4.5 (2.04)
Chinook			
CA	9.8 (4.45)	7.3 (3.31)	8.7 (3.95)
OR	9.6 (4.36)	8.2 (3.72)	8.5 (3.86)
WA	10.8 (4.90)	10.5 (4.77)	9.4 (4.27)

CA = California; OR = Oregon; WA = Washington
Source: Pacific Fisheries Management Council, 1989

most likely explanation would be some sort of depression of food supply at the necessary time for adult salmon growth. Comparable weight data do not appear to be available from previous El Niño periods.

Poor coho smolt survival

Considerable interest in understanding ocean survival of salmon has led to the systematic sampling of juvenile salmon in the OPI area. Because of a continuous series of survey data on salmon abundance, location, and growth, it is possible to determine that survival of juvenile coho was extremely poor in 1983 and 1984 (Fisher & Pearcy, 1988). Because the young salmon grew normally in these years despite below-average primary productivity in the ocean, the suspected cause of mortality is considered to be predation by other species of fish and possibly sea birds. Predators may find juvenile coho an alternative prey when natural abundance of normal prey species is decreased by the lower productivity of ocean waters during El Niño periods. Increased rates of predation on juveniles from hatchery stocks may also be a factor (Walters, 1988). No time series data on juvenile survival are available from previous El Niño periods in the northeast Pacific.

Diversion of migratory route – Fraser River sockeye

Under normal conditions, the majority of sockeye salmon returning to the Fraser River enter the Strait of Juan de Fuca where they are first intercepted by US fishers. They then turn to the northeast into Canadian waters where they become available to Canadian fishers. A small percentage of the run usually passes around the northern end of Vancouver Island through Johnstone Strait, migrating southward to the Fraser estuary, and remaining entirely in Canadian waters (see Fig. 4.6).

During the El Niño event of 1957–58, there was a major increase in the diversion of Fraser River sockeye through Johnstone Strait with approximately 35 percent of the run following that route (IPSFC, 1959). During the 1982–83 El Niño event, the diversion reached a record high, with over 80 percent of the sockeye run using the northern migration route through Johnstone Strait. The diversion rate appears to have no discernable effect on the abundance of Fraser River sockeye, but it makes a large difference in terms of which nation's fishers find the salmon most accessible.

As long as this diversion has been recognized, efforts have been made to understand its mechanisms and to predict its occurrence (Barber, 1983). Speculations on the mechanisms include temperature, salinity, olfactory cues, and sea level. Recent work points to temperature anomalies off northern Vancouver Island as being an important factor (Hamilton, 1985). Warm years in the area northeast of Vancouver Island generally result in higher diversion rates than cool years. Quantitative forecasts made using a diversion model based on temperature appear to predict this phenomenon fairly well (Xie & Hsieh, 1989). Major El Niño periods seem to result in diversion rates higher than other locally warm years.

Other effects of El Niño events on salmon have been mentioned in the literature – for example, reduction in fecundity, timing of runs, condition of the salmon, etc. – but these are not as well studied as those noted above (PFMC, 1984; Pearcy & Schoener, 1987). Thus, much remains unknown about the biological effects of El Niño on salmon in the northeastern Pacific. In summary, El Niño years can be said to provide a limited but useful set of insights on possible sources of variation in salmon catches but considerably more work is needed to advance this line of inquiry.

Long-term variations in ocean environment and salmon abundance

If El Niño events provide limited "experimental" evidence of the influence of ocean environment, are there other extremes that can provide insights? Some clues may be provided by the period of generally high harvests of coho in Washington, Oregon, and California and of chinook in British Columbia, Washington, and Oregon in the mid-1970s, as well as by the record high total salmon catches in Alaska in the early to mid-1980s. This period coincided with decreased harvests of coho and chinook to the south. Generally, it is thought that the cool temperatures and strong upwelling that were experienced off Washington and Oregon in the mid-1970s provided favorable conditions for salmon survival and growth. In contrast, anomalously warm water in the northeastern Pacific is thought to have contributed to favorable conditions for salmon survival off Alaska in the early 1980s (Rogers, 1984).

Fairly long time series of sea surface temperature measurements (averaged) are available for the northeastern Pacific (Fig. 4.7).

While these long-term temperature records confirm the cooling off the coasts of Oregon and Washington, and the warming off Alaska, year-to-year variations in temperatures do not appear to correspond closely with variations in harvests, and therefore do not explain all of the variability in catches. This is to be expected because of the multiyear life-cycle of salmon. Thus, the underlying patterns of variation are only partially understood.

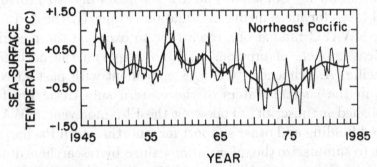

Fig. 4.7 Time series of sea surface temperature averaged over the northeastern quadrant of the North Pacific Ocean. Thin line = 3-month running average; heavy line = 25-month running average (Chelton, 1984).

Research and management response to the 1982–83 El Niño

Scientific research

At the beginning of 1983, the prospect of a large-scale El Niño event was seen as both an opportunity and a problem by the scientific community. On the one hand, given the rarity of the impacts of El Niño events manifested in the northeastern Pacific, it was felt that every effort should be made to learn as much as possible from it. On the other hand, scientific research agendas had already been set and it would not have been reasonable to abandon planned scientific activities to follow events during the El Niño period. In addition, there were basic problems of funding availability and the lack of a sufficient base of monitoring data on fisheries oceanography with which to compare the El Niño event to normal conditions. Finally, relatively little was known about the impacts of previous El Niño events in the northeastern Pacific – especially from the coast of Oregon northward. The most comprehensive documentation of the impacts in the region of an El Niño event came from the proceedings of a conference following the 1957–58 El Niño event (CalCOFI, 1960). It was therefore difficult to mount even a modest scientific monitoring effort of the event in 1982–83.

The Northwest and Alaska Fisheries Center (NWAFC), under the National Marine Fisheries Service, allocated funds for an El Niño Task Force to coordinate the timely acquisition and exchange of information by Canadian and US scientists and to provide a central focus for mass media information (Fluharty, 1984). In addition, NWAFC funded one researcher to examine the record of biological impacts of previous El Niño events in the northeastern Pacific (A. Schoener, "Changes coincident with past El Niño events in the coastal waters of the eastern sub-Arctic Pacific," unpublished manuscript). Following the El Niño event, NWAFC provided funding and other support for a northeastern Pacific conference to summarize the information gained by researchers during this period (Wooster & Fluharty, 1985). Coincidentally, a major workshop on the influence of ocean conditions on the production of salmonids in the North Pacific was jointly sponsored by the Cooperative Institute for Marine Resources Studies and the Sea Grant Program at Oregon State University (Pearcy, 1984).

The research effort coordinated through the Task Force funded by NWAFC was quickly disbanded as the oceanographic signal of El Niño diminished. Some of the scientists went on with other research activities, giving relatively little new emphasis to long-term studies of the effects of the ocean environment on fisheries. However, for the few long-term research and monitoring efforts on salmon, the El Niño period is now starting to yield considerable insights (see Nickelson, 1986; Fisher & Pearcy, 1988; Pearcy, 1991).

Salmon fishery management response

Given that the impacts of El Niño events are very infrequent and difficult to predict along the northwest coast of North America, fisheries resource managers do not regularly plan for El Niño effects. Prior to the 1982–83 El Niño event, the perception by most salmon managers of El Niño events and of their reputed effects were colored by vague recollection of low salmon years off the US Pacific Northwest during and following the 1957–58 El Niño. Institutional memory of such events is weak given the rapid turnover of personnel, the lack of mechanisms to record agency experience, and given the quarter of a century that had passed since the area had been most recently affected by a previous El Niño (R. Hilborn, "Institutional memory of fisheries agencies," unpublished lecture notes, January 1989). Therefore, no explicit contingency planning for El Niño events was drawn up as part of the management process and no guidelines were carried over in administrative protocols (Hilborn, 1987).

In the remainder of this section, focus is generally restricted to examination of: (1) management under the Pacific Fisheries Management Council (PFMC) which has jurisdiction over the salmon fisheries off the Washington, Oregon, and California coasts outside the three-mile limit of state jurisdiction, and (2) management under the International Pacific Salmon Fisheries Commission (IPSFC) established by treaty between Canada and the US and having, among other things, principal responsibility for managing the Fraser River sockeye fishery.

The Pacific Fishery Management Council

The PMFC has had an annual salmon management plan since 1978. This plan is normally developed by a team of biologists and other disciplinary specialists, including individuals from a variety of management perspectives. It is based on an assessment of stocks presented to the Council by the plan development team late in the year prior to the season for which its recommendations apply. Before its adoption by the Council, generally in late spring, the plan undergoes a period of comment and review.

Late in 1982, at the time the plan development team was gathering information and starting its deliberations, the first evidence of the impacts of the El Niño event was noted off Oregon (Huyer & Smith, 1985). By February of 1983 there was conclusive evidence that the El Niño event was affecting the PFMC management area. The plan development team did not take this information into consideration in its presentation to the Council, and it appears that no discussion of the possible impact of the El Niño event was held by the PMFC. Thus, no adjustments were made in the 1983 salmon management plan for possible El Niño effects and no provisions were made for in-season adjustments (PFMC, 1983). Procedural difficulties in plan development, unrelated to the El Niño, meant that the 1983 Council plan had to be implemented under emergency regulations (again, unrelated to El Niño). The plan itself was not finalized until mid-September – long after the bulk of the salmon fishing season had passed.

With the advantage of hindsight, it is possible to question why the Council did not devote any attention to the El Niño event and its effect on the plan. Perhaps the PFMC did not act because the strength of the El Niño event could not be realistically estimated at the time – would it be small and disappear or would it increase in intensity? However, even if the Council had wished to respond, it lacked the required scientific basis for the adoption of measures to adjust harvest rates, locations, seasons, or gear types. In addition, the procedural difficulties of hearings and advance notification may have prevented revisions of the plan. The opportunity still existed to use emergency regulations to adjust for El Niño effects if these were deemed necessary based on the regular in-season monitoring of the fishery but, even here, no actions were taken. As a result of the small run and the Council's inability to make appropriate management adjustments, 1983 escapements to many spawning areas were well below targeted levels.

Actual 1983 harvests and escapements demonstrated that ocean conditions affecting the survival, catchability and condition of salmonids in the northeastern Pacific differed significantly from those of "normal" years. For coho and chinook stocks, the PFMC's pre-season stock assessments considerably overestimated the abundance of actual runs, in virtually every instance (Table 4.2) (Hayes & Henry, 1985). It should be noted that the North Pacific Fishery Management Council pre-season estimates for Alaskan stocks erred in the opposite direction by considerably underestimating the record returns of many stocks.

Given the problems with the management of harvest and escapement in the 1983 season, it was decided that the 1984 plan should incorporate an "El Niño adjustment factor" to take into account the presumed higher mortality of salmon stocks due to the anomalous oceanic conditions (PFMC, 1984). As part of the development of the 1984 management plan (Hayes & Henry, 1985), the normal stock predictors for 1984 were reduced by an adjustment factor of approximately 30 percent.

In the light of experience, the adjustment was perhaps too conservative as it underestimated the actual run size in the OPI area

Table 4.2

Estimated adjustments for anticipated impacts of El Niño to 1984 stock abundance forecasts by area and stock

Area/stock	1983 Abundance estimates (× 1000 fish)		El Niño adjustment		1984 Abundance estimates (× 1000 fish)		
	Pre-season	Post-season	Ratio	factor	Forecasts Unadjusted	Adjusted	Post-season
Coho							
OPI	1553.6	657.9	0.42	0.69	806.6	556.6	658.7
Private Hatcheries	n/a	n/a	n/a	0.71	119.0	84.0	119.5
Willapa	70.0	27.8	0.40	0.61	52.3	31.9	88.6
Grays Harbor	103.3	56.3	0.55	0.72	56.4	40.6	n/a
Washington Coastal	40.8	32.7	0.80	0.85	44.4	37.7[a]	72.8
Puget Sound	1213.7	1154.2	0.95[b]	0.89	1187.8	1064.8	n/a
Chinook							
California							
Central Valley	756.6	350.6	0.46	0.72	651.9	469.4	504.3
Klamath	70.1	57.9	0.83	0.93	55.0	51.0	43.3
Columbia River							
Lower River							
Natural	26.4	18.3	0.69	1.00[c]	n/a	16.7	12.9
Hatchery	162.5	86.4	0.53	0.50[c]	n/a	69.6	81.5
Upriver Brights	77.8	81.5	1.05	1.00[c]	n/a	90.1	159.1
Bonneville Pool							
Hatchery	94.2	30.8	0.33	0.50[c]	n/a	21.3	34.9

[a] Grouped data: team analysis by five units
[b] Grouped data: adjusted for absence of Area 20 fishery
[c] Calculated by Columbia River Management Group
Source: Hayes and Henry (1985)

by approximately 20 percent. However, the unadjusted prediction would have overestimated the OPI run by a similar amount (Hayes & Henry, 1985). Regulations based on these predictions resulted in sharp reductions in commercial troll harvests and ocean sport harvests in Oregon and Washington, while the coho and chinook harvests of other commercial gears (the fisheries inside the three-mile limit and not under PFMC jurisdiction, including the Columbia River gillnet fishery), increased.

International management of Fraser River sockeye

Formal US–Canadian cooperation in the management of Fraser River sockeye salmon dates from the ratification of the International Pacific Salmon Fisheries Treaty in 1937 (IPSFC, 1939). A major management objective has been an equal division of the annual Fraser River sockeye harvest from Convention waters (i.e., the Strait of Juan de Fuca and southern Strait of Georgia) between US and Canadian harvesters. Another important objective has been assuring adequate escapement of each racial stock. When these management objectives were initially set, apparently little attention was given to the potential complication of large interannual changes in the Johnstone Strait diversion.

In most years, diversion rates ranged between 10 and 25 percent. An unusually high rate of diversion (35%) occurred during the 1957–58 El Niño. This led the IPSFC to make emergency amendments to its management measures for the 1958 season, including closings in Canadian waters and extra openings in US waters (IPSFC, 1959). Emergency regulations have been used during other high diversion years, such as in 1983 when the 1982–83 El Niño resulted in a record high diversion through the Johnstone Strait, exceeding 80 percent. High rates of diversion also occurred in 1972 (34%), 1978 (53%), 1980 (70%), and 1981 (67%) (Hamilton, 1985).

Over the years, efforts to balance the harvest between the two national fisheries have been relatively successful, given the occasional use of emergency regulations. However, this success began to erode as the Canadian troll fisheries expanded outside Convention areas, reducing the effective control of the IPSFC. In addition, both Canada and the US (Washington State) have had difficulty achieving desired allocations for native American bands (tribes) and among gear types. The inability of salmon managers to accurately forecast changing diversion rates during the late 1970s and early 1980s, meant that goals for the equitable sharing of the harvest were not achieved even with the use of emergency in-season adjustments. During 1983, for example, the US share of the harvest fell to 39 percent, its lowest level since 1947. In 1985–87, the US share was also well below 50 percent (Washington State Department of Fisheries, 1987).

In 1985, the IPFSC was replaced by the Pacific Salmon Commission (PSC) as a result of a new Convention between Canada and the US on the conservation, management, and optimum production of Pacific salmon (PSC, 1986). Management under the PSC has become increasingly complex. It involves pre-season estimates of stocks, and negotiation of allocations and escapement goals as well as in-season management involving monitoring and adjustment of fishing schedules to meet objectives for catches, escapement, allocations, and estimates of racial composition of the salmon stocks.

This type of management incorporates considerable flexibility which promotes the protection of the resource despite its inherent variability. However, this arrangement is also very costly. A "payback" policy instituted in 1988 calls for compensation during the following season of harvests diverging from an agreed division of the catch between the two nations. This may reduce the future cost of monitoring and of in-season management adjustments. The cost of management may also be reduced if variations in the Johnstone Strait diversion rate become more predictable in the future. Recent research suggests some progress toward that goal (Hamilton, 1985; Xie & Hsieh, 1989).

Sorting out the socioeconomic impacts of the 1982–83 El Niño

Just as it is difficult to sort out the biological effects of the 1982–83 El Niño along the Pacific Northwest coast, so it is difficult to determine the socioeconomic impacts. A cursory analysis, focusing only on the fortunes of commercial salmon trollers and charter-boat operators along the Washington, Oregon, and California coasts might conclude that the El Niño had nothing but deleterious impacts. Yet, Alaskan salmon harvesters and con-

sumers of salmon fared quite well during this period. While the
specific effect of the El Niño on the Alaskan harvest, as opposed to
other factors, is an unresolved question, it is clear that a different
picture is obtained by taking a broad rather than a narrow view of
the northeastern Pacific salmon fishery during the El Niño period.

Impacts on commercial harvesters

From the harvest figures presented in Table 4.3, it can be readily
inferred that the impacts of El Niño were highly uneven. While
Alaskan harvests continued to increase, harvesters targeting the
stocks of coho and chinook along the coasts of Washington, Oregon, and California were hard hit. In Washington, northern Puget
Sound purse seiners and gillnetters were also hurt by the reduced
availability of Fraser River sockeye.

For commercial trollers from Washington southward, 1983 was
a very bad year. Not only were harvests of their primary target
species (coho and chinook) down sharply from already depressed
levels, but the prices that they received for the harvest fell to their
lowest level in many years. The combination of a small harvest
and low prices caused the real ex-vessel value of the troll harvest
to plummet in all three states (see Table 4.3). While conditions
improved the following year in California, for trollers in Washington and Oregon, 1984 conditions were much worse. This was
the case despite the fact that, in both states, the harvests of coho
and chinook by other commercial gears increased slightly in 1984.
The change in the 1984 distribution of the harvest between trollers
and other commercial gears (see Figs. 4.8 and 4.9) can be traced
to very restrictive ocean fishing regulations put in place by the
PFMC in an effort to prevent overharvesting of stocks weakened
by the 1982–83 El Niño, one of the biggest in a century.

The fact that both prices and output could fall simultaneously
is a result of the highly competitive, international nature of the
market for salmon. Prices are determined by world supply and
demand, and are influenced by fluctuations in currency exchange
rates. There is also a high degree of substitutability between
salmon species, and between Alaskan and west coast salmon, particularly in the growing market for fresh and frozen salmon (Department of Agricultural and Resource Economics, Oregon State
University, 1978). In addition, chinook and coho harvested along

Table 4.3

Commercial troll salmon harvest, real value of harvest, and number of vessels

	Washington			Oregon			California		
Year	Total catch[a] (troll)	Real value[d] of harvest[b]	Number of vessels[c]	Total catch[a] (troll)	Real value[d] of harvest[b]	Number of vessels[c]	Total catch[a] (troll)	Real value[d] of harvest[b]	Number of vessels[c]
1979	7,786	23,430	2,778	7,273	26,377	3,114	8,746	30,523	4,593
1980	3,409	10,071	2,626	4,294	11,587	3,875	6,019	18,615	4,738
1981	3,591	7,657	2,439	5,216	12,380	3,615	6,042	18,522	4,102
1982	2,674	8,190	2,253	5,057	12,042	3,269	7,996	23,718	4,013
1983	936	1,721	2,045	1,753	2,698	2,951	2,410	5,414	3,223
1984	201	462	381	621	1,799	771	2,970	8,529	2,569
1985	962	1,748	1,259	2,773	6,302	2,050	4,630	12,590	2,308
1986	658	1,249	1,252	5,276	8,454	2,288	7,598	16,062	2,582
1987	763	2,030	883	7,142	17,362	2,111	9,296	26,539	2,442
1988*	796	2,337	650	7,622	21,536	2,053	14,671	41,629	2,562

[a]Thousand pounds dressed weight
[b]Ex-vessel 1988 US$ in thousands
[c]Number of vessels landing salmon
[d]Does not include pink salmon
*Preliminary
Source: Pacific Fishery Management Council (1989)

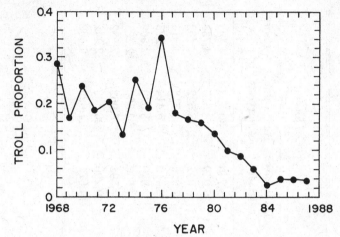

Fig. 4.8 Troll as a proportion of Washington commercial salmon harvest.
(Data sources: INPFC, 1952–85; Pacific Marine Fisheries Commission, unpublished data [PACFIN data set, 1986–87].)

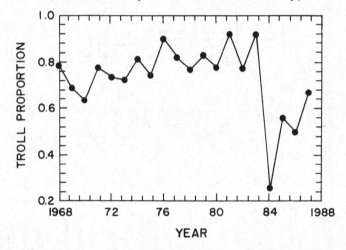

Fig. 4.9 Troll as a proportion of Oregon commercial salmon harvest. (Data sources: INPFC, 1952–85; Pacific Marine Fisheries Commission, unpublished data [PACFIN data set, 1986–87].)

the Washington, Oregon, and California coasts constitute only a small share of the North American commercial salmon harvest (approximately 3% in terms of weight in 1985, and 11% in 1976 at the peak of this fishery and prior to the recent explosive growth of the Alaskan harvest). There also appears to be a tendency for countervailing movements in the harvest of salmon at other locations.*

* For total salmon harvest the Pearson correlation coefficent between Alaskan and Oregon harvests = −0.315; Alaskan and California harvests = −0.584. For

For these reasons, variations in the harvest of coho or chinook in Washington, Oregon, or California, such as those occasionally associated with the El Niño, appear to have virtually no influence on the average prices received by the harvesters.†

Since the prices received by west coast trollers are not influenced by fluctuations in local harvests, the harvesters are vulnerable to declines in externally determined prices occurring in conjunction with poor harvests. This is exactly what happened in 1983. Measured in constant 1988 US dollars, the real ex-vessel price of coho harvested in the Oregon troll fishery fell to its lowest level since 1971, while the price for chinook was the lowest since 1975 (PFMC, 1989). The low prices in Oregon may have been partly a result of the small size and poor condition of the salmon harvested there, but the US average prices of coho, chinook, and of the total salmon harvest were also unusually low that year (National Marine Fisheries Service, 1988, 1989). However, the problem of depressed earnings in the coastal salmon troll fishery did not begin in 1983.

Participation rates and depressed earnings

As the total harvest of coho and chinook in these three states expanded over the course of the 1960s and early 1970s, so did participation in the troll fishery. This unrestricted increase in effort resulted in declining earnings per vessel despite large harvests and increasing real salmon prices. In California, for example, the number of commercial salmon trollers peaked in 1978 at 4,919 (PFMC, 1989). At that time, the size of the fleet had more than doubled since 1972, when 2,392 troll vessels landed commercial salmon in California. In 1960, there had been only 1,365 vessels in the fleet. When California's total commercial salmon harvest reached its all-time peak in 1976, the number of pounds landed was double that

coho: Alaskan and Oregon harvests = −0.348; Alaskan and California harvests = −0.389. All of these coefficients are significant at the 95% level of confidence or better.

† Regression analysis performed by authors. It was found, for example, that variations in real ex-vessel coho prices received by Oregon trollers bore no statistically significant relationship to variations in either Oregon's coho harvest or to coho harvests elsewhere along the coast. This suggests that other products, including other salmon species, are viewed by the market as good substitutes for coho.

landed in 1960. Even so, and even though the real average price
of the fish had increased by US$.64/lb – US$1.41/kg (in 1988 US
dollars), real gross annual earnings per vessel had dropped from
US$9,048 to US$5,829. The declining coho runs and the collapse
of the chinook run in 1983 caused real gross earnings per vessel to
drop further to US$1,680. A large number of vessels dropped out
of the California troll fishery that year (see Table 4.3), and there
were many foreclosures on mortgaged vessels (Bathgate, 1984). In
the wake of the El Niño, active participation in California's commercial troll fishery appears to have been permanently reduced.

In Washington and Oregon, the troll harvest peaked both in
terms of numbers of salmon and real value in 1976. As in California, participation in the troll fishery expanded in response to the
high total value of the harvest. After 1979, the earnings of Washington and Oregon trollers decreased rapidly due to a combination
of falling prices and reduced harvests. When the effects of the El
Niño were felt in 1983, they had already experienced three straight
years of relatively disappointing earnings. Recent entrants to the
industry who had borrowed money to acquire their vessels may,
therefore, already have been in a vulnerable financial position.

Concern about "overcapitalization" and resulting low harvester
income was a major driving force behind the establishment of
limited-entry programs in all of the coastal states. The goal of pro-
moting more effective biological management may also have been
a very important driving force. During the late 1970s, the Pacific
Fishery Management Council issued a recommendation urging all
Pacific Coast states to limit entry into the salmon fishery (Huppert
& Odemar, 1986).

Washington's limited-entry program had the additional goal of
assisting the state in enforcing the terms of the 1974 Boldt decision, which guaranteed Indian treaty rights to take half the salmon
that would have naturally returned to the traditional Indian fishing grounds in the absence of interception by other harvesters.
Since trollers are generally fishing on mixed stocks and "in-front"
of the traditional Indian fishing gears and areas, trolling was targeted early for reduction as part of the effort to comply with the
Boldt decree. A moratorium on new troll licenses was instituted
in 1976, followed by increasingly stringent gear restrictions and a
vessel buy-back program, that ended in 1986 (Jelvik, 1986).

The division of the salmon harvest between Washington's Indian and non-Indian commercial harvesters seems to have been
worked out just immediately prior to the El Niño. In terms of
total numbers of salmon, the Indian harvest first reached parity
with the non-Indian commercial harvest in 1980, and from 1982
onward, the division appears to be virtually 50–50 (see Fig. 4.10).

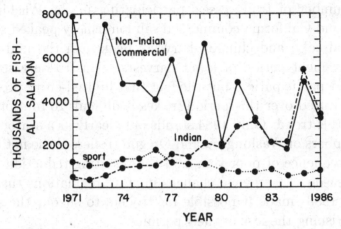

Fig. 4.10 Washington salmon harvest by user group. (Data source: Washington State Department of Fisheries, 1987.)

After peaking in 1976, the share of Washington's salmon harvest
taken in the troll fishery declined steadily to a low in 1984 (see
Fig. 4.8), when the combined effects of the El Niño and of ongoing
efforts to reduce the troll fishery resulted in a troll harvest of only
97,000 salmon (down from more than 1.7 million salmon in 1976).
It is somewhat ironic that coho and chinook were the species most
seriously affected by the El Niño, since they are the most important to the already stressed troll fishery. For Washington trollers,
the distress caused by the El Niño was apparently compounded by
ongoing regulatory changes.

In California, a moratorium on salmon troll permits went into
effect in 1980 followed by a limited-entry program in 1982. This
program allowed a large number of historically active vessels to remain licensed and included no buy-back or other directed fleet reduction program. Therefore, the subsequent reductions in fleet size
can be attributed to the effects of reduced run sizes, low salmon
prices, and increasingly restrictive fishing regulations (Huppert &
Odemar, 1986).

In Oregon, participation in the troll fishery peaked in 1980 when,
according to the PFMC (1989), "The establishment of a restricted

vessel permit system drew a number of historically active vessels back into the fishery...". As in California and Washington, the number of active salmon trollers in Oregon has been much smaller in the post-El Niño period than just prior to this event (see Fig. 4.9 for the proportion of the Oregon commercial harvest taken by trollers).

The number of troll vessels participating in the Washington, Oregon, and California commercial salmon fishery peaked shortly after total coho and chinook harvests peaked in the mid-1970s. The subsequent period of lower harvests seems to be associated with lower participation, but the fact that limited-entry programs were instituted over this period makes it difficult to interpret the cause of this trend. Only 1984 stands out clearly as a season when large numbers of Washington and Oregon trollers chose not to fish as a consequence of poor runs and the resulting reduction in the fishing season. Emergency changes in state regulations that year (PFMC, 1989) made it possible for trollers to sit out the season without risking the loss of their permits.

The movement of vessels into and out of the coastal salmon troll fishery in response to fluctuating returns can be seen as a socioeconomic response to the inherent variability of this fishery. This response has perhaps been facilitated by the fact that the capital requirements for commercial trolling are quite modest, and since most of the coastal spawning runs occur in the summer, trollers have traditionally included many part-time participants who pursue other occupations as their primary source of income. Those trollers who derive most of their income from fishing may also fish for crab, tuna, or bottom-fish.

The available data on participation indicate that the coastwide reduction in the number of troll vessels over the past decade cuts fairly evenly across all vessel size classes (PFMC, 1989). This suggests that owners of small vessels, many of whom may have been part-timers, were no more likely to be "marginal," in an economic sense, than the owners of larger vessels.

Fisherman relief – declaration of disaster

Although the total coastwide commercial salmon harvest was at near record levels in 1983 and 1984, from the narrower perspective of salmon fishing interests in Washington, Oregon and California, those years were viewed as disastrous. As has been previously shown, the El Niño-related reductions in salmon availability were not the sole cause of the distress experienced by salmon harvesters and related enterprises. Nevertheless, the possibility of obtaining federal disaster assistance from the US Small Business Administration's (SBA's) Economic Injury Disaster Loan Program led these interests to emphasize El Niño as the primary cause of their distress and to call on the governors of their states to declare the event a natural disaster.

Governors of the Pacific coast states made disaster declarations and requested federal assistance in late 1983. Washington Governor Spellman's declaration stated that: "Economic hardship has been severe among the salmon and crab fishermen and processors, but a broad range of marine related businesses such as equipment suppliers, shipyards, repair shops, retail stores and hotels also face economic problems due to the devastated fishing industry which normally supports their business" (letter from Governor Spellman of Washington to Regional Administrator, Small Business Administration, S.J. Hall, 23 November 1983).

SBA recognized that salmon harvests were substantially below average. However, the SBA initially denied the designations of emergency, arguing that: "The atmospheric or weather condition known as "El Niño" does not of itself constitute a physical disaster" (J.C. Sanders, Administrator of Disaster Assistance Division of the Small Business Administration, telegraphic message to Governor Spellman, 8 December 1983). El Niño is an "atmospheric condition ... not conducive to the operations of a certain type of business," but the possibility of such conditions "constitute[s] part of the risk of being in business" (*Rocky Mountain News*, 1984).

Fishing interests were dissatisfied with this response, and questioned why farmers were eligible for disaster assistance during droughts but not fishing interests affected by a similar short-term condition. Pointing to the fact that the US Agency for International Development had provided approximately US$100 million in disaster aid to South American countries affected by El Niño, they turned to their Congressional delegations for assistance.

Under continued pressure from Congress, the SBA reversed its decision on the disaster declaration in late September 1984, making low interest loans (4% for up to US$500,000) available to fishing-related businesses (*The Seattle Times*, 1984b). SBA was thus forced to provide assistance despite the agency's serious mis-

givings about the financial health of the industry and the ability
of the borrowers to repay the loans even under normal conditions.
The agency was also concerned about setting a precedent of becoming a banker for weather-sensitive businesses. For SBA, the
losses were part of doing business in fisheries, analogous to losses
incurred by a ski resort operator without snow or a beach resort
with no sun. Some years are bad, but others may be very good.

The program loaned approximately US$28 million, in a total of
319 loans, to west coast fishing firms. However, given the delay
in availability of funds, many individuals had already defaulted
on loans and lost vessels and collateral and had suffered other
personal damages (*The Seattle Times*, 1984a,b). Although salmon
harvesters had been instrumental in gaining implementation of
the program, its biggest beneficiaries were Bering Sea crab fishing
firms.

The final decision to provide assistance was not the result of
an objective analysis of the situation. Rather, it was the result of
pressure from local politicians responding to a powerful and highly
visible natural-resource-based interest group.

In retrospect, this episode provides a microcosm of issues surrounding efforts to come to grips with global climate change issues.
There is enormous uncertainty surrounding the assessment of climate impacts, particularly in the case of a fishery. SBA was not
prepared to perform such an assessment, since the agency was un-
accustomed to dealing with climate-related biological impacts on
fisheries and did not have a staff of oceanographers and biologists
to provide technical advice. As should be clear from the previous discussion, it would be extremely difficult and probably pro-
hibitively costly to determine accurately the level of the biological
impacts of an event like an El Niño and to further determine the
socioeconomic effects directly and solely attributable to those impacts. This suggests that governmental decisions to allocate public
funds for assistance are likely to continue to be made largely on a
political basis.

Impacts on other interests

Consumers of salmon were largely unaffected by the reduced
harvests of coho and chinook in Washington, Oregon, and California during 1983 and 1984. The Alaskan salmon industry had been

moving strongly away from canning and into the fresh and frozen market for several years before the El Niño event (Department of Agricultural and Resource Economics, Oregon State University, 1978). This fact, coupled with the rapid growth in Alaskan harvests simply overwhelmed the effects of reduced coastal coho and chinook harvests in the retail market for fresh and frozen salmon. There may, however, have been some adverse effects on consumers in certain sub-markets. For example, large troll-caught chinook and coho (as opposed to net-caught) are especially desired for the production of smoked salmon. The reduced availability of these particular salmon may therefore have affected consumers in that market.

While processors were reported to be adversely affected by the El Niño, their operations are often well designed to deal with fluctuating harvests. Many deal with the problem of biological variability by diversifying across a wide variety of species. As seasons may be short for any given species, diversification allows a more complete utilization of processing capital. Processors may also structure their buying arrangements to promote harvester loyalty and insure a more even utilization of processing capacity. For example, season-end bonus payments are commonly used, and a purchase agreement for salmon may also bind the harvester to sell that year's crab harvest to the same processor.

The salmon sport fishery was also affected by the El Niño and resulting changes in fishing regulations, particularly in 1984 in Washington. The Washington sport harvests of coho and chinook actually increased slightly in 1983, relative to 1982, but then dipped in 1984 when season restrictions caused a sharp drop in the number of charter-boat trips and, therefore, in revenues for those businesses.

Reduced harvester incomes and poor salmon sport fishing seasons undoubtedly had adverse economic impacts on a number of coastal communities. The magnitude of those impacts would, however, be very difficult to determine, particularly since this was a period of general economic recession, with high unemployment rates also in the locally important lumber industry.

Conclusions and lessons for global warming

Several conclusions can be drawn from this case study that shed light on the potential biological and socioeconomic impacts of global climate change. First, this case study demonstrates that both the biological and socioeconomic impacts of climatic variability are very complex and, at present, poorly understood. Salmon stocks in the northeastern Pacific appear to be affected by both short-term ocean–climate events such as El Niño and by longer-term variations in patterns of sea surface temperatures, upwelling, and primary productivity. However, little is known about the exact nature of the impacts of the ocean environment on salmon. Salmon abundance may also be affected by changes in runoff patterns in spawning streams, and the effects of changes in ocean circulation patterns and salinity levels are unknown. The interactions between climate-related ocean conditions and salmon abundance are so complex that there has been a tendency to view the ocean as a "black box." However, there has been recent research progress in this area. Although some observers debate the value of additional research and of improved monitoring of oceanic and other environmental conditions for annual management purposes (Walters & Collie, 1988), such efforts may improve our understanding of the effects of long-term climate changes.

The socioeconomic effects of variations in salmon abundance are also complex. The impacts in any given region are intertwined with the effects of changing salmon abundance elsewhere, as well as with the effects of changing market conditions and regulatory programs.

It has been shown that the apparent biological effects of the 1982–83 El Niño were highly uneven, and that this unevenness exacerbated the impacts of the event on those harvesters relying on adversely affected stocks. This suggests that climate change will, likewise, have uneven impacts. While individuals in one region may lose as a result of climate change, there may be winners elsewhere. It is easy to misconstrue the net socioeconomic effects of climatic events or of projected climate changes if one focuses only on adversely affected regions, or if the market effects of expanding output in other regions are misunderstood.

This complexity suggests that, even as they occur, the impacts of global climate change on fisheries may be far from obvious. This should lead us to be wary of facile conclusions drawn about the anticipated impacts of global climate change. This can also be seen as an argument in favor of improving our base of research on climate impacts. We can have little hope of correctly anticipating the effects of potential climate changes without an improved understanding of the interactions between climate, biological processes, and the socioeconomic activities dependent on those processes.

The El Niño period described in this case study gives us a glimpse of the kinds of biological and socioeconomic effects that may accompany the increased frequency of previously unusual climatic conditions. A change in the relative frequency of various climatic extremes is one of the potential effects of CO_2/trace gas-induced climate change (Mearns et al., 1984).

With limited scientific and managerial resources, little attention is generally given to the potential effects of rare events until such an event occurs. At that time, both the scientific and managerial responses may, of necessity, be *ad hoc*. A conclusion that can be drawn from the El Niño experience is that the scientific understanding of the impacts of an El Niño event on salmon was so fragmentary and uncertain that there was little basis upon which to make stock predictions for management purposes. In 1983, this left salmon management largely unprepared for the reduced abundance and poor condition of Washington, Oregon, and California coho and chinook stocks and for the altered migration pattern of Fraser River sockeye. In 1984, the conservative strategy adopted by the Pacific Fisheries Management Council resulted in a severely diminished coastal troll fishery, while harvests improved for operators of other gears. This suggests that efforts to equitably balance in-season harvest allocations among competing gears are unlikely to succeed when managers are confronted with conditions outside their range of recent experience.

Since climate change may increase the frequency with which resource managers confront unusual climate-related conditions, this case suggests that it may be valuable to devote increased attention to anticipating the effects of currently unusual climate-related conditions and to planning response strategies. While it is certainly possible to "muddle through" such periods, the El Niño experience suggests that reliance on this approach may have undesirable consequences such as an increased potential for resource damage or inequity in harvest allocation.

The longer-term effects of interannual variability in marine conditions on the salmon fishery and socioeconomic adaptations to this variability are also of interest to those concerned with the potential impacts of global warming. Variability will continue to be an inherent feature of climate, although ranges of variability may change.

In the case of the salmon fishery, harvesters, processors, managers, and markets have already adapted to an enormous amount of interannual variability. These adaptations to current variability imply that similar socioeconomic adjustments will be made to the effects of a changing climate. Since adaptations to current variability may provide the avenues for adjustments to climate change, it will be increasingly valuable to understand where they do and do not work well.

References

Barber, F.G. (1983). Inshore migration of adult Fraser sockeye, a "speculation." Canadian Technical Report of Fisheries and Aquatic Science No. 1162. Ottawa: Department of Fisheries and Oceans.

Barber, R. & Chavez, F. (1983). Biological consequences of El Niño. *Science*, **222**, 1203–10.

Bathgate, D.L. (1984). *Fishermen's Response to Reduced-Season Management in the Northern California Commercial Salmon Fishery*. Doctoral thesis. Boulder: Department of Anthropology, University of Colorado.

Bell, F.H. & Pruter, A.T. (1958). Climatic temperature changes and commercial yields of some marine fishes. *Journal of the Fisheries Research Board of Canada*, **15**, 625–83.

Bevan, D. (1988). Problems of managing mixed-stock salmon fisheries. In *Salmon Production, Management and Allocation*, ed. W.J. McNiel, pp. 103–7. Corvallis: Oregon State University Press.

CalCOFI (California Cooperative Oceanic Fisheries Investigations) (1960). *Reports, January 1958 to June 1959*. Vol. 7. Monterey: California Cooperative Oceanic Fisheries Investigations.

Cane, M. (1983). Oceanographic events during El Niño. *Science*, **222**, 1189–95.

Cannon, G.A., Reed, R.K. & Pullen, P.E. (1985). Comparison of El Niño events off the Pacific Northwest. In *El Niño North: El Niño Effect in the Eastern Subarctic Pacific Ocean*, ed. W. Wooster & D. Fluharty, pp. 75–84. Seattle: Washington Sea Grant Program, University of Washington.

Chelton, D.B. (1984). Commentary: Short-term climatic variability in the northeast Pacific Ocean. In *The Influence of Ocean Conditions on the Production of Salmonids in the North Pacific*, ed. W.G. Pearcy, pp. 87–99.

Corvallis: Cooperative Institute for Marine Resources Studies and Oregon State University Sea Grant College Program.

Cooley, R. (1963). *Politics and Conservation: The Decline of the Alaska Salmon.* New York: Harper and Row.

Crutchfield, J. & Pontecorvo, G. (1969). *The Pacific Salmon Fisheries: A Study of Irrational Conservation.* Resources for the Future. Baltimore: Johns Hopkins University Press.

Department of Agricultural and Resource Economics, Oregon State University (1978). *Socio-Economics of the Idaho, Washington, Oregon and California Coho and Chinook Salmon Industry.* Final report to the Pacific Fishery Management Council. Corvallis: Oregon State University Press.

Dodimead, A.J., Favorite, F. & Hirano, T. (1963). *Salmon of the North Pacific Ocean – Part II.* Bulletin No. 13. Vancouver: International North Pacific Fisheries Commission.

Favorite, F., Dodimead, A.J. & Nasu, K. (1976). *Oceanography of the Subarctic Pacific Region, 1960–1971.* Bulletin No. 33. Vancouver: International North Pacific Fisheries Commission.

Fiedler, P. (1984). Satellite observations of the 1982–1983 El Niño along the U.S. Pacific coast. *Science,* **224**, 1251–4.

Fisher, J.P. & Pearcy, W.G. (1988). Growth of juvenile coho salmon (*Oncorhynchus kisutch*) off Oregon and Washington, USA, in years of differing coastal upwelling. *Canadian Journal of Aquatic and Fishery Science,* **45**, 1036–44.

Fluharty, D. (1984). *1982–1983 El Niño Task Force Summary.* Seattle: Institute for Marine Studies, University of Washington.

French, R. (1977). *Incidence of Salmon in Japanese, Polish, and USSR Trawl Catches off California, Oregon, Washington and Southern British Columbia 1976.* Northwest and Alaska Fisheries Center Processed Report. Seattle: US Department of Commerce, National Oceanic and Atmospheric Administration, National Marine Fisheries Service.

Fredin, R., Major, R., Bakkala, R. & Tanonaka, G. (1977). *Pacific Salmon and the High Seas Salmon Fisheries of Japan.* Northwest and Alaska Fisheries Center Processed Report. Seattle: US Department of Commerce, National Oceanic and Atmospheric Administration, National Marine Fisheries Service.

Gross, M., Coleman, R. & McDowall, R. (1988). Aquatic productivity and the evolution of diadromous fish migration. *Science,* **239**, 1291–3.

Hamilton, K. (1985). A study of the variability of the return migration route of Fraser River sockeye salmon (*Oncorhynchus nerka*). *Canadian Journal of Zoology,* **63**, 1930–43.

Harris, C.K. (1988). Recent changes in the pattern of catch of North American salmonids by the Japanese high seas salmon fisheries. In *Salmon Production, Management, and Allocation,* ed. W.J. McNeil, pp. 41–65. Corvallis: Oregon State University Press.

Hayes, M. & Henry, K. (1985). Salmon management in response to the 1982–83 El Niño event. In *El Niño North: El Niño Effect in the Eastern Subarctic Pacific Ocean,* ed. W. Wooster & D. Fluharty, pp. 226–36. Seattle: Washington Sea Grant Program, University of Washington.

Hilborn, R. (1987). Living with uncertainty in resource management. *North American Journal of Fisheries Management*, **7**, 1–5.

Huppert, D. & Odemar, M. (1986). A review of California's limited entry programs. In *Fishery Access Control Programs Worldwide: Proceedings of the Workshop on Management Options for the North Pacific Longline Fisheries*, ed. N. Mollett, pp. 301–2. Alaska Sea Grant Report No. 86-4. Fairbanks: University of Alaska Press.

Huyer, A. & Smith, R. (1985). The signature of El Niño off Oregon, 1982–1983. *Journal of Geophysical Research*, **90**, C4, 7133–42.

INPFC (International North Pacific Fisheries Commission) (1979). *Historical Catch Statistics for Salmon of the North Pacific Ocean*. Bulletin No. 39. Vancouver: International North Pacific Fisheries Commission.

INPFC (1952–85). *Statistical Yearbook, 1952–1985*. Annual series. Vancouver: International North Pacific Fisheries Commission.

IPSFC (International Pacific Salmon Fisheries Commission) (1939). *Annual Report 1938*. New Westminster, Canada: IPSFC.

IPSFC (1959). *Annual Report 1958*. New Westminster: IPSFC.

Jackson, R. & Royce, W. (1987). *Ocean Forum*. Surrey, England: Fishing News Books Ltd.

Jelvik, M. (1986). Washington State's experience with limited entry. In *Fishery Access Control Programs Worldwide: Proceedings of the Workshop on Management Options for the North Pacific Longline Fisheries*, ed. N. Mollett, pp. 313–6. Alaska Sea Grant Report No. 86-4. Fairbanks: University of Alaska Press.

McLain, D.R. (1984). Coastal ocean warming in the Northeast Pacific. In *The Influence of Ocean Conditions on the Production of Salmonids in the North Pacific*, ed. W.G. Pearcy, pp. 61–86. Corvallis: Cooperative Institute for Marine Resources Studies and Oregon State University Sea Grant College Program.

Mearns, L., Katz, R. & Schneider, S. (1984). Extreme high-temperature events: changes in their probabilities with changes in mean temperature. *Journal of Climate and Applied Meteorology*, **23**, 1601–13.

Myers, K., Harris, C., Knudsen, C., Walker, R., Davis, N. & Rogers, D. (1987). Stock origins of chinook salmon in the area of the Japanese mothership salmon fishery. *North American Journal of Fisheries Management*, **7**, 459–74.

Namias, J. & Cayan, D. (1981). Large-scale air–sea interactions and short-period climatic fluctuations. *Science*, **214**, 869–76.

National Marine Fisheries Service, Fisheries Statistics Division (1988). *Fisheries of the United States 1987*. Current Fishery Statistics No. 8700. Washington, DC: US Government Printing Office.

National Marine Fisheries Service, Fisheries Statistics Division (1989). *Fisheries of the United States 1988*. Current Fishery Statistics No. 8800. Washington, DC: US Government Printing Office.

Netboy, A. (1980). *Salmon: The World's Most Harassed Fish*. Tulsa, OK: Winchester Press.

Nickelson T.E. (1986). Influences of upwelling, ocean temperature, and smolt abundance on marine survival of coho salmon (*Oncorhynchus kisutch*) in the Oregon production area. *Canadian Journal of Fisheries and Aquatic Sciences*, **42**, 527–35.

Nickelson, T.E. & Lichatowich, J.A. (1984). The influence of the marine environment on the interannual variation in coho salmon abundance: An overview. In *The Influence of Ocean Conditions on the Production of Salmonids in the North Pacific*, ed. W.G. Pearcy, pp. 24–36. Corvallis: Cooperative Institute for Marine Resources Studies and Oregon State University Sea Grant College Program.

Northwest Power Planning Council (1986). *Council Staff Compilation of Information on Salmon and Steelhead Losses in the Columbia River Basin*. Portland: Northwest Power Planning Council.

PFMC (Pacific Fishery Management Council) (1983). *Proposed Plan for Managing the 1983 Salmon Fisheries off the Coasts of California, Oregon, and Washington*. Portland: Pacific Fishery Management Council.

PFMC (1984). *A Review of the 1983 Ocean Salmon Fisheries and Status of Stocks and Management Goals for the 1984 Salmon Season off the Coasts of California, Oregon, and Washington*. Portland: Pacific Fishery Management Council.

PFMC (1989). *Review of the 1988 Ocean Salmon Fisheries*. Portland: Pacific Fishery Management Council.

PSC (Pacific Salmon Commission) (1986). *Annual Report 1986*. Vancouver: Pacific Salmon Commission.

Pearcy, W.G. (ed.) (1984). *The Influence of Ocean Conditions on the Production of Salmonids in the North Pacific*. Proceedings of workshop 8–10 November 1983, Newport, OR. ORESU-W-83-001. Corvallis: Cooperative Institute for Marine Resources Studies and Oregon State University Sea Grant College Program.

Pearcy, W.G. (1988). Factors affecting survival of coho salmon off Oregon and Washington. In *Salmon Production, Management, and Allocation*, ed. W.J. McNeil, pp. 67–73. Corvallis: Oregon State University Press.

Pearcy, W.G. (1991). *Ocean Ecology of North Pacific Salmonids*. Seattle: Washington Sea Grant Program.

Pearcy, W. & Schoener, A. (1987). Changes in the marine biota coincident with the 1982–1983 El Niño in the northeastern subarctic Pacific Ocean. *Journal of Geophysical Resarch*, **92**, C13, 14,417–8.

Philander, S. (1983). El Niño–Southern Oscillation phenomena. *Nature*, **302**, 295–301.

Quinn, T. & Groot, C. (1984). Pacific salmon (*Oncorhynchus*) migrations: orientation versus random movement. *Canadian Journal of Aquatic and Fisheries Science*, **41**, 1319–24.

Rasmusson, E. (1985). El Niño and variations in climate. *American Scientist*, **73**, 168–77.

Rasmusson, E. & Wallace, J. (1983). Meteorological aspects of the El Niño–Southern Oscillation. *Science*, **222**, 1195–202.

Rocky Mountain News (Denver, CO) (1984). Veto threatened of El Niño disaster aid. Tuesday, 5 June 1984, p. 33.

Rogers, D.E. (1984). Trends in abundance of northeastern Pacific stocks of salmon. In *The Influence of Ocean Conditions on the Production of*

Salmonids in the North Pacific, ed. W.G. Pearcy, pp. 100–27. Corvallis: Cooperative Institute for Marine Resources Studies and Oregon Sea Grant Program.

Royce, W.F. (1989). Managing Alaskan salmon fisheries for a prosperous future. *Fisheries*, **14**, 8–13.

Schelle, K. & Muse, B. (1986). Efficiency and distributional aspects of Alaska's limited entry program. In *Fishery Access Control Programs Worldwide: Proceedings of the Workshop on Management Options for the North Pacific Longline Fisheries*, ed. N. Mollett, pp. 317–52. Alaska Sea Grant Report No. 86-4. Fairbanks: University of Alaska Press.

Seattle Times (Seattle, WA) (1984a). El Niño victims struggle for aid. 24 March 1984, B1.

Seattle Times (Seattle, WA) (1984b). NW fishermen declared eligible for low-interest disaster loans. 28 September 1984, C5.

Tabata, T. (1984a). Anomalously warm water off the Pacific coast of Canada during the 1982–83 El Niño. *Tropical Ocean–Atmosphere Newsletter*, **24**, 7–9.

Tabata, T. (1984b). Oceanographic factors influencing the distribution, migration, and survival of salmonids in the northeastern Pacific Ocean: a review. In *The Influence of Ocean Conditions on the Productions of Salmonids in the North Pacific*, ed. W.G. Pearcy, pp 128–60. Corvallis: Oregon State University Sea Grant College Program.

United States v. Washington (1974). 384 F. Supp. 312 (W.D. Wash. 1974) (Boldt, J.), *affirmed*, 520 F. 2d 676 (9th Cir. 1975), *cert. denied*, 423 US 1086 (1976), discussed in *Washington v. Washington State Commercial Passenger Fishing Vessel Association*, 443 US 658, 1979.

Walters, C. (1988). Mixed-stock fisheries and the sustainability of enhancement production for chinook and coho salmon. In *Salmon Production, Management, and Allocation*, ed. W.J. McNiel, pp. 109–15. Corvallis: Oregon State University Press.

Walters, C., Hilborn, R., Peterman, R. & Stanley, M. (1978). Model for examining early ocean limitation of Pacific salmon production. *Journal of Fisheries Research Board of Canada*, **35**, 1303–15.

Walters, C.J. & Collie, J.S. (1988). Is research on environmental factors useful to fisheries management? *Canadian Journal of Fisheries and Aquatic Sciences*, **45**, 1848–54.

Washington State Department of Fisheries (1987). *1987 Fisheries Statistical Report*. Olympia: Department of Fisheries.

Wooster, W. & Fluharty, D. (eds.) (1985). *El Niño North: El Niño Effect in the Eastern Subarctic Pacific Ocean*. Seattle: Washington Sea Grant Program, University of Washington.

Xie, L. & Hsieh, W. (1989). Predicting the return migration routes of the Fraser River sockeye salmon (*Oncorhynchus nerka*). *Canadian Journal of Fisheries and Aquatic Sciences*, **46**, 1287–92.

Young, O. (1977). *Resource Management at the International Level: The Case of the North Pacific*. New York: Nichols Publishing Co.

5

The US Gulf shrimp fishery

RICHARD CONDREY and DEBORAH FULLER

Coastal Fisheries Institute
Louisiana State University
Baton Rouge, LA 70803, USA

Introduction

The US Gulf of Mexico shrimp fishery is one of the most diverse and valuable in the nation. Presently it is mainly dependent upon the harvest of three closely related, estuarine-dependent species: brown, white and pink shrimp (*Penaeus aztecus*, *P. setiferus*, and *P. duroraum*, respectively). The present-day fishery is a classic example of an open access fishery which has been allowed and, in some cases, encouraged to expand well beyond the point of maximum net economic return.

The fishery finds itself embroiled in a number of heated controversies especially over the incidental capture of sea turtles and finfish, with red snapper being the current example. Given the sheer size of the industry and the low marginal returns the average shrimper receives, it would be difficult enough for the industry to respond to these charges. Furthermore, recent massive imports of pond-raised shrimp, especially from China, have greatly eroded the shrimpers' already limited economic flexibility. Added to this is the possibility or likelihood of precipitous declines in yields associated with loss of productive estuarine habitats and the release into the marine environment of unspecified amounts of stored toxic wastes.

Nothing in the history of the fishery until the mid-1970s prepared the shrimpers to expect anything more than a larger cumulative harvest. During the past 300 years the fishery has undergone a mostly unplanned expansion with little or no regard for the future of the resource.

Today, a vocal component of the industry is embroiled in opposition to the required use of devices designed to release endangered sea turtles trapped in shrimp trawls: turtle excluder devices (TEDs). As emotional as the TED issue is, it is dwarfed by the implications of the unchecked, continued loss of habitat. While the TED controversy continues, and will likely rekindle with mounting concerns over finfish by-catch, regional management is beginning to deal with the finite, fragile nature of shrimp resources at a time when habitat degradation is anticipated by some to result in declines in yields which may be precipitous.

Background

Settlement in a rich but fragile system

The US Gulf shrimp industry has its origin in the seventeenth and eighteenth century colonization of the New World, and in the seafarers and trade practices of that time. The industry's history centers around the early development of New Orleans (Louisiana) and Biloxi (Mississippi) and the settlement of the surrounding cypress swamps, grassy marshes, and barrier islands. It is a period in which Europeans, Africans, and Asians settled among native Americans in a rich, wild, and fragile environment of tremendous productivity, beauty, and hardship. Buffalo, bear, panther, wolf, and parakeets were abundant as were "crabs, lobsters, scallops, shrimp, and oysters" (Dumont de Montiguy, 1753; Lowery, 1974a,b). The "fish on its shores [were] in such abundance that the noise they made at night, wakened us several times ... Grande Ecaille [tarpon], Red Drum, and very large Catfish, and some Gars" (Cathcart, 1819). Smallpox, yellow fever, floods, and hurricanes were frequent visitors, as were pirates and buccaneers.

The Mississippi River flowed wild and sweet when the Europeans first arrived. Its mouth was not confined to the present narrow bird-foot delta of some 25 miles (40 km). Rather it extended for more than 160 miles (250 km) along the Louisiana coast through a vast series of bayous and bays characterized by oak-lined natural levees (cheniers) (Iberville, 1699 in Brasseaux, 1979; Du Ru, 1700; Collot, 1826). The natural flow of the River into its rich marshes and bays was such that, when in 1785 Don Jose de Evia explored the lower reaches of Barataria Bay (a fingerling bay to the west of the Mississippi delta), he reported that "on the bay, which is a large one, one always encounters a strong current." That strong current, absent today, was the flow of the

Mississippi River through its vast delta (Stielow, 1975). With European settlement, a process of leveeing was begun by at least 1722 (Kniffen & Hilliard, 1988). As for the quality of these waters which drained the heartland of America, Stoddard (1812, p. 164) noted that "The people who live on the banks of the Mississippi prefer its waters to all others. When filtrated, it is transparent, light, soft, pleasant, and wholesome."

French and Spanish interest in the region was minimal, because the colony lacked the "golden plunder" obtained in Central and South America. While trade was officially limited to the sovereign nation or its commercial designee, smuggling was a socially accepted practice, which was partially condoned by the Spanish governors out of a necessity to obtain supplies. A quasi-official smuggling route occurred between British/US-held Baton Rouge and New Orleans, while the bayous, swamps, and marshes south of New Orleans provided the smuggler an endless array of hiding places (Saxon, 1940). This early stamp of self-reliance and at times an almost casual disregard for regulations imposed from outside the colony was to persist.

The region around New Orleans was settled by a rich and diverse ethnic mixture: first the French (Creole and, later, Acadian) settled among the native Americans. They were followed by "Spaniards, Italians, Irish, Chinese, Portuguese, Danes, Greeks, Swedes, and Eastern Europeans – peasants driven from their native lands by poverty or repression, sailors who had jumped ship, adventurers of every kind." Fishing, trapping, and abundant household gardens became a way of life in these isolated communities each of which "clung jealously to its own customs and way of life" (Crete, 1981, p. 283). This isolation was broken by periodic, if not frequent, trips to the urban centers to market their fresh catch or contraband, and sometimes to bury their dead.

A white shrimp fishery

Hearn (1883) provides one of the oldest written accounts of a commercial fishing village in the region. The village was located some 25 miles (40 km) from New Orleans on the southern shore of Lake Borgne. It had been founded more than 50 years earlier by Malay fishermen, some of whom had left their forced participation in the Spanish Crown's Mexico–Philippine trade route. Raised on stilts of cypress, the village of thirty residents supplied dried fish and shrimp to the New Orleans market (Kane, 1944). The practice of shrimp drying was expanded by others, and enriched by the addition of Chinese nationals, who exported large quantities of dried shrimp to China.

In 1810, Jean Lafitte, a young New Orleans blacksmith, joined a band of pirates or buccaneers who had settled in the existing communities on the barrier islands of Grand Terre and Grand Isle. Sailing under the flags of Central and South American nations, which were in revolt against Spain, Lafitte's band raided both Spanish and neutral ships for goods and slaves. Lafitte brought a harsh new dimension to the culture of these isolated communities, by some accounts taking no prisoners with the exception of African slaves. Lafitte held auctions in New Orleans and Grand Terre which were well attended by wealthy, respectable merchants and planters, eager for well discounted goods and slaves, especially given that the importation of new slaves was not allowed (Saxon, 1940).

The native self-sufficiency and fierce independence of those who inhabited this region (including Lafitte's band) were evidenced by the crucial role they played in helping to defeat the British at the Battle of New Orleans in 1814 (Frantz, 1937).

In 1875 the Dubois brothers refined the existing process of canning shrimp, and as a result this method, together with drying, became a way to export shrimp outside the local markets (Kniffen & Hilliard, 1988). In 1889, the first year for which complete estimated catch statistics are available, the Gulf shrimp catch was 8.3 million pounds (3.8×10^6 kg) with an average ex-vessel price of US$0.015/lb ($0.033/kg).

Under sail

Scientific inquiries into shrimp and shrimping did not begin until near the turn of the twentieth century. At that time, the fishery was still mainly limited to the estuaries and shallow bays surrounding New Orleans and Biloxi. Harvesting was accomplished by crews of up to 20 men, pulling seines which were sometimes in excess of 2,000 feet (600 m) long. Harvests were limited by water temperature to the warmer months of the year and by access to market. Vessels were large row boats, fitted with sails. Principal products were canned and dried shrimp. There were two seasonal closures in Louisiana, which had evidently been enacted as a result of industry's concerns. Both were intended to prevent the harvest of "small" white shrimp less than 4 inches (10 cm) in length (Louisiana Conservation Commission, 1920).

The fishery was predominately dependent on white shrimp and occasionally used a smaller "sea-bob" shrimp (*Xiphopenaeus kroyeri*) in the production of dried shrimp. The brown shrimp, which is currently a major portion of the US Gulf harvest, was "never as abundant as either the sea-bob or lake (white) shrimp, and consequently are almost negligible as a commercial proposition" (Tulian, 1920, p. 108). By 1908 the reported Gulf catch had increased by 50 percent (from 1889) to 12.6 million pounds (5.72 million kg) and the ex-vessel value had jumped to US$0.021/lb ($0.046/kg).

The fishery operated differently earlier in the twentieth century than it does today. Schools of white shrimp were hunted by the use of a cast net or by sightings of "white boils on the water surface" or "muddy boils" indicating feeding shrimp (Julius Collins, personal communication, 1990).

Frank Schoonover (1911) reached a shrimp and fish drying platform (Manila Village, Fig. 5.1) some 25 miles (40 km) south of New Orleans after "a day's journey and more by waterway through a great swamp." His account is instructive because the great swamp, the great schools of white shrimp, and the platform he found do not exist today and soon the remnant of the marsh will be gone.

> As we drew near there spread before our eyes a great fleet of sailing boats with red sails drying in the sun; dugouts, painted green and red, were tied to a wharf that ran back to a huge platform ... . [A]nd extending back along a narrow bayou were twenty or more houses, all raised high above the water on posts of cypress ... . There were French, Creole, Mexicans, Spaniards, half-tamed men of the Manila Islands, dark-skinned Indians ... (p. 81).

> We could see the old Captain ... as he made cast after cast with a small net ... . After a time the schooner drops the peak of her sail and a seine fifteen hundred feet or more is played out. The Captain has found a great

*school of shrimp ... . Presently a lot of men, maybe
a dozen, are splashing about, tugging and pulling in the
marsh at a long rope .... [N]othing but their heads can be
seen above the tall grass ... . With long-handled dip-nets
the live shrimp are lifted and dumped into the schooner
... . The Captain and the crew are fortunate, ... the
catch is estimated to be a hundred baskets [7,000 pounds]
... (pp. 84–5).*

Fig. 5.1 A drawing by Frank Schoonover (1911) showing a traditional turkey-
red sail boat setting out from the drying platform (Manila village)
in the early morning hours to hunt for great schools of white shrimp.

Under power

As would be repeated, the advent of scientific inquiries brought
about an expansion of the industry. Atlantic coast fishermen observed scientists from the US Bureau of Fisheries using trawls. The
fishermen modified the gear for their purposes and began pulling
trawls in 1912 (Johnson & Lindner, 1934).

The trawl entered the Gulf in 1917 and the fishery began a
period of further expansion and mechanization. The roar of gaso-
line engines replaced the ruffle of sails and ripple of oars and were
themselves replaced by diesel engines. Open skiffs were fitted with
cabins. The fishery was no longer limited to the height of a man
pulling a seine or by a scarcity of manpower brought on by World

War I. By 1920, in Louisiana at least, the fishery had tested the Gulf waters out to 18 miles (29 km) and found, according to management, "an immense fishing ground where a boundless supply of adult (white) shrimp always exist, with endless possibilities for the future of the shrimp industry" (Tulian, 1920). In comparison to 1908, the 1918 reported catch of 29 million pounds (13 million kg) indicates a near tripling of the catch. Ex-vessel value also rose by more than 50 percent to US$0.034/lb ($0.075/kg).

Management considered the trawl more advantageous than the seine, as it was "generally operated in deeper water," "usually took only bottom species" which were generally "second class fish" in comparison to "the important shore and surface feeders which were taken incidentally in seining" (Tulian, 1924).

Scientific management also became concerned with growth overfishing and saw a necessity to limit the harvest of shrimp less than 6 inches (15 cm) via closed seasons (Viosca, 1924). Management was largely unconcerned with recruitment overfishing, feeling that adult shrimp in the Gulf were "largely protected by natural conditions" and suggested that laws be enacted which would allow fishermen to harvest adults "whenever and wherever they may" (Viosca, 1928).

After 1900, the northern Gulf coast entered into an era of commercial industrial expansion based upon the exploration for and exploitation of oil and gas. Refineries and chemical plants were set up near major oil and gas fields, many being built along the Mississippi and other major rivers and bays (Louisiana Writers' Project, 1941). By 1989 these activities would play a major role in making Louisiana and Texas the leaders in emissions of toxic gases and liquid wastes in the nation.

The highly destructive 1927 flood of the Mississippi River ended the somewhat patchwork-like system of discontinuous public and private levees. The US Army Corps of Engineers was charged with developing and maintaining a series of continuous levees south of Baton Rouge, Louisiana to the mouth of the Mississippi. It would construct a system which would deprive the surrounding marshes of most of the Mississippi's freshwater and mineral-rich silt, while sending much of the latter cascading down the continental shelf. Bays such as Barataria would no longer be characterized by constant currents that resulted from the vital flow of the Mississippi River.

In the 1930s, oil and gas exploration activities, including seismic blasts, entered the marshes. The account of Louisiana's Lafitte Oil Field, located in the estuarine heart of the Gulf shrimp fishery, is instructive. "Additional canals, essential because the ground will not support the weight of a man are being built. Storage tanks and field buildings are set up on the edge of the canals." The canals and marsh buggy scars will speed the inflow of saltwater into marshes already deprived of much of the Mississippi's flow (Louisiana's Writers' Project, 1941).

The landings and value of shrimp continued to increase. By 1928 the reported catch was 79 million pounds (36 million kg) and ex-vessel value was US$0.038/lb ($0.084/kg). This was 2.5 times the 1918 catch, though ex-vessel value was similar [US$0.034/lb ($0.075/kg) in 1918].

A major expansion

In 1931 a state–federal cooperative shrimp investigation was initiated between the US Bureau of Fisheries and the natural resource departments of the states of Georgia, Louisiana, and Texas. These efforts were initially designed to develop yield models consistent with management's concerns about growth overfishing. They represent the first period of scientific concern over the finite nature of the resource.

Writing in the mid-1930s, the chief federal government scientist, Milton Lindner, noted the recent dramatic expansion of the Atlantic shrimp fishery off the southeastern coast of the US, and called for all the states to implement a program for obtaining daily catch statistics (Lindner, 1938). Lindner, like other federal scientists, was concerned about the possibility of spawner-recruit overfishing. They noted that if the annual shrimp spawn was dependent upon a single year class, then the fishery lacked stability and that the results of recruit overfishing could be sudden and severe (Weymouth et al., 1933). "Because of the constantly increasing drain on the shrimp population," Lindner (1938) pointed out the necessity of knowing "whether or not there is a reserve supply of shrimp available beyond the range of the commercial fishery" and for taking appropriate actions if such a reserve did not exist.

Ironically, Lindner's desire to determine the extent of a spawning reserve resulted in a rapid expansion of the fishery. Fishermen from Florida's east coast migrated with their large Florida-style vessels to Morgan City, Louisiana in 1937, once they learned from Lindner, that same year of his initial efforts to map it, the extent of the large schools of white shrimp off the Louisiana coast (Lindner, 1940). In that single year, the Louisiana annual reported catch jumped from 60 to 76 million pounds (27 to 34 million kg), a 27 percent increase which was credited to the new offshore fishery (Louisiana Department of Conservation, 1944). At the same time the ex-vessel value increased by 28 percent.

"A year later (1938) another invasion (the Morgan City vessels being the first), this time by oil drillers, took place" and the continental shelf of the north central Gulf eventually became one of the most important oil exploitation regions in the nation (Kniffen & Hilliard, 1988).

In 1938 and 1939 Lindner mapped the shrimp concentrations of the US Gulf "between the beach and the one hundred fathom contour from the Mexican border to Carrabelle, Florida." He found that shrimp were so abundant that "a small nine-foot trawl towed at full speed ... [took] ... as much as eight gallons of shrimp ... in a half-hour." These studies revealed no additional concentrations of shrimp comparable to those being exploited off the central Louisiana coast and concluded "that there appears to be little likelihood of other offshore areas being [similarly] developed" (Lindner, 1940, p. 391, p. 393). He mentioned no commercial concentrations of brown shrimp, although he was well aware of this species. It is noteworthy that great concentrations of brown shrimp would be reported later in a similar survey in 1950.

A summary of Lindner's findings was published for the general public in Walford's (1947) *Fishery Resources of the United States.* Walford's pictorial description on the extent of the fishery clearly defines it as a nearshore white shrimp fishery with the Louisiana coast as its geographical center (Fig. 5.2).

The advent of World War II essentially ended these scientific studies and any efforts Lindner might have contemplated to protect the stock of white shrimp whose spawning grounds were now completely covered by the fishery.

The offshore component of the shrimp fishery continued a rapid rate of growth and expansion over the period from 1938 to 1948, with a majority of the fleet retaining Morgan City (Louisiana) as its home port. White shrimp continued to account for 95 percent

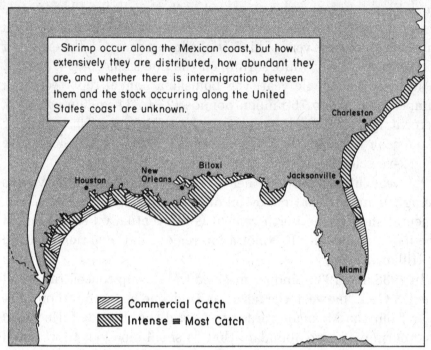

Fig. 5.2 Prior to 1948, the US Gulf and South Atlantic shrimp fishery
         was mainly restricted to the locally known concentrations of white
         shrimp. This illustration, taken from Walford (1947), shows where
         those concentrations were greatest.

of the catch and Louisiana continued as the center of production
(Anderson et al., 1949). Beginning in 1946, US shrimpers began
fishing off the Mexican coast, and in 1947 at least 48 vessels transferred from US to Mexican registry, so as to shrimp legally in these
waters. Others continued to fish outside the Mexican territorial
sea. By 1950 the reported US Gulf catch was 143 million pounds
(65 million kg) and the reported Mexican Gulf catch was 47 million
pounds (21 million kg).

A new fishery: shifts in species dominance and abundance

In 1948 some shrimpers began to notice increasing catches of
brown shrimp. These were difficult to market in Texas, though
catches were small, but not in New Orleans where they found a
ready market (Denham, 1948a,b).

The dominance of white shrimp in the fishery and fishing
grounds appears to have ended abruptly in the late 1940s and

early 1950s (Burkenroad, 1949; Viosca, 1958). Available literature (e.g., Werlla, 1954) from that time suggests that it was a
cataclysmic drop in abundance of white shrimp which coincided
with a sudden increase in abundance of the nontraditional brown
shrimp (Fig. 5.3). Toward the end of the decline (in 1957), the
white shrimp harvest in Louisiana was reportedly 10 percent of
the pre-1952 average (Viosca, 1958), representing a 70 million
pound (32 million kg) reduction. The decline was felt by Viosca
to be associated with spawner-recruit overfishing and the severe
drought of 1952–57 – the impact of the drought was magnified by
the greatly restricted flow of the Mississippi River into the marshes
(Viosca, 1958). However, growth-overfishing, the spraying of DDT
on nearby sugar cane fields (Charles Lyles, personal communication, 1989), and early oil and gas exploration, and manufacturing
activities should not be discounted as having had adverse effects
on the fishery.

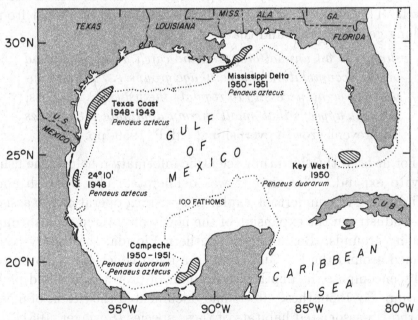

Fig. 5.3 New shrimping grounds for pink and brown shrimp discovered by
shrimpers and the US Bureau of Fisheries during the decline in abundance of white shrimp. (After Springer, 1951.)

Despite the then-national pre-eminence of the Gulf shrimp, it
appears that no major national research efforts were undertaken
to assess the extent of this decline in white shrimp production or
to measure the magnitude of this possible species shift. Rather,

national attention focused on the two new fisheries for large brown and pink shrimp.

The last expansion

In 1948, two vessels from northeastern Florida found fishable concentrations of pink shrimp on the coral-rich bottoms of southeast Florida's Dry Tortugas and a fishery developed almost overnight. Two boats fished the grounds in January; 125 to 175 boats fished in February; and by March 1, there were 250 to 300 boats from all of the US South Atlantic and Gulf states, except Louisiana (Idyll, 1950). The total catch for that first year, 17 million pounds (7.7 million kg) (tails), was the maximum ever recorded for these grounds.

Despite what must have been a "gold-rush" development, no caution was suggested by management, at least at the national level. The view was that "much of the area is protected from fishing gear by coral" and that

> shrimp is an annual crop and the catch does not depend
> on the accumulation of several age groups.... [T]here ap-
> pears to be no necessity to regulate the fishery ... [unless]
> later, it appears that small shrimp dominate the catches
> [to prevent growth overfishing] (Idyll, 1950, pp. 15–6).

For at least the third time scientific information helped the fishery to expand. This time it was deliberate. The US Fish and Wildlife Service undertook exploratory fishing operations to assist the industry in the expansion of the newly discovered pink shrimp fishing grounds. Beginning in southern Florida, the fishery continued around the US Gulf coast in a band which later proved to fully encompass the depth distributions of brown, white and pink shrimp. By 1950, these efforts had documented the extent of most of the US-associated habitats of these species (Springer, 1951).

Lindner's suggestion that a spawning reserve was needed, or Viosca's concern that white shrimp had been spawner-recruit overfished, did not surface again to any appreciable extent. The scientific management of these species entered a period where it was apparently believed that it was essentially impossible to overfish these resources, for a variety of reasons, including high individual egg production rates and areas of untrawlable bottoms. Lindner's concerns over the potential of a cataclysmic decline in such fisheries and Viosca's dramatic depiction of such a decline for white shrimp became buried in the literature. Until this writing, they have remained as muffled calls of concern from the past.

The period from 1950 to 1976 is marked by continued growth of the fleet, full use of the domestic fishing grounds, and continued expansion of the foreign fishing grounds into Central and South America (Fig. 5.4). Reported US landings increased to 210 million pounds (95 million kg) with an average ex-vessel value of US$1.31/lb ($2.89/kg) in 1976, as compared to 143 million pounds (65 million kg) and US$0.06 to US$0.28 ($0.13 to $0.62/kg) ex-vessel value in 1950.

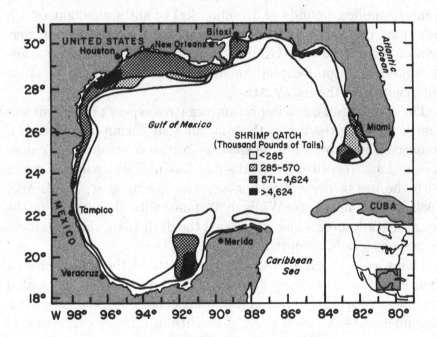

Fig. 5.4 Average annual offshore harvest of brown, white, and pink shrimp for the period 1959 to 1963. Catch composition was 52 percent brown shrimp, 26 percent white shrimp, and 22 percent pink shrimp. (After Osborn et al., 1969.)

In 1976 the Fishery Conservation and Management Act (FCMA) was passed, and the Gulf shrimp fishery entered a new era with the adoption of a federal fishery management plan for US Gulf shrimp.

An analogy – the federal shrimp plan and its impact on the fishery

Management by action: pros and cons of the action portion of the plan

There were at least two negative aspects associated with the federal shrimp plan. First, the expansion into foreign fishing grounds was reversed. Unilateral extension of the US exclusive fishing zone out to 200 miles resulted in similar, retaliatory extensions by Central and South American countries. This was not unexpected. Indeed, shrimpers who had traditionally fished these waters had lobbied unsuccessfully to have shrimp excluded from the FCMA. The almost immediate impact of the FCMA, then, was a loss of some 10 million pounds (4.5 million kg) of shrimp caught by US fishermen operating in foreign waters, the equivalent of 10 percent of the US harvest (Griffin & Beattie, 1978). Efforts by a few shrimpers to illegally exploit Mexican waters were met by vigorous enforcement of the Lacey Act.

The second and more important negative aspect of the plan was that it gave legitimacy to the concept that shrimp constituted a resource which could not be spawner-recruit overfished. The plan states that "recruitment overfishing has not and will not occur with the use of present technology and fishing gear" (GMFMC, 1980, pp. 6-2 and 6-3). While at variance with the opinion of the scientific task force which developed the draft plan, this statement became generally accepted.

There were at least two positive aspects of the plan. First it identified that the "major concern for future stocks is related to concern for adequate habitat, particularly for the estuarine-dependent brown, white, and pink shrimp." In an expression of that concern, the Gulf of Mexico Fisheries Management Council (GMFMC) established a Habitat Committee which has actively addressed a number of concerns, largely through its advisory role.

Second, the plan resulted in focused, Gulf-wide attention on shrimp population dynamics and the Council provided for an annual scientific forum for review of selected portions of the plan through its Scientific and Statistical Committee (SSC). While much of this attention was focused by the Council on two narrowly constructed management measures intended to reduce growth-overfishing, scientists continued to draw their, and the Council's, attention to the critical link between shrimp production and estuarine habitat and to the finite nature of the resource (e.g., Condrey, 1991).

*Forcing management to wrestle with
the finite nature of the resource*

At the January 1988 meeting of the Council's SSC, National Marine Fisheries Service (NMFS) researchers from the Galveston (Texas) Laboratory (Dr. Ed Klima and Dr. Jim Nance) discussed a possible conceptual link between three well recognized phenomena. The first was a direct measure of the role that flooded marsh surface has in providing juvenile shrimp with food and protection from predation (e.g., Browder et al., 1989; Minello et al., 1989; Minello & Zimmerman, 1991). The second was that marshes in the north central Gulf were in the process of deteriorating, subsiding, and converting to open water. The third was a pattern, as yet unexplained, of increased annual recruitment of juvenile white and brown shrimp and menhaden to their respective fisheries from 1960 to the present.

The Galveston researchers eloquently pointed out that if flooded marshes serve as an important habitat for shrimp and the marshes are undergoing subsidence, then the recent increases in brown and white shrimp landings may be the result of a temporary stimulation of the carrying capacity of marshes for shrimp. If so, and this possible stimulation declines, the fishery and stock could go into a state of collapse.

A motion was made that the Council should recognize this possibility and establish a stock assessment panel to address the issue. Though that motion failed for lack of a second, a similar motion, broadened to include pink shrimp, passed unanimously at the next January (1989) meeting (GMFMC, 1988, 1989). The Council, acting on this advice, requested that NMFS convene a shrimp stock assessment panel.

That panel met in June 1989 and concluded that

> *The possibility that current recruitment of brown and
> white shrimp is linked to expansion of shrimp nursery
> habitats due to sea level rise, subsidence and marsh de-
> terioration, and that this process has stimulated a short-*

term increase in recruitment from 1960 to the present, should be recognized. If this link exists, and if critical habitat is not maintained, gradual or precipitous declines in recruitment and yields should be anticipated.

Pink shrimp production in the Dry Tortugas is correlated with freshwater discharge into the Everglades and with seagrass bed acreage in that area. Recently, pink shrimp landings have undergone an unpredicted 5-year decline in the Tortugas fishery from a previously stable average of 10 million pounds (1960–80) to a low of 6 million pounds (1986). During this same time period, there have been nontraditional manipulations of the patterns of freshwater discharge and wide-scale, and apparently unprecedented, seagrass bed die backs in the Everglades. If these modifications of the nursery area are causal and uncorrected, further reductions can be anticipated in the Tortugas fishery (Nance et al., 1989, pp. 34–6).

The recognition by the SSC and the Council that habitat-related recruitment-overfishing was a possibility for white, brown, and pink shrimp necessitated a process of plan amendment. As part of that process, NMFS required that the Council define overfishing.

The general issue of habitat vulnerability dominated the discussions at the January 1990 meeting of the SSC. During the first part of that meeting the SSC was asked to comment on a proposed seven-million-dollar (US), five-year research plan to assess whether habitat alteration or degradation was responsible for recent drastic reductions in the pink shrimp fishery.

The proposal was not favorably received by the SSC, because it felt that the research plan was too costly, given the limited monies available. In addition, it focused attention on pink shrimp, which are of minor economic importance in relation to brown and white shrimp, which are also threatened by habitat loss. One member of the SSC (Condrey, the senior author) argued that there seemed little likelihood that pink shrimp landings would be restored, since Florida did not appear committed to addressing the likely reason for the decline. He noted that the research plan stated that it is "unlikely that natural water flow regimes [into the Gulf adjacent to South Florida] are ever going to be restored" (Anon., 1989) and that statistical relationships exist positively correlating pink shrimp landings with freshwater runoff (e.g., Browder, 1985). Other members of the SSC pointed out that Louisiana citizens had made a commitment to the environment via their votes to establish a wetland trust fund and freshwater diversion projects.

The SSC adopted a motion suggesting that

> *Further research which quantifies the relationship between shrimp recruitment and habitat should be conducted, specifically in the northern Gulf of Mexico. This research would include, but not be restricted to, studies of the effect of subsidence on marshland in all nursery areas and studies of the effects of reduced seagrass acreage, increased freshwater discharges, and increased use of pesticides* (GMFMC, 1990, p. 3).

That motion was adopted by the Council with the interpretation that the northern Gulf of Mexico encompassed pink, as well as brown and white shrimp (Terry Leary, personal communication, 1990).

The SSC was also asked to comment on NMFS's efforts to develop a definition of recruitment overfishing. While no specific recommendation was made, the Vice Chairman of the Council's Shrimp Committee asked that the following concerns expressed by one of the scientists be transcribed verbatim for the Council's consideration. These concerns are relevant to managers attempting to deal with global warming, if one replaces the phrase recruitment "overfishing" with "global warming." They are that

> *Detection of recruitment overfishing in any of the Council's fisheries, especially shrimp, is going to be very difficult to ascertain given the changing environmental conditions. ... The shrimp fishery is not stable.... . It is based ... at least on habitat ... which is continuing to undergo degradation, despite the best efforts of the Council's Habitat Committee, and there are clear indications in the literature that degradation may in the near future (for brown and white shrimp), or may currently (with pink shrimp), be impacting the fishery.... For the Council to expect ... any group of scientists to come up with a plan of action which will allow the Council to identify recruitment failure ... and ... then to reverse (it) is beyond our capability. So I would urge the Council to proceed*

cautiously and to be very concerned about the combined effect of high levels of harvest and a resource which is based on an unstable estuarine environment. That's not any number, that's not any procedure, but it's a difference in approach – a radical difference in approach from the unmodified shrimp plan (GMFMC, 1990, p. 5).

At its January 1990 meeting the Council incorporated these remarks into its new efforts. A draft definition of overfishing was developed by April 1990. It reflects caution in the face of instability and uncertainty, and it is a radical departure from past management practices, one which may hopefully help to heal the environment and better the life of the average shrimper.

Management by inaction and reaction

Turtle excluder devices (TEDs)

In the early 1970s turtle biologists began to recognize that incidental capture was one of the major identifiable sources of sea turtle mortality and could have a serious impact on sea turtle populations, especially the endangered Kemp's ridley. In 1947 approximately 40,000 adult female Kemp's were counted during a single nesting arribada (concentrated emergence of turtles). In 1988 approximately 600 females nested during the entire season. In spite of nearly 100 percent protection of the nesting beach since 1978, the number of nesting adults continues to decline. This problem was at a low level of recognition among fishery scientists, as evidenced by the fact that the Gulf States Marine Fisheries Commission Shrimp Plan did not address sea turtle by-catch (Christmas & Etzold, 1977). Shortly thereafter, in 1978, the Gulf of Mexico Fishery Management Council developed a Shrimp Management Plan for the Gulf of Mexico (GMFMC, 1980). The Council's plan recognized that there were five species of endangered or threatened sea turtles that were caught and, in many cases, drowned in shrimp trawls. Of these, the Kemp's ridley was believed to be the "most" in danger of extinction.

The specific management objective of the plan regarding turtles was to promote consistency with the Endangered Species Act. This was to be addressed through the following two measures: (1)

encourage research on and the development of shrimping gear to reduce or eliminate the incidental capture of sea turtles, and (2) develop and implement an educational program to inform shrimpers of the current status of sea turtles and of proper methods of resuscitation and return to the sea of incidentally captured turtles (GMFMC, 1980). As more data began to be collected on estimates of the magnitude of incidental capture, these measures became outdated.

The appearance of a record number of dead sea turtles on the southeastern coasts shortly after the onset of the 1980 shrimping season, prompted an emergency meeting of government officials, environmentalists and leaders of the shrimping industry. At this meeting NMFS introduced its TED as a technological solution that would save sea turtles, increase the shrimper's shrimp catch, and reduce finfish by-catch by 50 to 70 percent.

In 1982 a TED Voluntary Use Committee of conservationists and industry representatives was formed and set a goal of a 50 percent TED use in the Gulf and South Atlantic shrimp fleet by November 1985. Progress toward that goal was minimal. With a gear they felt could reduce incidental capture, environmentalists and scientists now began to press for a more rigid adherence to the Council's objective of consistency with the Endangered Species Act.

Late in 1985, the US Fish and Wildlife Service (which has primary responsibility for sea turtles), presented NMFS (which has responsibility for sea turtles at sea) with a formal recommendation for mandatory TED use in inshore and offshore waters of the Gulf and South Atlantic by March 1986. NMFS rejected the recommendation and the US Fish and Wildlife Service approached the Gulf Council in January 1986 to act on the treatment of sea turtles in its plan. The Council declined to act.

In August 1986, the Center for Environmental Education (CEE) formally informed the US Department of Commerce that it was violating the Endangered Species Act because it had not closed the shrimp fishery nor required the use of TEDs. Faced with this inherent threat of litigation, the National Oceanic and Atmospheric Administration (NOAA) and NMFS sponsored a series of four mediation meetings from October to December 1986 between the relevant federal agencies, the major offshore and nearshore shrimping organizations, and concerned environmental groups. The local fishermen's awareness of the TED controversy at this time was varied. In the south Atlantic, where much of the voluntary TED program occurred, TEDs were a familiar issue. However, it became apparent during a series of meetings held by the Concerned Shrimpers of Louisiana during the summer of 1986 that individual fishermen (at least in Louisiana) were mostly unaware of impending TED regulations.

When the mediations concluded, one member of the group, the President of Concerned Shrimpers of Louisiana, did not sign the report and the process began to unravel. The unravelling was highlighted at a meeting on 5 March 1989 of the Concerned Shrimpers of Louisiana, held in a small, rather isolated town in south Louisiana. It was attended by over 5,600 shrimpers, with major elected state officials pledging their support for the shrimpers. Shortly thereafter, the Concerned Shrimpers of Louisiana gained support from shrimpers along the entire Gulf and south Atlantic coast and became the Concerned Shrimpers of America (CSA). This group then took up the leading role in fighting the imposition of TED regulations.

In February 1987, NOAA published its proposed rules on mandatory TED use based upon the mediation agreement. These rules called for a three-year phase-in to begin 15 July 1987 and would eventually require TED use on all vessels. The single exception was a 90 minute/tow limit for vessels pulling a single trawl with a headrope length less than 30 feet (9 m).

Thousands of shrimpers attended the public hearings which were held on the proposed rules. Federal legislators from some southern states pressed for less restrictive measures.

On 29 June 1987, NOAA issued its final rules just two weeks prior to the proposed initial implementation date of 15 July. The final rules were substantially less encompassing than the proposed rules. Specifically, all inshore shrimpers could opt to trawl for 90 minutes or less instead of pulling TEDs. The 90-minute option also applied to any offshore shrimpers whose vessels measured less than 25 feet (7.6 m) in length. Implementation for the remaining component of the offshore fleet was delayed until January 1988 for southwestern Florida and until March 1988 for the remainder of the Gulf. A Louisiana Congressman claimed an "87.5 percent" victory, since that was the percentage of Louisiana vessels now exempt from mandatory TED use.

While much attention was paid to publicly fighting mandatory TED use during this time, little attention was paid to the lack of infrastructure to support the proposed massive gear modification. In January 1987, the senior author of this chapter undertook a small study designed to introduce TEDs into the Louisiana fishery only to find out that NMFS did not expect the devices to work off some portions of the Louisiana coast because they had not been adapted to those regional conditions. The federal expertise necessary to make these modifications was unavailable. Many of the manufacturers listed in the draft regulations were either not making the approved TEDs or were only making one or two types of TEDs (Condrey & Day, 1987).

In October 1987, the Attorney General for the State of Louisiana filed suit against the US Department of Commerce. He was joined in that suit in early 1988 by the Concerned Shrimpers of America, following the failure of a US House of Representatives bill, attached as an amendment to the Endangered Species Act, to delay the requirement of TEDs for two years (or until 1989). The State of Louisiana and CSA lost their lawsuit but did get an injunction against the enforcement of the regulations until the final resolution of their legal appeal. This injunction was granted for two reasons: (1) the testimonies of shrimpers as to the unavailability of TEDs and their inexperience in using them, and (2) the fact that NMFS had been studying turtles for 10 years already and if no action were taken the species would not become extinct for 20 years and that programs such as hatcheries could extend the time before the species' extinction. It was decided that a delay of several months would not substantially hurt the species. Given the highly endangered status of Kemp's ridleys and the questionable success of hatchery programs, this was not a view endorsed by the scientific community. In early April 1988, the appeals court ruled that TED regulations had to be enforced, but because of the general unavailability of TEDs, enforcement would start on 1 September 1988.

During the summer of 1988, the US Congress was in the process of reauthorizing the Endangered Species Act. However, this reauthorization was being delayed by several southern senators looking for concessions on the TED controversy. The Act was at last reauthorized in October 1988 but with one important amendment. The Heflin–Mitchell Amendment delayed the starting date for the enforcement of TED regulations until May 1989 and also set up a National Academy of Sciences (NAS) panel to conduct a study of the status of sea turtles. That study would commence in April 1989.

Once again the CSA filed a law suit to block the regulations, but that suit failed in April 1989 and beginning 1 May 1989, TEDs were required. However, a 60-day grace period was granted during which time no legal citations would be issued. On 1 July 1989 TED regulations began to be enforced with citations being issued.

From this time on the sequence of events proceeded quite rapidly. An unusual condition existed in the Gulf of Mexico resulting from a change in two major currents which brought large amounts of sargassum into the Gulf. This brought a great outcry from shrimpers who claimed that it was impossible to pull a TED through the debris, and resulted in a suspension of TED regulations on July 10, pending study of the situation. Ten days later the regulations were reinstated and, during the following weekend, shrimpers in Texas and Louisiana (most notably in Galveston and Cameron, respectively) staged a blockade of major ship channels. While no one was injured during these blockades, shots were fired, TEDs were thrown overboard or burned, and tempers were at dangerously high levels. The US Secretary of Commerce issued a 45-day suspension of the regulations to allow for a cooling-off period.

Conservation groups immediately filed suit claiming that the Secretary had no authority to suspend enforcement of the Endangered Species Act. On August 1, a federal judge ruled the suspension illegal and gave NMFS three days to come up with regulations to protect turtles. NMFS did come up with regulations in the form of reduced tow time. Shrimpers who were not pulling TEDs would be allowed to trawl for no longer than 105 minutes. In order to enforce the regulations, a 24-hour trawling schedule was set up detailing when the designated trawling periods would begin and end. These regulations displeased both the shrimpers and the environmentalists.

On September 8, environmentalists sued once again claiming that the regulations did not protect turtles and that in fact many turtles would not survive the 90-minute tow time in the original regulations. They won this suit and on September 9, the original TED regulations were once again in effect. This was decreed to be the final word on the regulations. However, a grace period was established until September 22, during which no fines would be issued. From September 22 until October 15, the fines would be reduced contingent on the immediate installation of a TED. With the reinstatement of the regulations on September 9, another series of blockades were conducted. At this same time, however, President Bush happened to be in New Orleans and representatives for the shrimpers were successful in meeting with him. This diffused the blockade situation and in October the shrimpers received a reply from the President's chief of staff that alternatives to the TED would be looked at but only if supported with hard evidence. There was hope that the results of the NAS study would shed some light on this issue.

Concerned shrimpers once again sought a temporary restraining order on the TED regulation in February 1990, but were unsuccessful. The TED regulations are in effect and citations have been issued. Shrimpers have been urged not to pay their fines in the hope of getting the TED regulations soon overturned.

In April 1990, the NAS study was released. It identified incidental capture as the major human-caused mortality and upheld the use of TEDs as the way to reduce it, essentially leaving the TED regulations as is. It remains to be seen if the federal commitment to enforcement will bring the majority of shrimpers into compliance or fan the fires of opposition. Meanwhile, in the four years since 1986, when debate on this issue began, there has been no observable recruitment of adult nesting Kemp's ridleys. The average number of turtles in a nesting arribada has declined to less than one-half of one percent of a 1940 arribada, and the number of nests has continued to decrease by three to four percent annually.

By-catch and red snapper

The Council's plan recognized that the incidental catch of finfish in trawls was "a matter of concern" and that "most of the incidental catch die ... before they are discarded." While one objective of the plan calls for the by-catch to be minimized "when appropriate," the only management measure in the plan which addresses the issue encourages research on gear to reduce the by-catch.

Despite an estimated annual take of 1.1 billion pounds (0.5 billion kg) (Nichols et al., 1987), the impact of the by-catch on a stock's ability to replenish itself has only recently begun to receive attention.

The March 1990 findings of the Council's Reeffish Assessment Panel on the status of red snapper are instructive as to the impacts of shrimping on other stocks and fisheries. The Panel noted that although red snapper comprised only a small fraction of the by-catch, the by-catch comprised a large percentage of the red snapper landings allowed under the Council's red snapper management criteria. Those criteria call for the implementation of a harvest schedule which will allow for a rebuilding of the spawning biomass of red snapper to a level in the year 2000 that is 20 percent of that expected in the long-term absence of the fishery. The incidental mortality of zero- and one-year-old red snapper in shrimp trawls reduces the potential spawning level to 25 percent. This mortality occurs before the fish enter the directed fishery, further reducing the red snapper spawning biomass to the 0.5–1.0 percent range.

The Panel observed that there were potential management regimes which dramatically increased red snapper yield while attaining the Council's spawning biomass goal, because of the excessive growth overfished nature of the fishery. These options, however, required across-the-board reductions in fishing mortality. The Panel was told that such measures could not be recommended because the Council's shrimp plan did not contain a measure to reduce the by-catch. The only viable option the Panel was told it could recommend was a zero catch in the directed fishery, until such time, if ever, that the shrimp plan was amended.

So, while the Council was able to avoid the TED issue, it finds itself faced with by-catch problems. The difference is that if it turns to trawling efficiency devices, it will find that these are now a heated emotional issue in many coastal communities and that the available technical solutions are not as simple as was the case with turtles.

Summary of the analogy

The analogy for those dealing with global warming is simple. The Gulf shrimp fishery was allowed to expand without regard to its Achilles' heel, the estuarine habitat, while that habitat was encountering a massive adverse impact – drastic restrictions in freshwater flows, industrial pollution, canalization, bulkheading, and construction. That expansion of human impacts left the natural environment impoverished and the average shrimper with few options for lateral movement. It was facilitated by scientists (the senior author included) who had been taught that they must measure a decline and document the cause of that decline before they suggested caution.

Management is now attempting to deal with the issue of habitat degradation and by-catch. This will not be easy. A portion of the fishery (pink shrimp) may already be in a state of decline and the major portion (brown and white shrimp) may soon follow suit. Conflicts, confusion, and mixed national signals over the required use of TEDs have rocked the industry and hindered an orderly introduction of the gear into many communities. In addition, recent massive imports of pond-raised shrimp from China (ironically one of the first major international markets for US shrimp) have undercut the prices fishermen receive, while greatly increasing the demand for quality shrimp. The important point, however, is that the Council now seems to be moving on the issue and is hopefully not waiting for concrete evidence that irreparable harm has been done.

Impacts

The impacts of global warming on the shrimp fishery are likely to be devastating. If the current rate of sea level rise is accelerated, it will hasten the flooding and loss of productive habitat for brown, white, and pink shrimp. If estuarine temperatures rise 4°C, summer temperatures will likely exceed the lethal limit for white shrimp during their period of peak estuarine abundance. That same 4°C rise could increase the survivability of brown shrimp in the spring, but the eventual loss of estuarine habitat would make any such gain short term.

If global warming results in a severe and prolonged drought in the Great Plains, it may result in a partial drying of Louisiana's estuaries. Such dryings have been associated with large losses of marsh lands. As such, the drought could further accelerate the marsh loss and the eventual loss of shrimp yields. In addition, severe droughts in the early 1950s may have resulted in a dramatic shift in species dominance and abundance on the main fishing grounds: going in a step function from 95 percent white shrimp to 50 percent white and 50 percent brown shrimp (with a large decrease in white shrimp abundance). It is conceivable that future droughts could cause similar, but unpredictable shifts.

If at the least global warming results in a protracted flooding of only the Louisiana coastal zone below the Intracoastal Canal, it will allow the release of an unknown but not insignificant amount of toxic substances from numerous registered and unregistered waste deposits and storage sites. Given the apparent poor condition of the computer files on the priority Superfund sites (Magistro, 1987) and the slow progress being made in site cleanup, it is conceivable that the major sites will not be cleaned up before the area becomes inundated. Given the limited circulation and somewhat confined nature of the Gulf of Mexico, it is also conceivable that such a release could render shrimp and other valuable living marine resources unusable or unfit for human consumption. Regardless, debris left by flooded communities and oil refineries would make trawling difficult. Finally, if global warming were to result in only a 25 percent reduction in shrimp yields, an uncertainty over the timing of shrimp harvest, and a decrease in basic human comfort (i.e., hotter, more humid summers; increased residential flooding) it would most likely exceed the already strained capacity of the coastal shrimp communities to adapt. The social chaos brought on in the summer of 1989 through the imposition of TEDs can serve as a case in point.

Lessons

There are clear and concise messages to be learned from the management of shrimp by fisheries managers contemplating how they might deal with the effects of global warming.

- The system has its limits: plan to live within those limits or plan to abandon the fishery. The critical threat to this fishery – the ultimate dependence on the resources of a threatened and deteriorating environment – was long recognized. Efforts to restore that environment have been largely superficial and unproductive. The adverse consequences are either now being felt (in pink shrimp) or may be felt in the near future (brown and white shrimp). The industry is not yet prepared for a reduction in yield.

- Be prepared for the unexpected; do not expect to explain it by hindcasting. A dramatic species dominance shift appears to have occurred in the dominant portion of this fishery in the late 1940s and early 1950s. By the time a continuous program of scientific investigation was implemented, researchers found themselves perhaps more interested in the new dominant species, but at least more involved in a study of the new system. The limited account of the species shift became buried in the gray literature, and to this date has never received investigation as a scientific hypothesis.

- Seek a scientific consensus across a broad array of backgrounds and experience and provide forums for constructive scientific debate. It is possible that the recent scientific consensus on the critical link between shrimp production and habitat will develop in time to allow managers and the industry to deal with an impending crisis constructively.

- Where there is scientific debate, weigh carefully the possible consequences of following non-conservation (for the resource) advice and for funding programs of "continued study" as a justification for inaction. The existing program for continuous monitoring of the shrimp fishery may only alert us when "its heart has stopped beating."

- Know the history, culture, and practices of those you are dealing with. Before the signing of the TED mediation report, little work had been conducted toward introducing the device into Louisiana, despite the single fact that Louisiana has the largest number of shrimpers. These shrimpers represent a wide array of ethnic backgrounds, with strong family ties and a rich heritage of individualism and isolation. It is from this group that the first organized opposition arose, and it arose with a speed and intensity that inflamed the Gulf and South Atlantic fisheries.

- The GMFMC initiated a progressive program to deal rationally with the national concerns about sea turtle and finfish incidental capture and death in shrimp trawls in its original shrimp plan. The Council (until recently) never implemented that portion of its plan. Many, if not most, of the shrimpers who blockaded Gulf ports in the summer of 1989 to protest the mandatory use of TEDs had only recently been introduced to the device by a handful of individuals who had very limited experience (e.g., Condrey & Day, 1987).

References

Anderson, W.W., Lindner, M.J. & King, J.E. (1949). The shrimp fishery of the Southern United States. *Commercial Fisheries Review*, **11**, 1–17.

Anon. (1989). *Research Needs for South Florida Fisheries and Habitats*. Galveston: National Marine Fisheries Service.

Brasseaux, C.A. (1979). *A Comparative View of French in Louisiana, 1699 and 1762*. Lafayette: The Center for Louisiana Studies.

Browder, J.A. (1985). Relationship between pink shrimp production on the Tortugas grounds and water flow patterns in the Florida Everglades. *Bulletin of Marine Sciences*, **37**, 839–56.

Browder, J.A., May, L.N. Jr., Rosenthal, A., Gosselink, J.D. & Baumann, R.H. (1989). Modelling future trends in wetland loss and brown shrimp production in Louisiana using thematic mapper imagery. *Remote Sensing of the Environment*, **28**, 45–59.

Burkenroad, M.D. (1949). Occurrence and life histories of commercial shrimp. *Science*, **110**, 688–9.

Cathcart, J.L. (1819). Southern Louisiana and southern Alabama in 1819: the journal of James Leander Cathcart, ed. W. Prichard, F.B. Kniffen & C.A. Brown. *Louisiana Historical Quarterly*, **28**.

Christmas, J.Y. & Etzold, D.J. (1977). *The Shrimp Fishery of the Gulf of Mexico United States: A Regional Management Plan*. Technical Report No. 2. Ocean Springs: Gulf Coast Research Laboratory.

Collot, V. (1826). *A Journey in North America*. Paris: Arthur Bertrand, Bookseller.

Condrey, R.E. (1991). Shrimp population models and management strategies: potential for enhancing yields. In *Frontiers in Shrimp Research*, ed. P. DeLoach, W. Dougherty & M. Davidson, pp. 33–45. Amsterdam: Elsevier.

Condrey, R.E. & Day, R. (1987). *Results of a Program for Introducting Turtle Excluder Devices (TEDs) into the Louisiana Shrimp Fishery*. Baton Rouge: Center for Wetland Resources, Louisiana State University.

Crete, L. (1981). *Early Life in Louisiana 1815–1830*. (Translated by P. Gregory.) Baton Rouge: Louisiana State University Press.

Denham, S.C. (1948a). Gulf coast shrimp fisheries, January–June, 1948. *Commercial Fisheries Review*, **10**, 11–6.

Denham, S.C. (1948b). Fisheries review: Gulf states, 1947. *Commercial Fisheries Review*, **10**, 1–8.

Du Ru, P. (1700). *Journal of Paul Du Ru*. (Translated by R.L. Butler.) Chicago: The Caxton Club.

Dumont de Montigny, L.F.B. (1753). Historical memoire on Louisiana, Part 1. In *Survey of Federal Archives in Louisiana*. (Translated by Olivia Blanchard.) Baton Rouge: Louisiana State University. (Copy housed in Hill Memorial Library.)

Frantz, H.W. (1937). Manila village is subject of official Filipino survey. *Tribune* (New Orleans, LA), 14 March 1937, p. 3.

GMFMC (Gulf of Mexico Fishery Management Council) (1980). Fishery management plan for the shrimp fishery of the Gulf of Mexico, United States waters. *Federal Register*, **45**, 74178–308.

GMFMC (1988). *Minutes of the January 1988 Shrimp Scientific and Statistical Committee Meeting*. Tampa: GMFMC.

GMFMC (1989). *Minutes of the January 1989 Shrimp Scientific and Statistical Committee Meeting*. Tampa: GMFMC.

GMFMC (1990). *Summary of Actions of the Shrimp Scientific and Statistical Committee, New Orleans, Louisiana, January 17, 1990*. January 1990 Briefing Book, Tab B, No. 5. Tampa: GMFMC.

Griffin, W.L. & Beattie, B.R. (1978). Economic impact of Mexico's 200-mile offshore fishing zone on the US Gulf of Mexico shrimp fishery. *Land Economics*, **54**, 27–38.

Hearn, L. (1883). Saint Malo: a lacustrine village in Louisiana. *Harper's Weekly*, **74**, 196–9.

Idyll, C.P. (1950). A new fishery for grooved shrimp in Southern Florida. *Commercial Fisheries Review*, **12**, 10–6.

Johnson, F.J. & Lindner, M.J. (1934). *Shrimp Industry of the South Atlantic and Gulf States*. Investigational Report No. 21. Washington DC: US Bureau of Fisheries.

Kane, H.T. (1944). *Deep Delta Country*. New York: Duell, Sloan & Pearce.

Kniffen, F. & Hilliard, S. (1988). *Louisiana: Its Land and People*. Baton Rouge: Louisiana State University Press.

Lindner, M.J. (1938). The cooperative shrimp investigations. In *Thirteenth Biennial Report of the Louisiana Department of Conservation, 1936–1937*, 447–9. New Orleans: Louisiana Department of Conservation.

Lindner, M.J. (1940). Biennial report, shrimp investigations. In *Fourteenth Biennial Report of the Louisiana Department of Conservation, 1938–1939*, 389–99. New Orleans: Louisiana Department of Conservation.

Louisiana Conservation Commission (1920). Close seasons. In *Fourth Report of the Conservation Commission of Louisiana*, 102–65. New Orleans: Conservation Commission.

Louisiana Department of Conservation (1944). Louisiana shrimp: one of the world's great fisheries. In *Sixteenth Biennial Report of the Department of Conservation, State of Louisiana*, 147. New Orleans: Department of Conservation.

Louisiana Writers' Project (1941). *Louisiana: A Guide to the State*. New York: Hastings House.

Lowery, G.H., Jr. (1974a). *Louisiana Birds*. Baton Rouge: Louisiana State University Press.

Lowery, G.H., Jr. (1974b). *The Mammals of Louisiana and Its Adjacent Waters*. Baton Rouge: Louisiana State University Press.

Magistro, J.L. (1987). *Association of Superfund Sites with Wetlands: A Pilot Study*. Washington, DC: Office of Wetlands Protection, Environmental Protection Agency.

Minello, T.J. & Zimmerman, R.J. (1991). The role of estuarine habitats in regulating growth and survival of juvenile *penaeid* shrimp. In *Frontiers of Shrimp Research*, ed. P. DeLoach, W. Dougherty & M. Davidson, pp. 1-16. Amsterdam: Elsevier.

Minello, T.J., Zimmerman, R.J. & Martinez, E.X. (1989). Mortality of young brown shrimp in estuarine nurseries. *Transactions of the American Fisheries Society*, **118**, 693–708.

Nance, J.M., Klima, E.F. & Czapla, T.E. (1989). *Gulf of Mexico Shrimp Stock Assessment Workshop*. Technical Memorandum SEFC-NMFS-239. Galveston: NOAA.

Nichols, S., Shah, A., Pellegun, Jr., G. & Mullin, K. (1987). *Estimates of Annual Shrimp Fleet Bycatch for Thirteen Finfish Species in the Offshore Waters of the Gulf of Mexico*. Pascagoula: National Marine Fisheries Service, Pascagoula Facility.

Osborn, K.W., Maghan, B.W. & Drummond, S.B. (1969). *Gulf of Mexico Atlas*. Bureau of Commercial Fisheries Circular 312. Washington, DC: US Department of the Interior.

Saxon, L. (1940). Their faces tell the story. *Jefferson Parish Yearly Review, 1940*, 34–55.

Schoonover, F.E. (1911). In the haunts of Jean Lafitte. *Harper's Monthly Magazine*, **124**, 79–91.

Springer, S. (1951). The *Oregon*'s fishery explorations in the Gulf of Mexico, 1950. *Commercial Fisheries Review*, **13**, 1–8.

Stielow, F.J. (1975). Isolation and development on a Louisiana gulf coast island: Grande Isle, 1881–1962. Ann Arbor: University Microfilms International.

Stoddard, A. (1812). *Sketches, Historical* [sic] *and Description of Louisiana*. Philadelphia: Mathew Carey.

Tulian, E.A. (1920). Louisiana: Greatest in the production of shrimp, *Penaeus setiferus*. In *Fourth Biennial Report of the Louisiana Department of Conservation*, 106–14. New Orleans: Louisiana Department of Conservation.

Tulian, E.A. (1924). The present state of the Louisiana shrimp industry. In *Sixth Biennial Report of the Louisiana Department of Conservation, 87–90*. New Orleans: Louisiana Department of Conservation.

Viosca, Jr., P. (1924). Life history and habits of the shrimp (*Penaeus setiferus*). In *Sixth Biennial Report of the Louisiana Department of Conservation*, 90–3. New Orleans: Louisiana Department of Conservation.

Viosca, Jr., P. (1928). Status of the shrimp industry. In *Eighth Biennial Report of the Louisiana Department of Conservation*, 212. New Orleans: Louisiana Department of Conservation.

Viosca, P. Jr. (1958). Shrimp studies. In *Seventh Biennial Report of the Louisiana Wildlife and Fisheries Commission, 1956–1957*, 135–42. New Orleans: Louisiana Wildlife and Fisheries Commission.

Walford, L.A. (1947). *Fishery Resources of the United States*. Washington, DC: Public Affairs Press.

Werlla, W.S. (1954). Commercial fresh and saltwater fisheries, shrimp, oysters, fur and hides, shell, sand, gravel, sportfishing and hunting. In *Fifth Biennial Report of the Louisiana Wildlife and Fisheries Commission, 1952–1953*, 147–64. New Orleans: Louisiana Wildlife and Fisheries Commission.

Weymouth, F.W., Lindner, M.J. & Anderson, W.W. (1933). Preliminary report on the life history of the common shrimp *Penaeus setiferus* (Linn.). *Bulletin of the US Bureau of Fisheries*, **14**, 1–26.

6

The menhaden fishery: interactions of climate, industry, and society

LUCY E. FEINGOLD*

College of Marine Studies
University of Delaware
Lewes, DE 19958, USA

Introduction

Menhaden are uncommon as a food fish, and thus are unfamiliar to the general public. There are two species of menhaden off the US coast, each of which supports a separate US menhaden fishery, the Atlantic coast fishery (*Brevoortia tyrannus*), and the Gulf of Mexico fishery (*B. patronus*). Together, they comprise one of the largest (by weight) commercial fisheries in the US (Henry, 1971; Rothschild, 1983). Prior to World War II, Atlantic menhaden dominated the total catch. Between World War II and 1962, larger and more efficient vessels led to a steady increase in Gulf landings (Vaughan et al., 1988). By 1963 Gulf menhaden landings exceeded those from the Atlantic (Henry, 1971).

The menhaden processing industry produces fish meal, fish oil, and fish solubles. During 1980, for example, more than 75 percent of the fish meal produced was used domestically, and over 90 percent of the fish oil was exported, primarily to European markets (Hu et al., 1983). Menhaden, therefore, are of value not only to the US economy and fishing industry, but also play a role in the international marketplace. Present-day development of new products and uses of menhaden by-products, such as medicinal uses of fish oil and surimi (minced fish), is bringing the menhaden fishery into a new prominence which will directly benefit the diet and health of citizens in the US and elsewhere.

* The author was a Visiting Scientist with the Environmental and Societal Impacts Group, National Center for Atmospheric Research, at the time this research was carried out.

This chapter focuses on historical changes in Atlantic menhaden. Interannual and interdecadal fluctuations in commercial fish catches are well known, although their causes are often poorly understood. Recruitment, or year-class strength, can be influenced by year-to-year regional and global-scale environmental changes, as well as by other population responses and differences attributable to stock density (Cushing, 1974). The understanding of the relationship between the early life history, juvenile and adult stages of menhaden and their environmental settings is crucial in assessing what impact a global climate change may have on their distributions and abundance (Fig. 6.1). These will affect the fisheries, and therefore society in general. Based on analogies drawn from past societal actions, lessons can be drawn and used as warnings or guidelines for the future.

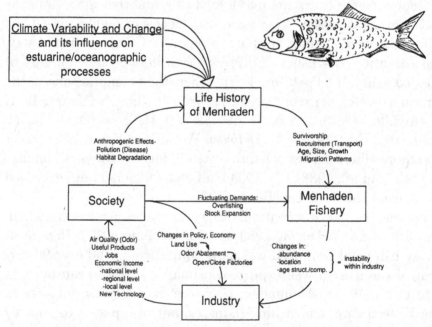

Fig. 6.1 Influencing effects and interrelationships of climate variability and change on menhaden life history, fishery and industry, and on society.

Life history and environmental setting

Atlantic menhaden are found from Nova Scotia (Canada) to Florida (USA) (Reintjes, 1969). They undergo extensive migration, encountering variable environmental conditions throughout

their life cycle. Migration is generally northward in the spring
and summer with adults stratifying by age and size (older and
larger fish are usually farther north), and southward in the fall
and winter (Nicholson, 1978).

Spawning most probably occurs during every month of the year
along the US Atlantic coast from Maine to Florida, but not at the
same time in all areas (Lewis et al., 1987) (Fig. 6.2). Dispersal of
eggs along the coast protects any single year class from complete

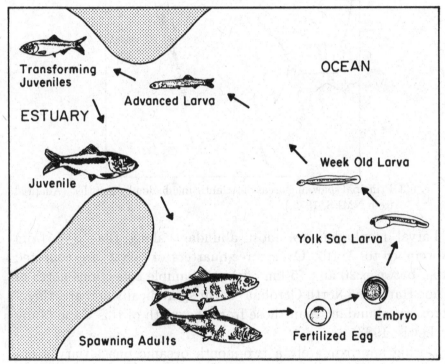

Fig. 6.2 Life cycle of Atlantic menhaden. (Adapted from Henry, 1971.)

mortality, although small parts of any brood may be destroyed by
adverse environmental conditions (Hoss et al., 1974). Spawning activity during the northward migration is limited, taking place near
shore in the bays, sounds, and estuaries along the coast (Higham
& Nicholson, 1964). Intensity increases to its maximum during the
fall/winter migration from North Carolina southward, and spawning occurs increasingly further offshore, nearer to the edge of the
continental shelf (Fig. 6.3). All age groups of migrating fish from
the north are represented during the fall/winter spawning migration (Nelson et al., 1977; Nicholson, 1978).

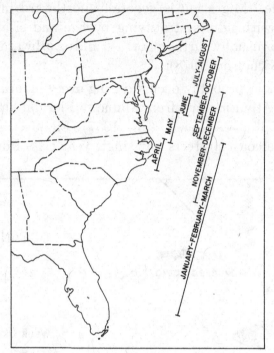

Fig. 6.3 Principal spawning areas of Atlantic menhaden by months. (Adapted from NMFS, 1973.)

Larval menhaden are most abundant along the coast from November to April. Over three-quarters of larvae are collected from between 20 and 75 km offshore. Sample collections north of Cape Hatteras (North Carolina) occur sporadically and are, therefore, less abundant than those from the south of this Cape (Judy & Lewis, 1983).

During approximately a two-month oceanic phase, larvae undergo a period of passive drift; movement is the result of ocean currents (Nelson et al., 1977) (Fig. 6.3). Atlantic menhaden larvae are then transported into estuaries by ocean currents, spending several months in the tributary streams almost reaching fresh water. During this time, they metamorphose into juveniles (Henry, 1971).

In late summer the juveniles leave the tributaries in schools, returning to the bays, sounds, and estuaries, and by late fall and early winter they migrate to the ocean (Henry, 1971). Schooling is a distinct characteristic of menhaden, beginning as early as the late larval/early juvenile stage and continuing through old age (Reintjes, 1969). The menhaden fishery is dependent on these abundant surface schools for exploitation.

Most Atlantic menhaden are mature at age two, and all are mature at age three (Lewis et al., 1987). Maximum reported age for menhaden is 10 years (Nicholson, 1975; Reintjes, 1969), and therefore they can potentially spawn for about seven years (Higham & Nicholson, 1964).

Menhaden feed primarily by filtering phytoplankton, zooplankton and other small particles in the water column ranging in size from 2 to 200 μm depending on fish size (Durbin & Durbin, 1975; Friedland et al., 1984; Lewis & Peters, 1984). Algae are also very important to the diet of older menhaden especially in non-marsh environments (Durbin & Durbin, 1975). Feeding studies indicate that juvenile menhaden derive much of their energy from sources other than phytoplankton-based food webs, implicating land-based detritus as an important food and energy source either directly (Lewis & Peters, 1984; Peters & Schaaf, 1981) or by aiding in the retention of other filtered particles (Friedland et al., 1984). *Spartina alterniflora*, or salt marsh cordgrass, is the major source of detritus in many marsh-dominated estuaries along the Atlantic coast, and therefore may be an important source of energy for menhaden (Lewis & Peters, 1984). The diet of menhaden changes as the fish increase in size (from being mainly herbivorous to omnivorous, and with the relative importance of detritus) and also varies geographically (Friedland et al., 1984; Peters & Schaaf, 1981).

Juvenile and adult menhaden are a major constituent in the diet of carnivorous fishes including abundant and recreationally important species such as bluefish (*Pomatomus saltatrix*), striped bass (*Morone saxatilis*), summer flounder (*Paralichthys dentatus*) and weakfish (*Cynoscion regalis*) (Reintjes, 1969). Primary predators of larval menhaden are copepods. The smaller copepod, *Centropages typicus*, is able to ingest yolk-sac larvae of menhaden, but not the larger first-feeding larvae. The larger copepod, *Anomalocera ornata*, however, can ingest both larval sizes at approximately equal rates for a given prey concentration (Turner et al., 1985).

Impact of climate

Atlantic menhaden are highly influenced by many marine environmental conditions, both oceanic and estuarine. This influence extends not only directly to survivorship in early life history, but also to food, predators, migration and spawning patterns, and recruitment into the fishery. Indirectly, influences are felt by the industry which depends on abundance of the resource and, in turn, by society.

Effects of climate on menhaden are most pronounced on their behavior and physiology and are associated with changing temperature and salinity. Changes in ocean temperatures will alter migration patterns of adults, thereby altering spawning patterns. Changes in migration patterns directly affect the fishery, possibly relocating its center of abundance. For example, processing plants that are active now could be forced to close and presently inactive ones could be reopened. With a global warming of the atmosphere, the center of the fishery would likely shift northward. However, processing plants located in the North and mid-Atlantic regions would have difficulty reopening because of social and economic changes in land use which had previously caused them to be closed. Catches in these areas could be hauled to plants in the South Atlantic region, but at some cost to the industry. In any event, with the increasing desire for developing coastal property, it may not be long before areas in North Carolina develop similar social and economic pressures (e.g., intensive coastal development and anti-odor legislation) that forced plants to close elsewhere.

Changes in spawning patterns could have a drastic effect on the fishery, limiting reproductive success and the survival of eggs and larvae. Eggs and larvae are present in the open ocean environment long enough to be affected by favorable as well as adverse oceanic conditions consequential to adult spawning migrations (Nelson et al., 1977). Changes in temperature and salinity would alter the development, movement, and survival of eggs and larvae. Changes in one parameter can affect the biological tolerance level of the organism to other parameters. More important, though, is the location of spawning in relationship to the shore combined with the effects that temperature has on growth and development. In the northern part of the menhaden range, spawning occurs nearshore and sometimes in bays, sounds and estuaries. In the southern part of their range menhaden spawn offshore. In both cases (although with greater weight in the southern part of the range) transport of passive oceanic larvae by onshore currents from the offshore spawning areas into estuarine nursery grounds is essential for survival of metamorphosing larvae. Winter storms, with their associated upwelling and cross-shelf circulation provide the physical conditions optimal for survival and shoreward transport of menhaden eggs and larvae (Checkley et al., 1988). The circulation pattern begins with warm nutrient-rich water upwelling at the Gulf Stream Front. This water mass, which moves shoreward at the surface, loses heat and nutrients, sinks at the Mid-Shelf Front, and finally flows offshore at depth. This storm/circulation phenomenon also enhances the production and vertical aggregation of microzooplankton (larval menhaden food). Thus, during winter, menhaden eggs and larvae drift shoreward in warm water with abundant food (Checkley et al., 1988).

Ocean circulation is driven by exchanges of energy involving such climatic variables as air temperature, evaporation, humidity and wind velocity, which interact at the air–sea interface. Climate change may, therefore, have significant impacts on ocean circulation resulting in a variety of long- and short-term effects (Frye, 1983). As an indirect climatic effect, a short-term change in currents or circulation can cause high mortality to the larvae in need of a transport mechanism into the estuarine nursery ground. Years with strong onshore larval transport result in good year classes, while years with weak onshore transport or offshore water movement increase the distance for larvae to travel to their nursery grounds, resulting in a poor year class (Nelson et al., 1977).

The length of time it takes larvae to reach the end of the passive oceanic phase is dependent on water temperature and ranges from 30–45 days (Nelson et al., 1977). Furthermore, an increase in temperature has been found to increase the rate of larval menhaden yolk utilization, to result in an earlier age at first feeding, and to decrease the ability of larvae to withstand the deprivation of food (Powell & Phonlor, 1986). Thus, environmental conditions (onshore currents and temperature), influenced by climate, can determine the survival of larvae.

Upon entering US estuaries and tributaries, late-stage larvae are subjected to various hydrographic conditions. Local variable currents are influenced by wind and runoff that result in large variations in salinity, from approximately 1 to 36‰ and low temperatures ranging from 4° to 10°C (Lewis, 1965; Reintjes & Pacheco, 1966; Reintjes, 1969). A high level of freshwater discharge from estuarine tributaries could also possibly hinder the entering larvae (Nelson et al., 1977). It is also feasible that localized hydrographic conditions maintain the larvae, as well as larval prey, in patches. Changes in these hydrographic conditions (currents, temperature, runoff, wind, and salinity) would occur as a result of climate variability and change, thus altering larval distribution, larval food, and survival. Although menhaden are highly tolerant euryhaline fish (for example, they are physiologically able to adapt temporarily to altered salinity regimes), drastic changes in coastal oceanic and estuarine waters may provide unacceptable conditions for metamorphosing larval and juvenile menhaden (Engel et al., 1987).

Because menhaden spend a crucial part of their early life in the estuaries and can be found in every major estuary along the US Atlantic coast, from Florida to Massachusetts during different times of the year (Pacheco & Grant, 1965), it is important to look at the overall role of the estuary and the effects of climate. An increase in sea level, related to climate change, will extend the habitat used for early life stage nursery areas in some regions of the menhaden's range – most probably toward the southern end. In the northern extreme of their range, however, nursery habitats as well as spawning grounds could be lost to flooding. The flooding of these areas could also release pollutants and other terrestial wastes into the marine environment that currently are filtered through marshes bordering the estuaries. This would have a deleterious effect on all estuarine organisms. On the other climatic extreme, severe droughts would have an adverse effect by drying up the marsh environments and greatly extending saline waters up-estuary. In both cases, local changes in species dominance as well as the introduction of new species would most likely occur. This would include menhaden and other organisms that utilize the estuary as a nursery area, and predators and prey of these species as well as disease-bearing species.

Juvenile menhaden play an important role in the transfer of energy, which they derive from marsh-generated detritus, from the marsh through the estuary to the ocean (Lewis & Peters, 1984).

The impact of climate variability and change on the estuarine and marsh environments would surely be evidenced in local changes in food webs, perhaps from plankton-based diets to increased carnivory, leading to changes in important energy transfers through the ecosystems and to the menhaden fishery.

A disease common to menhaden throughout its range is ulcer disease syndrome (UDS). This disease, represented by lesions, eventually causes death and is most probably aggravated by unfavorable, stressing environmental conditions (Hargis, 1985). It is also possible that predatory species dependent on menhaden (especially juveniles) are also affected. Climate-related environmental change could accelerate or spread disease among juvenile menhaden and to many predators via these juveniles, thereby greatly increasing mortality and greatly decreasing recruitment to the fishery.

The introduction of new species, plant or animal, or the increase in abundance of existing species, can cause competition for food and space in the estuary. Depletion of oxygen by plants has contributed to massive menhaden mortality. Under environmental stress, menhaden are also more susceptible to parasites. During the menhaden's stay in estuaries, parasites are much in evidence and could possibly flourish under the same changing conditions (Reintjes & Pacheco, 1966).

Anthropogenic impacts on the marine environment such as changes in land use, economic policies, water quality, and overfishing, compound environmental impacts on the fishery. For example, intense fishing pressure during the 1960s continued over a number of years, considerably altering the age structure of the spawning stock. Approximately 40 percent of the estimated spawning stock in 1958 was at least four years old. The number of age 4 and older fish in the 1969 spawning population decreased to about 9 percent, and the average number of eggs per spawning female was about 50,000 less than in 1958. Thus, spawning potential was greatly reduced as a result of fishing pressure. This reduction in spawning potential also reduced opportunities for a broad-scale response to favorable transport in the 1968–70 year classes (Nelson et al., 1977). Therefore, it is possible for environmental parameters to be favorable while the impacts of human activities counterbalance the situation, turning it into an unfavorable one. For example, the closing of menhaden industrial plants in the New England area because of local factory-odor abatement ordinances (Smith et al., 1987), the dumping of sewage offshore, or increased demand for menhaden by-products can adversely affect favorable environmental settings. Clearly, both climate change and anthropogenic factors can influence this fishery in many ways.

History of the fishery

Throughout its long history, the menhaden fishery has withstood many periods of fluctuation in the fish population (Fig. 6.4). Numerous explanations for these fluctuations involve the age composition of the fishery. Once menhaden enter the fishery, a single year class is harvested by the industry over a four- to five-year period. Because of the length of this harvesting, the failure of an entering year class may be unnoticed, as other year classes are harvested at the same time. This can lead to serious stock depletion (Nelson et al., 1977). Therefore, a large increase in fishing effort which continues after the year class has been harvested reduces the stock size in subsequent years. This can be devastating to the fishery and the industry, if a large year class is followed by several poor year classes (Nelson et al., 1977).

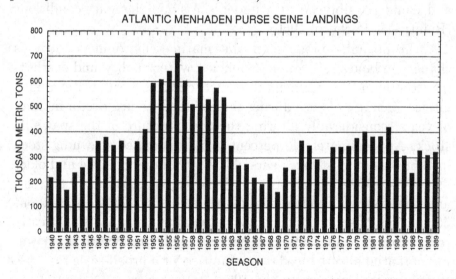

Fig. 6.4 Landings of Atlantic menhaden. (Partial data from Atlantic States Marine Fishing Commission, 1986.)

Catch statistics have been collected sporadically since the 1870s, and regularly since the 1940s (Ahrenholz et al., 1987; Blomo, 1987). Before 1940, landings showed an initial decline, with increasing catch throughout the 1940s (Ahrenholz et al., 1987) (Fig. 6.4).

Catch continued to rise through the late 1940s and early 1950s and peaked at a record high of 712,000 mt in 1956. The 1950s have been described as years of stock expansion with a broadened age structure and several dominant year classes (Vaughan et al., 1988). Peak catches in the 1950s were due to an increase in fishing pressure (despite a decrease in the number of vessels), improved harvesting technology, and for the first time, statistical sampling which covered the entire geographic range of the fishery (Ahrenholz et al., 1987; Blomo, 1987; Vaughan et al., 1988).

From 1955 to 1968 age 1 and 2 fish made up the bulk of the catch. The mean average age of catch during these years was 1.67 years (Henry, 1971). Large catches of the older age groups were evident only when a large year class, such as 1951 or 1958, was present.

The next decade witnessed a different trend. Landings fell in 1961 to 575,900 mt, and by 1962 the dominant year classes of the 1950s had disappeared (Smith et al., 1987; Vaughan et al., 1988). Landings continued to drop and by 1967 the age structure of the stock began to contract. The severe truncation of the stock resulted in a record low catch of 161,600 mt in 1969 (Blomo, 1987; Nicholson, 1975; Smith et al., 1987; Vaughan et al., 1988).

From 1964 until the early 1970s, fish four years and older had almost disappeared from catch samples (Nicholson, 1975). Age 1 and 2 fish continued to make up the bulk of the catch (and they still do). This decline in age structure and abundance was attributed to increased fishing pressure (Nicholson, 1971). Noticeable increases in growth rates occurred during this same time period as compared to growth rates in the late 1950s and early 1960s. These age-specific increases in length and weight are documented as the result of the decrease in the average age of the menhaden stock (Nicholson, 1975). Although an estimated 80 to 90 percent of the potential spawners were being taken by the fishery, the Atlantic menhaden supported a large commercial fishery during the 1970s (Ahrenholz et al., 1987). Speculation about the regrowth of the population under these conditions has centered on either of two views: (1) certain basic aspects of the reproductive biology of Atlantic menhaden were altered because of fishing pressure, resulting in reduced size and average age of the stock (no evidence for marked changes in reproductive biology was found, however); and (2) favorable environmental conditions for recruitment over a period of years allowed for better recruitment and a stable fishery, in spite of the heavy dependency on the age 2 year class by the spawning stock and the fishery (Lewis et al., 1987).

The mid-1970s represented a period of temporary decline (in the midst of an overall increasing trend) with landings fluctuating between 365,900 mt in 1972 and 250,000 mt in 1975. This decline was hypothesized to have resulted from climatic and environmental conditions which affected fishing operations and the menhaden stock, as well as from changes in market demand for menhaden products (Blomo, 1987). During the late 1970s, landings remained above the 340,000 mt level and continued to climb (Blomo, 1987; Smith et al., 1987; Vaughan et al., 1988).

In 1980 the stock showed signs of strengthening and significant expansion – 401,500 mt were landed. Landings in 1983, the highest since 1962, reached 418,600 mt. They fell, however, in 1986. Rather than a decline in stock abundance, this decline was more a reflection of socioeconomic conditions (Smith et al., 1987; Vaughan et al., 1988).

History of the industry

Menhaden have been used as fertilizer as far back as the early 1600s. The name menhaden is derived from the American Indian word *Munnawhatteaug*, meaning "fertilizer" (Frye, 1978). According to legend, the American Indians helped the European colonists to avoid starvation by teaching them how to use whole menhaden to fertilize their crops (Blomo, 1987; Hu et al., 1983). Documented evidence, however, supports the view that Squanto, a native New England Indian, simply introduced the use of fish as a fertilizer to the European settlers in Plymouth from his contact with other Europeans already using this agricultural practice (Ceci, 1975). In any event, the use of whole fish as fertilizer in the US became popular and widespread, when crops failed on Long Island and along the New England coast (Harrison, 1931). From colonial times into the 1800s, the principal value of menhaden was derived from its use as fertilizer.

New England fishermen also utilized great quantities of menhaden for bait, sparking a substantial trade (Ellison, 1951).

Growth of the fishery as a whole-fish fertilizer and bait trade, however, was limited. In the early 1800s, haul seines, and sometimes gill nets and traps were used by small companies to collect menhaden for fertilizer production (Harrison, 1931). It was not until the introduction of the purse seine in Rhode Island in 1845 (Frye, 1978) that the fishery had a chance for expansion. Successful usage of the purse seine spread down the coast to Virginia by 1869, and remains today as the most common gear used in the menhaden fishery.

It was, however, the discovery of the valuable menhaden by-products, such as fish oil, that actually turned the fishery into a major industry (Harrison, 1931). Centuries after the French and English produced oil and fertilizer from fish, the menhaden oil industry began in the US. In 1811, the first menhaden plant was opened in Rhode Island. Replacing whale oil, menhaden oil was used for tanning and curing leather, for the production of margarine and shortening for European markets, and as an ingredient in soap and paints (Frye, 1978; Hu et al., 1983).

Following the introduction of the purse seine, the use of the first curb press in 1856 added the ability to produce a greater yield of oil and a drier residue in the processing industry. The implementation of steam cooking at this time in combination with these technological advances prompted the development of steam factories as well as major expansion of the menhaden fishery along the Atlantic coast. By 1866 steam plants had been established from Maine (24 plants) to North Carolina (three plants) (Ellison, 1951; Harrison, 1931).

Not until the late 1800s, however, was the dried fish scrap utilized as a marketable commodity in the form of domestic farm animal feed (Blomo, 1987). By the turn of the twentieth century, the primary use of menhaden had changed from fertilizer to feed for poultry, swine, and cattle. The use of oil expanded to include the production of linoleum and waterproof fabrics (Hu et al., 1983).

The growth of the industry was concurrent with advancements in the technological aspects of the fish reduction process. Continuous steam cookers were replaced by steam cooking tanks; the press evolved from a lever press to a hydraulic curb press and eventually to a screw press. Other aspects of the reduction process included the replacement of solar evaporation drying techniques by a rotary flame drier, and the introduction of acidulation (commercial sulfuric acid) used in the preservation of the "chum" (scrap). As pressed fish putrefy rapidly, the introduction of an acid preservative was a major advancement overall in the industry and sparked the development of the fish-scrap industry. Steam-powered vessels, some built especially for menhaden fishing, increased fishing capacity by bringing fishing areas within easier reach and also enabling catches to arrive at the factory sooner, which meant that fish arrived in better condition. Together with the improved technology of the reduction equipment, powered vessels led to the construction of larger factories along the coast, many of which are still in existence today (Harrison, 1931; Hu et al., 1983).

By 1929, there were 37 factories in operation along the Atlantic coast utilizing an estimated 180 million kilograms of fish, with meal, oil, and scrap products valued at close to US$4 million. At this time the menhaden industry employed approximately 4,500 people earning over US$2.5 million (Harrison, 1931). The decline in menhaden catch during the end of the 1920s caused menhaden by-products to decline in the rank from first to third, giving way to the expansion of the freshwater mussel and pilchard by-products industries (Harrison, 1931).

Prior to World War II, the primary by-products of the fishery were dried fish scrap for fertilizers, and fish oil for fuel and industrial processes (June & Reintjes, 1976). After the War, the growth and rapid expansion of the poultry industry and the decline of the California (Pacific) sardine industry generated an increased demand for menhaden fish meal (McHugh, 1981; Vaughan et al., 1988).

Concurrently, the industry expanded and technology advanced. Larger, more efficient processing plants with greater capacities were supplied by faster, larger (>300 net mt) carrier vessels now being equipped with refrigerated fish holds. Spotter aircraft were used to locate schools of menhaden and new nylon purse seines were closed with hydraulic power blocks. Processing plants were established along the coast of the Gulf of Mexico by Atlantic coast companies (Vaughan et al., 1988).

In the early 1960s changes in the availability of menhaden off the coast of Maine (a change from great abundance to complete absence) caused temporary closings of plants along the North Atlantic coast (Fig. 6.5). The reappearance of significant numbers of

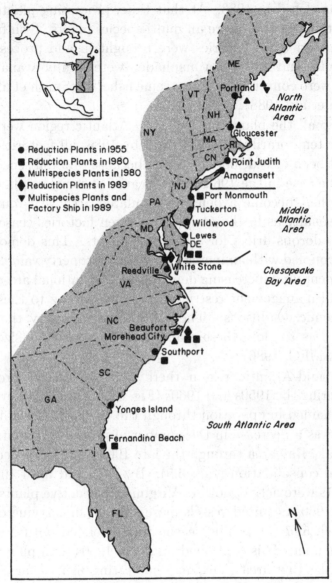

Fig. 6.5 Location of menhaden reduction plants in 1955, 1980, and 1989. Menhaden range from Nova Scotia to Florida. (Adapted from Atlantic States Marine Fisheries Commission, 1981, 1986.)

fish in the 1970s allowed some of these plants to reopen (Nicholson, 1975). In 1972, only two plants, one in Rhode Island and one in Massachusetts, were operating. By 1975, the Rhode Island plant had closed, and the Massachusetts plant was joined by two plants in Maine. Menhaden were still caught in Rhode Island but were trucked to Portland (Maine) for processing, until this plant was closed in 1983. At that time, processing plants in the North Atlantic region were all multi-species plants. Fish from the New England trawl fisheries were brought in and processed. In the summer and fall, when menhaden were locally available, the trawlers were converted to purse seine fisheries (Smith et al., 1987; Vaughan et al., 1988).

By 1986 all the plants in the North Atlantic region were either closed or temporarily inactive. And by 1988, all factories in this area had been closed, not, however, because of an absence of fish as was the case in the 1960s, but because of social conflicts. New England had become a densely populated area year-round, not only seasonally as in the early 1900s when factories ceased operation of odorous driers for summer residents. This demographic change, coupled with an increase in coastal property value, coastal development, and increasing demands for recreational areas, led to additional management restrictions corresponding to these shifting economic conditions. The resultant decision by the fishing industry was to close these plants and to reduce the size of the fleets (ASMFC, 1986).

In the mid-Atlantic region there were five active processing plants during the 1950s and 1960s (Fig. 6.5). Only one New Jersey plant remained in operation through the 1970s. It closed in 1981.

There was a decrease in the number of processing plants in the Chesapeake Bay area through the late 1960s and 1970s because of industrial consolidation (Fig. 6.5). By 1973 and until 1985 only two plants were active, both in Virginia. The active plants during the mid-1960s acquired vessels more efficient in carrying capacity and speed, and the fishing season was extended until November and December. This continued until the 1980s with plants in the Chesapeake Bay area (Virginia) harvesting fish as far south as Cape Hatteras (North Carolina) and, after 1983, north into the North and mid-Atlantic regions (Nicholson, 1975; Smith et al., 1987).

In the South Atlantic coastal area, the number of menhaden processing plants remained relatively stable from 1972 to 1984 (Fig. 6.5). During this period, two plants in Beaufort (North Carolina) operated during both the summer fishery and the North Carolina fall fishery season. A third plant in Beaufort and a plant in Southport (North Carolina) operated intermittently during the summer and North Carolina fall fishery season in this period. A single plant in Fernandina Beach (Florida) was active from 1972 to 1986 but suspended operation in mid-1987 (Smith et al., 1987; Vaughan et al., 1988).

For the entire Atlantic coast, the number of menhaden processing plants decreased steadily (Fig. 6.5), with little fluctuation, from 23 plants at 16 ports in 1955 to 11 plants at six ports in 1980 (seven of these 11 plants processed only menhaden; the others were multi-species plants). By 1984, only eight plants at five ports remained (ASMFC, 1981; Vaughan & Smith, 1988). Most of the plants were owned by four major companies (Hu et al., 1983). By 1989, only three plants remained at two ports (Reedville, Virginia, and Beaufort), in addition to a Russian factory ship operating off the coast of Maine (that is supplied by American vessels) and a multi-species plant and cannery off the coast of New Brunswick (Canada).

Economics of the industry

Through the mid-1900s, only sections of the menhaden industry along the coast were vertically integrated (Harrison, 1931). Today, the menhaden industry is the only fully integrated fishing industry in the US, meaning that companies own their own fishing vessels, processing plants, and marketing infrastructure (Hu et al., 1983).

The mechanization of menhaden plants has resulted in a less labor intensive industry in the processing phase than other seafood industries (Hu et al., 1983). In addition to processing plant jobs, the industry employs captains and crews for its vessels, airplane pilots, net repairers, administrative personnel, and other general support staff. Because the menhaden processing industry is a seasonal one, plants employ only maintenance and net-repair workers during the off-season (Hu et al., 1983). Employment in menhaden processing plants in 1982 ranged from 65 for large plants during peak production months to 15 for small plants during slow production months.

The estimated value of US menhaden products in 1980 was US$166.5 million, approximately 3.7 percent of all US processed fishery products (Hu et al., 1983). Menhaden constituted 76 percent of the fish meal produced in the US in 1980, with the balance made up by mackerel and tuna by-products. The US exported

77,000 metric tons of its fish meal (24%), primarily to West Germany. This quantity, however, was only 5 percent of the world's fish meal exports for that year. Approximately 94 percent of the fish oil produced in the US is derived from menhaden. In 1980, 130,000 kg of fish oil (91% of the total) was exported primarily to the Netherlands and UK. Japan is the largest US competitor in fish oil (Hu et al., 1983).

Markets for human consumption of menhaden existed in the 1700s and 1800s, when they were salted and eaten like herring. In the middle of the 1800s large quantities of salted menhaden were exported to the West Indies as well as being consumed in the US (Ellison, 1951; Lanier, 1985). However, the canning of small menhaden as "American sardines" (or "shadines") was abandoned around 1875 because of the superior quality of small herring caught off the coast of Maine (Ellison, 1951). Again, during World War II, menhaden was canned in fairly large quantities, most of which was exported. They were also possibly marketed for their roe; they are ripe when large quantities of roe from other fish are unavailable (Ellison, 1951).

Over the centuries menhaden has remained a highly abundant commercial fishery and is perhaps unequalled by any other known food in natural nutritive value at such a low cost (Ellison, 1951). Menhaden's high oil content and extremely bony structure has been the deterrent in its success as a direct food source. Today, however, enhanced use of menhaden for human consumption may result from increasing market demand for surimi. The protein-rich additive, surimi, is used in seafood analogue products and has potential for use in many processed meat products. Fillets and cuts also currently provide potential new markets for menhaden as a direct human food source. These are completely different items and markets from surimi.

New markets for menhaden oil products have also been identified. Extensive research at the Charleston Laboratory of the National Marine Fisheries Service focuses on the use of the oil as nutritional supplements and therapeutic agents. The US Food and Drug Administration has recently granted "safe status" to menhaden oil, which resulted in an increased demand by Americans (Anon., 1990). Refined menhaden oil contains high levels of omega-3 fatty acids; the oil product American food manufacturers favor. There are orders for the oil from manufactures of bread, pastries, cakes, cookies, imitation creams, and emulsifiers (Anon., 1990). Cooking oil produced from refined menhaden oil may also prove to be a new product line in America. Menhaden oil is already used in Europe as a cooking oil, and in margarine and baked goods.

Discussion

Human activities have direct effects on the menhaden fishery's success or failure; from the demands for by-products and location of factories to more complex human-induced degradation of habitats. After World War II, the demand for menhaden by-products increased, because of the decline of Californian sardine by-products. The present demand for menhaden is dependent upon the soybean products market, as well as on demands for polyunsaturated omega-3 fatty acid components and surimi. Market conditions for menhaden products determine the level of fishing pressure on the fishery. Presently, however, with the decrease in demand for menhaden fish meal as a result of the popularity of soybean-based feeds, the surimi market may fill the gap without generating an increase in fishing pressure. Approximately 20 percent of a fish is converted to surimi and the remainder is potentially available for meal production (Lanier, 1985). During times of high demand, menhaden have the potential to be over-exploited, but in this case changes in manufacturing techniques could prevent the undesired added pressure. The northern factories located along the coast, necessary because of proximity to fishing grounds, were overpowered by urban development, changes in land use from commercial to recreational, and the desire to end odor problems caused by the factories. Finally, society's role in the destruction of environments on which menhaden are dependent, by pollution and coastal development as well as potentially adverse changes in the ocean environment that might accompany a climate change, will certainly lead to changes in the fishery. For the US menhaden industry to remain viable, the interrelationships between the menhaden life history, fishery, and industry, society, and the driving forces of change in climate and estuarine and oceanographic processes, as well as management efforts must all be considered (Fig. 6.1).

Several comparisons have been made between the sardine (*Sardinops sagax*) fishery and the menhaden fishery (Clark, 1974; Henry, 1971; Nelson et al., 1977; McHugh, 1969). The bases of comparison lie in the parallels of life histories, spawner-recruit relationships and landings of the two clupeids. For sardine and for menhaden, a series of good year classes was followed by a series of years with poor survival. After the period of decline and restabilization of the sardine stocks, the fishery completely collapsed. There was concern that after the 1959–64 decline of the menhaden fishery, the stocks had been reduced to such a level that recruitment had been jeopardized (Henry, 1971).

The possibility exists that the Atlantic menhaden fishery could follow in the same path as the California Pacific sardine fishery. Extensive fishing pressure leads to unstable resource conditions and, therefore, leaves the fishery highly vulnerable to adverse environmental factors, which could also lead to collapse (Clark, 1974).

Overfishing of the Atlantic menhaden fishery resulted in the truncation of the age structure of the stock and the older, larger fish in the northern part of the range were fished out. This situation leaves the fishery economically vulnerable, as the resource abundance is concentrated in the southern part of the range where younger and smaller fish are harvested. Consequently, the overall spawning stock becomes depleted. Furthermore, as the area of fish concentration contracts, so too does the available fishing area; a greater number of vessels fish in a smaller area. This collapse of distribution also occurred in the sardine fishery. In order to control and stabilize this situation, the southern operations would have to be restricted to allow northern operations to expand. This presumes that an economic solution would be found to the problem of the processing plants being reopened in the north.

Improper management of the sardine fishery led to the disappearance of the fish meal and oil industry on the Pacific coast (McHugh, 1969). This could provide an important lesson to menhaden fishery managers.

Although the menhaden fishery still largely comprises prespawners, management of the menhaden fishery has prevented the fishery from exploitation to the point of complete collapse. The age structure of the stock is expanding as large quantities of fish are found once again in New England waters. With the closure of processing plants in this area, the resource has an even greater chance of expansion. Menhaden management consists of inputs from biological, fishery, industrial, and environmental sources. It appears that, although important research continues on menhaden life history and assessment of the stock, management is not waiting for irrefutable scientific evidence before taking action. This is important to the success and survival of the fishery.

In addition, regulations concerning estuarine habitats have increased since the 1960s in order to protect this environment as well as coastal marine environments from urbanization including the construction of new roads and other structures for the purpose of recreational use, and residential coastal development (Siry, 1984). New awareness in the US for protection of the environment may keep estuarine habitats from being degraded so that they may continue to provide nursery and spawning grounds for valuable marine life.

Knowledge and research on the effects of changes in water quality continues to be an important priority, whether these changes are introduced by society (pollutants) or by nature (environmental changes leading to changes in introduction, distribution, and abundance of species) via organisms which are parasitic or cause disease.

Much attention is placed on the possibility of a collapse of the fishery. Menhaden may, however, outlive this fishery simply because of socioeconomic changes along the Atlantic coast.

The history of the menhaden fishery provides some insight into the management of a living marine resource in the face of environmental uncertainty. Given the existing speculation about what global climate change might mean for living marine resources at the regional level, lessons drawn from the menhaden case study might offer insights into the problems associated with living marine resource management under conditions of societal as well as environmental change. The following lessons suggest the kinds of actions that might be considered to prepare better the fisheries facing the consequences of a potential global warming of a few degrees Celsius in the next few decades.

Lessons

- Overfishing in combination with adverse environmental conditions can lead to a serious decline in a resource and, consequently, to economic hardships in the industry dependent upon that resource.

- An industry must be able to adapt to changes in availability or abundance of resources and be able to create new products and uses for the resource in order to meet changes in market conditions and to ensure resilience of the industry.

- The use of newly available technologies, in the absence of new compensating management restrictions, puts a tremendous burden on an exploited living marine resource, resulting in declining economic viability of the industry.

- Changes in land use can affect the overall efficiency of fishing industries.

- Final scientific corroboration is not always necessary before effective management policies can be established.

- Analogies to similar fisheries which have undergone extreme fluctuation and collapse can provide abundant insight into necessary management decisions.

- Environments on which fishery resources are dependent must be protected, and care taken to monitor their changes in response to climatic changes.

- An industry which depends on a highly migratory resource with an extensive geographical range needs to be managed over the total range, in order to ensure sustainability of the resource.

Acknowledgments

The Southeast Regional NMFS Laboratory in Beaufort, North Carolina, conducts the majority of menhaden research. I thank the scientists at this laboratory for their research and for the time they spent discussing menhaden with me. I owe a great deal to Joseph E. Smith for his correspondence and discussions and, along with John V. Merriner, valuable comments and suggestions on the manuscript.

References

Ahrenholz, D.W., Nelson, W.R., & Epperly, S.P. (1987). Population and fishery characteristics of Atlantic menhaden, *Brevoortia tyrannus*. *Fishery Bulletin*, **85**, 569–600.

Anon. (1990). Menhaden oil in demand. *Fishing News International,* **29**(6), 91.

ASMFC (Atlantic States Marine Fisheries Commission) (1981). *Fishery Management Plan for Atlantic Menhaden,* Brevoortia tyrannus *(Latrove).* Fisheries Management Report No. 2. Washington, DC: Atlantic Menhaden Management Board.

ASMFC (Atlantic States Marine Fisheries Commission) (1986). *Supplement to Atlantic Menhaden Fishery Management Plan.* Fisheries Management Report No. 8. Washington, DC: Atlantic Menhaden Management Board.

Blomo, V.J. (1987). Distribution of economic impacts from proposed conservation measures in the US Atlantic menhaden fishery. *Fisheries Research,* **5**, 23–38.

Ceci, L. (1975). Fish fertilizer: a native North American practice? *Science,* **188**, 26–30.

Checkley, D.M., Raman, S., Maillet, G.L. & Mason, K.M. (1988). Winter storm effects on the spawning and larval drift of a pelagic fish. *Nature,* **335**, 346–8.

Clark, C.W. (1974). Possible effects of schooling on the dynamics of exploited fish populations. *Journal du Conseil,* **36**, 7–14.

Cushing, D.H. (1974). The natural regulation of fish populations. In *Sea Fisheries Research* ed. F.R. Jones, pp. 319–412. London: Elek Science.

Durbin, A.G. & Durbin, E.G. (1975). Grazing rates of Atlantic menhaden *Brevoortia tyrannus* as a function of particle size and concentration. *Marine Biology,* **33**, 265–77.

Ellison, W.A., Jr. (1951). The menhaden. In *Survey of Marine Fisheries of North Carolina,* ed. H.F. Taylor & Associates, pp. 85–107. Chapel Hill: University of North Carolina Press.

Engel, D.W., Hettler, W.F., Coston–Clements, L. & Hoss, D.E. (1987). The effect of abrupt salinity changes on the osmoregulatory abilities of the Atlantic menhaden, *Brevoortia tyrannus. Comparative Biochemistry and Physiology,* **86A**, 723–7.

Friedland, K.D., Haas, L.W. & Merriner, J.V. (1984). Filtering rates of the juvenile Atlantic menhaden *Brevoortia tyrannus* (Pisces: Clupeidae), with consideration of the effects of detritus and swimming speed. *Marine Biology,* **84**, 109–17.

Frye, J. (1978). *The Men All Singing.* Norfolk/Virginia Beach: Donning.

Frye, R. (1983). Climate change and fisheries management. *Natural Resources Journal,* **23**, 77–96.

Hargis, W.J., Jr. (1985). Quantitative effects of marine diseases on fish and shellfish populations. *Transactions of the North American Wildlife and Natural Resources Conference,* **50**, 608–40.

Harrison, R.W. (1931). *The Menhaden Industry.* Washington, DC: Bureau of Fisheries, US Department of Commerce.

Henry, K.A. (1971). *Atlantic Menhaden* (Brevoortia tyrannus) *Resource and Fishery – Analysis of Decline.* Technical Report NMFS SSRF-642. Washington, DC: NOAA.

Higham, J.R. & Nicholson, W.R. (1964). Sexual maturation and spawning of Atlantic menhaden. *Fishery Bulletin,* **63**, 255–71.

Hoss, D.E., Hettler, W.F. & Coston, L.C. (1974). Effects of thermal shock on larval estuarine fish: ecological implications with respect to entrainment in power plant cooling systems. In *The Early Life History of Fish*, ed. J.H.S. Blaxter, pp. 357–71. New York: Springer-Verlag.

Hu, T.–W., Whitaker, D.R. & Kaltreider, D.L. (1983). *The US Menhaden Industry: An Economic Profile for Policy and Regulatory Analysts.* National Technical Information Service No. PB83–165720. Springfield: NTIS.

Judy, M.H. & Lewis, R.M. (1983). *Distribution of Eggs and Larvae of Atlantic Menhaden*, Brevoortia tyrannus, *Along the Atlantic Coast of the United States.* Technical Report NMFS SSRF–774. Washington, DC: NOAA.

June, F.C. & Reintjes, J.W. (1976). The menhaden fishery. In *Industrial Fishery Technology*, ed. M.E. Stansby, pp. 146–59. New York: Krieger Publishing Company.

Lanier, T. (1985). *Menhaden: Soybean of the Sea.* Publication UNC–SG–85–02. Raleigh: University of North Carolina Sea Grant College.

Lewis, R.M. (1965). The effect of minimum temperature on the survival of larval Atlantic menhaden, *Brevoortia tyrannus. Transactions of American Fisheries Society*, **94**, 409–12.

Lewis, R.M., Ahrenholz, D.W. & Epperly, S.P. (1987). Fecundity of Atlantic menhaden, *Brevoortia tyrannus. Estuaries*, **10**, 347–50.

Lewis, V.P. & Peters, D.S. (1984). Menhaden – A single step from vascular plant to fishery harvest. *Journal of Experimental Marine Biology and Ecology*, **84**, 95–100.

McHugh, J.L. (1969). Comparison of Pacific sardine and Atlantic menhaden fisheries. *Fiskeridirektoratets Skrifter, Serie Havundersøkelser*, **15**, 356–67.

McHugh, J.L. (1981). Marine fisheries of Delaware. *Fishery Bulletin*, **79**, 575–99.

NMFS (National Marine Fisheries Service) (1973). *Status of the International Fisheries off the Middle Atlantic Coast.* Laboratory Ref. No. 73-2, 102–5. Woods Hole: Northeast Fisheries Center.

Nelson, W.R., Ingham, M.C. & Schaaf, W.E. (1977). Larval transport and year-class strength of Atlantic menhaden, *Brevoortia tyrannus. Fishery Bulletin*, **75**, 23–41.

Nicholson, W.R. (1971). Changes in catch and effort in the Atlantic menhaden purse-seine fishery 1940–68. *Fishery Bulletin*, **69**, 765–81.

Nicholson, W.R. (1975). *Age and Size Composition of the Atlantic Menhaden*, Brevoortia tyrannus, *Purse Seine Catch, 1963–71, With a Brief Discussion of the Fishery.* US Department of Commerce Technical Report NMFS SSRF-684. Seattle: NOAA.

Nicholson, W.R. (1978). Movements and population structure of Atlantic menhaden indicated by tag returns. *Estuaries*, **1**, 141–50.

Pacheco, A.L. & Grant, G.C. (1965). *Studies of the Early Life History of Atlantic Menhaden in Estuarine Nurseries. Part I. Seasonal Occurrence of Juvenile Menhaden and Other Small Fishes in a Tributary Creek of Indian River, Delaware, 1957–58.* Special Scientific Report, Fisheries No. 504. Washington, DC: US Fish & Wildlife Service.

Peters, D.S. & Schaaf, W.E. (1981). Food requirements and sources for juvenile Atlantic menhaden. *Transactions of the American Fisheries Society*, **110**, 317-24.

Powell, A.B. & Phonlor, G. (1986). Early life history of Atlantic menhaden, *Brevoortia tyrannus*, and Gulf menhaden, *B. patronus*. *Fishery Bulletin*, **84**, 935-46.

Reintjes, J.W. (1969). *Synopsis of Biological Data on the Atlantic Menhaden*, Brevoortia tyrannus. FAO Species Synopsis No. 42, Circ. 320. Washington, DC: US Fish & Wildlife Service.

Reintjes, J.W. & Pacheco, A.L. (1966). The relationship of menhaden to estuaries. In *A Symposium on Estuarine Fisheries*, ed. R.F. Smith, A.H. Swartz & W.H. Massman, pp. 50-8. Special Publication 3. Washington, DC: American Fisheries Society.

Rothschild, B. (Ed.) (1983). *Global Fisheries: Perspectives for the 1980s*. New York: Springer–Verlag.

Siry, J.V. (1984). *Marshes of the Ocean Shore: Development of an Ecological Ethic*. College Station: Texas A&M University Press.

Smith, J.W., Nicholson, W.R., Vaughan, D.S., Dudley, D.L. & Hall, E.A. (1987). *Atlantic Menhaden*, Brevoortia tyrannus, *Purse Seine Fishery, 1972–84, With a Brief Discussion of Age and Size Composition of the Landings*. Technical Report NMFS 59. Washington, DC: NOAA.

Turner, J.T., Tester, P.A. & Hettler, W.F. (1985). A laboratory study of predation on fish eggs and larvae by the copepods *Anomalocera ornata* and *Centropages typicus*. *Marine Biology*, **90**, 1-8.

Vaughan, D.S., Merriner, J.V. & Smith, J.W. (1988). The US menhaden fishery: current status and utilization. In *Fatty Fish Utilization: Upgrading from Feed to Food*, ed. N. Davis, pp. 15-38. Sea Grant Publication 88-04. Raleigh: University of North Carolina.

Vaughan, D.S. & Smith, J.W. (1988). *Stock Assessment of the Atlantic Menhaden*, Brevoortia tyrannus, *Fishery*. Technical Report NMFS 63. Washington, DC: NOAA.

7

Maine lobster industry

JAMES M. ACHESON

Department of Anthropology
University of Maine
Orono, ME 04473, USA

Maine lobstering: general information

The American lobster (*Homarus americanus*, see Fig. 7.1) is found in the waters off the Atlantic coast of North America from Newfoundland to Virginia. Concentrations of lobsters are greatest in waters less than 55 meters deep. Although lobsters are found on all kinds of bottom types, they prefer rocky areas, especially where there is a good deal of kelp in which to hide.

Lobsters eat a wide variety of foods, both living and dead organisms. Their preferred foods are fish, mollusks and small crustaceans. They can also filter plankton from the water, and thus can live in untended traps for considerable periods. They are also cannibalistic and will eat small lobsters and soft shelled lobsters regardless of size. For this reason fishermen immobilize the lobster's claws, usually by placing a thick rubber band around each claw, making it impossible to open.

When lobsters have outgrown the capacity of their shells, molting occurs. During molting the lobster wiggles out of its shell, after which the lobster is soft, weak, and highly vulnerable. Its only defense is to hide for a few weeks until its shell has hardened again. Although lobsters can molt in any month, a very large proportion molt from mid-June to mid-August. For this reason, fishing is bad during mid-summer, since so many are in the rocks and not feeding. Small lobsters molt several times a year, but commercial size lobsters molt only once.

Lobsters mate after the female has molted. The female can exude as many as 50,000 eggs, which remain attached to her abdomen until they hatch, usually during the following summer. Female lobsters do not mature sexually until they are at least 80 mm on the carapace; and 50 percent are not mature until they reach 90 to 95 mm (Krouse, 1972, 1973).

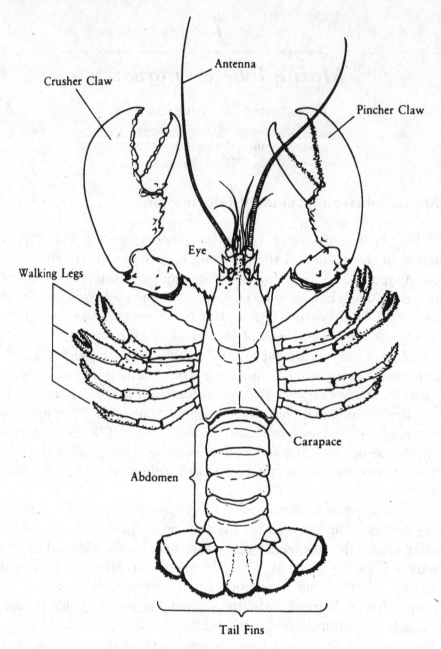

Fig. 7.1 The American lobster (*Homarus americanus*).

Lobsters are not found in rivers or estuarine areas as they require highly saline water. Recent studies indicate that lobsters migrate locally. They tend to migrate in towards the shore during the warm months and into deeper water in the cold winter months. Water temperature affects migration, growth rates, mortality, and

catches.   Catches are maximized when mean annual sea water
temperature is between 9° and 11°C (Dow, 1969, pp. 61–63).

Maine consistently produces far more lobsters than any other
state in the US (Fig. 7.2). In 1989, approximately 9,000 lobster
licenses were granted. About 2,400 of these fishermen can be considered as "full time." Only 9 percent of the licenses are held currently by women, and the majority of these go lobstering on a
part-time basis, if at all; many serve as helpers for their husbands
or boyfriends (Acheson, 1988b, p. 3).

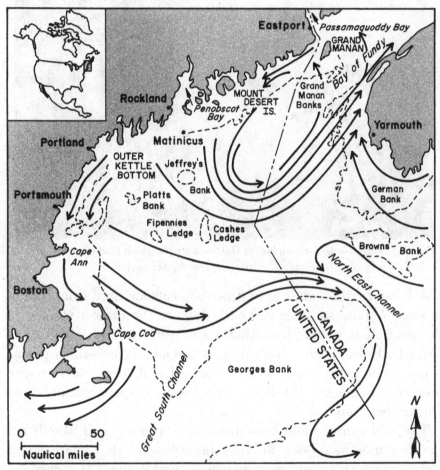

Fig. 7.2 Major fishing grounds and currents of the Gulf of Maine.

Most full-time lobster fishermen fish from a diesel or gas-
powered boat between 28 and 38 feet (8.5 and 11.5 m) long,
equipped with radio, hydraulic pot hauler, electronic sounding
gear, and perhaps a radar set. Lobsters are caught in three- or
four-foot traps covered with wooden lathes or wire (Fig. 7.3). Fish

remnants from processing plants generally serve as bait. The traps are attached to a styrofoam buoy. All of the traps of each fisherman are painted with a distinctive set of colors, which are registered with the State of Maine.

Fig. 7.3 A lobster fisherman leaves the dock with a small load of traps. (Courtesy of *The Times Record*, Brunswick, Maine.)

A lobster fisherman's activities vary considerably from season to season. Late summer and early fall are the busiest months of the year. It is at this time that a new year class of lobsters has molted into legal size. From August to October fishermen put as many traps in the water as possible and pull them every chance they get. Some two-thirds of the total annual catch is obtained during these months.

The mid-winter months of January, February, and March are the slow months of year for trapping lobsters. Bad weather and storms make the work more dangerous and difficult. Catches decline because lobsters are sluggish and not inclined to crawl into traps. Also, during these months, many fishermen pull their traps only a few times a month; others cease fishing entirely and use their time to build traps.

Any sizable harbor has at least one dealer who buys lobsters from local fishermen and sells them to tourists or to one of the wholesale firms in Portland (Maine) and Boston (Massachusetts), among other cities, distributing lobsters to the nation and foreign countries.

Fishermen ordinarily sell lobsters each day they go fishing to a single dealer with whom they have developed a long-term relationship (Acheson, 1985a). In recent years fishermen had formed cooperatives in 17 coastal communities (Acheson, 1988b, p. 129).

The lobster industry is highly territorial. To go lobster fishing, one must have a license from the State of Maine. In addition, one must be accepted as a member of a "harbor gang" – the group of fishermen who fish from a given harbor. Once entry has been gained, one can only fish in the traditional territory of that harbor gang. A typical fishing area might contain 130 to 260 km^2 of ocean. Repeated violations of territorial boundaries usually result in surreptitious trap cutting; that is, the offender's traps are pulled up, the buoys are cut off, and traps are thrown into deep water where they cannot be found. Although destruction of another's traps is illegal, it is a standard method of defending boundaries. Sometimes small-scale incidents can escalate into full-fledged "cut wars" (Acheson, 1988b, pp. 71–83).

From 1920 to the present, lobstering has changed relatively little. In the 1920s, boats and traps were essentially the same as those used today. The boats were smaller, cockpits were largely open and there was no electronic equipment. At that time, traps were all wood; the wire traps were not used in numbers until the 1970s. Only a very few important innovations have been adopted since the 1920s. The most important is the hydraulic trap hauler, which came into general use in the early 1950s. This innovation made it possible for fishermen to pull many more traps in a day. Another important technological change was the introduction of nylon string, which in the early 1960s replaced hemp and other vegetable fibers as the material used to make "heads" – the funnel-shaped net used to retain lobsters that have entered the trap. Nylon lasts much longer than hemp or other such fibers and its use has greatly reduced trap maintenance costs. A third innovation has been the electronic sounder which has made it much easier to get data about the ocean floor. While radios, radar and Loran (an electronic navigational device) all helped to make lobstering safer, these innovations did not change the nature of the industry much.

In the early decades of the twentieth century, the territorial system was in full bloom. However, in the past half century, territories have become larger, as abandoned island territories were incorporated into other areas, and the amount of area where mixed fishing has been allowed (i.e., an area fished by two harbor gangs) has increased because of competition among harbor gangs (Acheson, 1979, pp. 265–75).

On the marketing scene, the major change has been the advent of cooperatives, which began in the late 1940s. Although there are only 17 cooperatives at present (1990), they serve as a check on the power of the private dealers. Should a dealer's behavior become impossible, fishermen can always form a cooperative (Acheson, 1985b, p. 121–8).

Causes of the disaster of the 1920s and 1930s

During the entire inter-war period, the lobster industry was in the worst condition in its entire history. Catches fell dramatically from their pre-World War I levels, and remained low until World War II. Between 1880 and 1900 the lobster catch ranged from 14 million pounds (6.4 million kg) to a high of 24.5 million pounds (11.1 million kg) in 1889 (Dow, 1967, p. 4; Martin & Lipfert, 1985, p. 51). It appeared to be about the same until World War I, although the data from this period admittedly are poor. Just after the war, the catch dropped dramatically and remained at low levels until World War II. In 1919 the catch was only 5.8 million pounds (2.63 million kg); in 1924 it had fallen to 5.1 million pounds (2.32 million kg). In the 1930s the catch ranged from 5.5 million pounds (2.5 million kg) in 1934 to a high of 7.7 million pounds (3.5 million kg) in 1935 (Dow, 1967, p. 5).

From 1947 to the present, catches have been much higher, ranging from 15.9 million pounds (7.22 million kg) in 1948 to 24.4 million pounds (11 million kg) in 1957 (Townsend & Briggs, 1982, p. 10). It would appear that in the past half century, catches have returned to the pre-World War I level, which at the present time appears to be some kind of norm for the Maine lobster industry.

Five hypotheses have been suggested as to what caused catches to drop so drastically in the inter-war period. Each view has different champions.

Hypothesis 1

Changes in water temperature may have reduced the stocks and the catches. This hypothesis, which was seriously argued in the 1950s and 1960s, appears to have little support. Accurate records of sea water temperature changes have been kept at Boothbay Harbor (Maine) since 1905. There is no difference in the temperature range in the pre-World War I and the inter-war years. Robert Dow, who has thoroughly studied the effect of temperature changes on lobsters, concludes "that some other factor than temperature had limited lobster landings during the period between wars" (Dow, 1967, p. 4).

Hypothesis 2

This hypothesis suggests that prices were so low that fishermen could not make a living in lobstering. Thus, many fishermen went out of business and the total number of traps was reduced, which in turn reduced the catch. This hypothesis has more support than hypothesis 1. The argument for it is best phrased by Dow, who concludes that "The 'real' landed value of Maine lobster averaged US$0.023 to US$0.101 per pound below the 1915 and 1916 levels throughout the 1920s and the 1930s, and probably accounts for the low fishing effort, which averaged only 203,000 traps per year for the period, well below the 234,000 trap minimum effort of pre-World War I years" (Dow, 1967, p. 4).

Closer inspection of the data gives reason to pause, however. Certainly, average prices were very low by present-day standards, but they were not that low in terms of constant dollars. In the depths of the Depression, ex-vessel prices averaged about US$0.18 to US$0.19/lb ($0.39 to $0.42/kg), and in 1935 it was US$0.23/lb ($0.51/kg). A dollar was worth much more in the early 1930s (19 cents per pound then is equivalent to US$1.80 in 1989 dollars; and US$0.23 is worth US$2.17 today).* These are relatively

* Information on the consumer price index (CPI) was taken from McConnell (1963). These figures should be taken "with a grain of salt," since figures on the CPI are estimates at best and are calculated from the prices of a set of typical goods. The goods in the "market basket" change over time. There was nothing like personal computers in 1930, for example. Given the problems with

low prices by today's standards, but not by much. In the summer of 1989 the ex-vessel price for lobsters dropped to US$1.75/lb ($3.85/kg) in late August and early September. This was low enough to cause all fishermen to complain bitterly; many of them went on strike for higher prices. Prices are apt to be at their annual low in early September, since this is the time when the catch is largest. In November of 1989, the price had climbed to US$2.40/lb ($5.28/kg). This was only a bit lower than the US$2.56 ($5.63/kg) average price for the year, which was not much higher than US$2.17/lb ($4.77/kg) – the 1933 average ex-vessel price corrected for inflation. Thus, there is reason to suspect that low prices alone were not the cause of people dropping out of the industry.

The income of fishermen, of course, is determined both by the size of the catch and the price received for it. Low prices can be compensated for by increases in catch. Those who assert that low price alone caused fishermen to drop out of the industry are assuming that catches were not high enough to compensate. Is it possible that catches and catch per unit of effort (CPUE) were low because the stock was low? There is some reason to think this might have been the case. Older fishermen report that in the 1930s fishermen often came home with only 20 or 30 pounds (9 or 13.5 kg) for a day's work. Moreover, some of the statistical evidence points to the same conclusion. We know that in 1940, about 3,700 fishermen took approximately 7.5 million pounds (3.4 million kg) of lobster (Martin & Lipfert, 1985, p. 85). Assume that 1,500 of these men were full-time fishermen. (Large numbers of license holders do not fish full time; and it is the full-time fishermen who take virtually all the lobsters.) Under these conditions, the average fisherman, who remained in the industry, would catch approximately 5,000 pounds (2,270 kg) of lobster using about 100 traps. Thus, each trap would produce about 50 pounds (22.7 kg) a year. In the 1980s a typical full time fisherman might catch 20,000 pounds (9,080 kg) of lobster using about 400 traps. Today, each trap would still produce about 50 pounds a year. But now there are many more traps in the water. If indeed the average full-time fisherman in the 1930s caught 5,000 pounds of lobster annually

calculating the CPI over time, it would be reasonable to assume that a 1989 US dollar is worth about 1/10 of a 1933 dollar.

with this number of traps, CPUE was not high. This gives reason to suspect that all was not well with the stock.

Hypothesis 3

The third hypothesis is favored by many fishermen. They point out that the legal catch size of lobsters from 1907 to 1932 was 4.75 inches (12 cm) on the carapace (Judd, 1988, p. 598) – it is currently 3.25 inches (8.26 cm). This meant that a very large percentage of the lobsters which are currently legal, were illegal at that time. Fishermen called this large gauge the "poverty gauge," since it ostensibly impoverished so many fishermen. Many believe that the gauge itself was the basic cause for the low catches. As Phil Davis of Pleasant Point (Maine) once told me, "you could pull a lot of traps before you could get one you could keep."

Nevertheless, the data do not support these assertions as well as one might have expected. Certainly, a large gauge makes it illegal to take lobsters that could be taken at a smaller gauge size, but those smaller lobsters presumably would stay in the water for another year, and if there would be increased natural mortality in those years, those that remained would weigh a good deal more when caught. The idea that a larger gauge automatically leads to a smaller catch is not true. The only study that has been done concerning the economic effect of raising the minimum legal size argues that total pounds landed and the value of the catch would be increased by about 6 percent, if the gauge were increased in the 1980s up to 3.5 inches (8.9 cm) on the carapace (Acheson & Reidman, 1982, pp. 10–11). Moreover, the large measure was in force between 1907 and 1914, and catches during those years were in excess of 14 million pounds (6.36 million kg). Catches did decline during World War I, but this was probably due to a decline in the number of fishermen (Dow, 1967, p. 4).

In addition, if the large gauge had been the source of the decline in catches, then one would expect that when the legal minimum size was reduced, catches should have increased. This does not seem to have been the case. The legal minimum size was reduced from $4\frac{3}{4}$ to $3\frac{1}{16}$ inches (12 to 7.8 cm) in 1933. In 1932, the catch was 6.2 million pounds (2.81 million kg); in 1933, it fell to 5.8 million pounds (2.63 million kg); and in 1934 it fell again to 5.5 million pounds (2.5 million kg). The catch increased to 7.7 million pounds (3.5 million kg) in 1935, but declined once again in 1936 to 5.2 million pounds (Dow, 1967). It was not until 1943 that catches exceeded 10 million pounds (4.54 million kg) where they have remained to the present. If the big gauge were the source of the catch problem, it took 10 years after the minimum size was reduced for the benefits to show up.

Hypothesis 4

The fourth hypothesis is based on the belief that there was a tremendous amount of illegal activity in the lobster industry, which damaged the breeding stock. Indeed, many older fishermen can relate incidents which indicate widespread disregard for conservation laws concerning lobsters. One man told the author that his father and uncle, in time of need, used to scrub the eggs off female lobsters with a stiff brush and sell the lobsters. This was strictly illegal then, as it is now. According to some observers, there was a large trade in undersized lobsters. One older Bristol (Maine) fisherman recalls the time when two fishermen would leave large numbers of short lobsters in sunken crates in the middle of John's Bay (Maine), which would then be retrieved in the middle of the night by a smack (a vessel with flowing seawater tanks). They would then be transported to Boston. This one illegal operation alone accounted for untold hundreds of thousands of pounds of small lobsters per year. At that time it was also standard practice for fishermen to take short lobsters home to feed their families.

Such stories are reinforced by all observers of the industry. Martin and Lipfert report that violations of the legal size limit were so severe in 1920 that the Commissioner "suspended lobstering in mid-coast Maine because of repeated violations" (Martin & Lipfert, 1985, p. 55). Large numbers of illegal Maine lobsters also found their way to restaurants in neighboring New England states in the 1920s and 1930s (Judd, 1988, p. 612).

In the eyes of some fishermen this hypothesis is buttressed by the fact that the lobster catch recovered during World War II, when the numbers of fishermen and boats were reduced drastically because of a decline in the number of available fishermen (and the level of effort) brought about by the military draft and the war effort. It is also substantiated by the fact that today there is clearly much more support for the conservation laws, and, therefore, far fewer violations of them (Acheson, 1988b, pp. 137–41).

Hypothesis 5

The fifth hypothesis, relating to the drastic decline in lobster catches in the 1920s and 1930s, is referred to as the "culturing" hypothesis. It is argued that the lobster population is dependent in part on the food obtained from the bait they eat from the hundreds of thousands of traps placed in Maine's waters. In the 1920s and 1930s there were not enough traps in the water to provide enough food to sustain a large stock size. Dr. Robert Stennick, one of the very best and most imaginative lobster biologists, seriously argues that the amount of bait in traps adds considerably to the food supply for lobsters and that the lobster population is positively associated with the number of traps fished. He is currently gathering data to evaluate this hypothesis.

Some available data, however, tend to negate the culturing hypothesis. Before World War I, the number of traps was about 234,000, while in the 1920s and 1930s, the number of traps averaged 203,000 (13% less) (Dow, 1967). Can a 50 to 75 percent decline in lobster catches really be explained in terms of a 13 percent decline in the number of traps? If these figures on traps are accurate, this will be a difficult hypothesis to support.

Summary

Of the five explanations offered, those that appear to have the most validity are: (a) the idea that the low price resulted in fewer fishermen and reduced effort, which reduced the catch and (b) the massive violation of the conservation and size laws, probably reduced the breeding stock. The culturing hypothesis still remains to be demonstrated. The other two hypotheses have such serious flaws that they should not be taken seriously.

Alternative economic options

To assess how people in the fishing industry adapted to the serious and long-term decline in lobster catches, one needs to know about the economic options available in the region. As can be seen from the following discussion, they were not very good.

Communities along the coast of Maine witnessed their economic heyday in the years before the American Civil War in the early 1860s. At that time, the economy was based on fishing, lumbering, manufacturing, and – most importantly – shipbuilding. In the 1850s, Maine built more than a third of all the ships constructed in the US and more than any other state (Rowe, n.d., p. 144). Many of these were small fishing vessels or were built for the coastal trade, but a good percentage were very large vessels, well over 2,000 tons, which ranged to every deep water port in the world. Many of these ships, including the famous clipper ships of the 1850s, were built for owners in other states or nations. A sizable proportion, however, were locally owned and manned by Maine crews (Rowe, n.d., p. 144).

Shipping and commerce became big business even in relatively small towns like Boothbay. Farming was at its height in New England as a whole, and it is clear that farms in coastal Maine shared in the general prosperity (Russell, 1976, p. 258). In addition to brickyards and tanneries, manufacturing plants and sawmills were established, usually along streams where machinery could be operated by water power. In these years, Maine and Massachusetts were the top fishing states in the nation. Cod and mackerel were the two key species (McFarland, 1911, p. 169). The lobster industry was very small, because there was no way to transport lobsters to distant markets.

The closing decades of the nineteenth century were a period of economic decline for most of Maine's coastal towns. As railroads and steamships replaced sailing vessels, the shipbuilding and shipping industry – the economic lifeblood of these communities – ebbed rapidly. By the 1870s, large-scale lumbering had apparently ceased. Fishing went into a period of eclipse as the cod, menhaden, and mackerel industries declined.

The picture was not completely bleak, however, for a few industries developed to take up the slack. Quarries were established in many coastal communities and islands, and granite from them was shipped all over the US and beyond for the construction of public buildings (Stahl, 1956, p. 418). The ice industry was also at its peak. At that time, ice was cut from freshwater ponds and rivers and was shipped (packed in sawdust) to the hotter American cities and as far away as Brazil where demand was enormous (Clifford, 1961, pp. 46–9; Stahl, 1956, p. 362). Moreover, agricultural production was very high. Some figures suggest that agriculture reached its peak about 1900 and then declined. The lobstering and herring industries had begun to expand. In many towns, canneries had been established to put up lobsters, clams, salmon, and the like. In the 1870s tourism had already begun. Every little town had several guest houses. However, these enterprises were not enough to offset the decline of the older industries. During these years, people began to leave the area to seek better opportunities elsewhere.

The downward spiral continued well into the twentieth century. The granite industry, which had come to its peak after the American Civil War, was almost extinct by World War I; and the shipping of ice to places such as New York, New Orleans, and Brazil had ended by 1918 (Clifford, 1961, p. 91). The coastal schooner trade was almost dead by World War II. Moreover, after the turn of the century, agriculture declined rapidly. By the 1930s there were very few working farms in coastal Maine. Clifford's (1961) excellent history of Boothbay, which began its chronology in 1906, does not even mention agriculture.

In the inter-war years, there were few growth industries. One was tourism, although the tourist industry had changed considerably since its inception in the late 1800s. Before World War I, tourists brought trunks and tended to stay for weeks in the coastal hotels or cottage colonies; after World War I, increasing numbers of tourists built their own cottages or were on an auto tour and simply stayed overnight in a coastal motel. The importance of the tourist industry became apparent in the Depression years. In an era when unemployment was high, and local services and wages of public employees were reduced, and fish prices were low, it was one of the few industries that continued to expand.

Another growth industry in Maine was shipbuilding. During both World War I and World War II, shipbuilding boomed. Not only did the Bath Iron Works produce a large number of destroyers, but smaller firms in Thomaston, Boothbay, South Bristol, and Waldoboro (Maine) built smaller vessels for use in minesweeping and anti-submarine activities (Clifford, 1961, p. 263). In the middle of the Depression of the 1930s, however, few warships were built and the market for yachts and fishing boats declined dramatically. Nevertheless, during the Depression the shipyards stayed open.

Seen in this perspective, the decline of the lobster industry in the 1920s and 1930s was more like a crowning blow to an already faltering state economy than an isolated tragedy.

Adaptation to industrial decline

How did people adapt to the decline in catches in the lobster fishing industry? What choices did they make to solve the problems this decline caused? Four response strategies need to be considered.

First, Maine fishermen have always done a good deal of switching among fisheries (Acheson, 1988a, p. 49ff). One strategy that might have been resorted to was to switch out of lobstering into another fishery. In the study area, this strategy does not seem to have been a favored one. Certainly, there was no marked trend out of lobstering and into other fisheries. The reason was that if lobstering was bad, other fisheries were likely to be as bad if not worse. From World War I to the end of the 1920s, catches of groundfish declined due to a decrease in catches of cod. In addition, the herring catch was generally low in the 1930s in this part of Maine. The general decline in fish landings was offset to some extent by increases in haddock catches, which by 1935 had surpassed cod catches (Ackerman, 1941, p. 16). In the late 1930s there was some switching by fishermen into the mackerel industry. In the 1920s the annual mackerel catch averaged about 25 million pounds (11.35 million kg); between 1935 and 1940, catches increased to about 50 million pounds (22.7 million kg) (Ackerman, 1941, p. 31). Moreover, after 1930 and the advent of freezer technology, the catch of redfish increased considerably. This meant an increase in employment on two dozen offshore vessels operating out of Rockland, and a few other Maine towns.

One of the few towns in the region that seemed to have responded to these changes in species availability was New Harbor (Maine). In the 1930s there were seven or eight seiners working out of New Harbor, devoting most of their time to the capture of herring and mackerel. In addition, several boats in the town were gillnetting for groundfish (Ackerman, 1941). At the time, New

Harbor was known as a seiner port – not as a lobstering harbor (Acheson et al., 1978, p. 39).

Conditions in other fisheries may have been so bad that the movement during these years may have been into lobstering, not out of it. In this regard, Clifford (1961, p. 204) says that "fishing in general, as a means of livelihood, failed to advance in the 1920s, but lobstering moved still farther ahead of other elements in the industry."

Moreover, the number of people in the lobstering industry does not seem to have declined, as the lobster catch went down. In 1906, there were 2,562 fishermen in Maine; in 1940, after 20 years of low catches, there were 3,717 (Martin & Lipfert, 1985, pp. 76–77). To be sure, these figures almost certainly were obtained from the license list and many people who had licenses did not do much lobster fishing. Nevertheless, the figures do not suggest any general movement out of lobstering as catches declined; quite the contrary.

Second, it has been suggested that people may have entered other industries in the same area. This does not seem to have been the case either. If one compares the lists of occupations in the 1920 census to those in the 1940 census, one notes that no industry grew appreciably as fishing declined – with the exception of the hotel and restaurant industry.

A third response strategy to a decline in lobster catches was that people may have moved away from the local area. Evidence suggests that there was a good deal of out-migration. The town of Bristol is rather typical in the region. The population of Bristol reached its high point in 1860 with 3,019 permanent, year-round inhabitants. By 1920 that number had declined to 1,419. The population reached a low point in 1940, when only 1,355 people lived in town. It has increased since 1940, primarily because of an influx of retirees after 1960. The population changes in Bristol clearly mirror the economic fortunes of the town. In 1860, the fortunes of the town were favorable with large shipping, shipbuilding, and fishing industries. In the latter part of the nineteenth century and the early decades of the twentieth century, its population declined as literally every industry went into eclipse. The biggest drop in population (from 2,415 to 1,419) occurred between 1910 and 1920.

The same pattern can be seen in other nearby towns. In Bremen (Maine), for example, the population was at its height between

1860 and 1890 and then dropped sharply until 1920. The population of Bremen reached its low point again in 1940 (Acheson et al., 1978, p. 45). In Southport, "The population in 1860 was 708; ... the population in 1920 was 272 ..." (Clifford, 1961, p. 209).

Out-migration from the region may have been greater than these figures alone might suggest. An untold number of people left their families in Maine and sought jobs in other parts of the country. The author's grandfather was one of those. In 1931 and 1932, he went to work in the coal mines of Pennsylvania for US$0.25 an hour. Virtually all the money he earned was sent home. How many other people did similar things is impossible to know, since no figures on temporary migration were collected.

We cannot attribute all of this out-migration to a decline in the lobster fishing industry in the inter-war period. Much of the decline in population occurred between 1890 and World War I, when lobster catches were high. People were obviously responding to an overall lack of economic options in the region as well as to better opportunities elsewhere; but certainly low lobster catches were part of a general pattern of decline. Low catches did not provide people with any strong incentive to remain in Maine's mid-coast region.

Fourth, people might have responded to low lobster catches and poor economic options by intensifying their subsistence activities. This was apparently a general response to economic problems during the Depression era of the 1930s. A very large percentage of people in small towns and rural areas of Maine heated their homes with wood (many still do). Most did the cutting and splitting themselves. Many people in these areas operated small subsistence farms. It was very common for people to have large gardens (many still have them). In the 1920s and 1930s, it was not uncommon for a family to have one or two acres under cultivation. Many women did a good deal of canning, using mason jars to preserve for winter everything from jams to beans and corn. Many people kept chickens, a pig or two, and perhaps a cow. Many hunted. Such activities did not decline in the 1930s. In fact several people interviewed for this and earlier studies believe they had increased. The author's grandfather, who was rather typical, had a huge garden well into the 1970s when he was in his eighties; he had kept a flock of chickens for meat and eggs until the 1950s; he was an avid hunter all his life; and he ran a small trap line until 1960.

In the coastal areas, the opportunities to undertake subsistence activities were perhaps even greater. For example, during the Depression people ate more than their share of clams (Clifford, 1961, p. 217); and the "trash fish" from fishing operations found a market in such towns.

Some people in coastal areas, and especially on the islands, were certainly less dependent on the industrial world than city people who were forced to pay cash at market prices for everything they needed. If one had a house and a boat that was paid for, all one would need to survive would be food, clothing, heat, and a small amount of cash to buy those items considered necessities. Presumably, a lobster fisherman faced with a low cash income, might make out far better in a Depression than an underemployed worker in a large industrial city, who could only obtain the necessities with hard cash. The coastal rural fisherman was able to put much of his spare time to good use in the subsistence sector.

Nevertheless, life in coastal Maine in the inter-war years was far from pleasant for lobster fishermen or for others with very low incomes. People really could not live off the land, unless they were willing to subsist at a very low standard of living. By 1930 people wanted to go to the movies, and have a radio, electricity, and store-bought shoes. They, too, were tied to the cash economy. When there were no jobs to be had during the Depression, they suffered acutely. The fact that they could cut wood and dig clams only partially alleviated their sense of deprivation.

World War II and recovery

With the onset of World War II, the lobster industry began to recover; a recovery which has continued to the present. Since 1947, the lobster catch has averaged about 18 million pounds (8.2 million kg) and at no time in this period has it fallen below 15 million pounds (6.8 million kg). By any measure, it is the most important fishery in Maine today. It is no exaggeration to say that it is a mainstay of the coastal economy. In 1978, for example, there were 440 employed males in Bristol (Maine), which is a typical community in many respects. Of those, 161 were fishermen – by far the largest single occupational group. Of these 161 fishermen, no fewer that 108 were lobster fishermen (Acheson et al., 1978,

p. 54). Literally, more than one-fourth of all adult men in this community were employed in this one fishery.

Exactly what role the decline of the lobster industry had on the total economy of these communities in the 1920s and 1930s is still difficult to say. It is fair to say, however, that had that fishery been in better condition, the impacts of the Depression would not have been as severe as they were, and fewer people would have had to migrate from the region.

Conclusions

Several conclusions stem from an historical review of Maine's lobster fishery.

- The cause of the decline in the 1920s and 1930s in the lobster industry is still not certain. Overexploitation and economic factors are possibly the major reasons for the reduction in lobster catches.
- Although changes in water temperature probably did not play a primary role in the decline of the industry, the kinds of changes we observed in the 1920s and 1930s and the societal responses to them might be similar to those which would be observed if climatic change damages the lobster stock or lowers the catch.
- All major industries in Maine's central coast were economically depressed at this time. Although lobstering was in the doldrums, it may still have been a major employer.
- People adapted to the lack of economic opportunities by intensifying their subsistence activities or by migrating from the area. Although people in coastal Maine are very attached to their natal communities, they moved in large numbers rather than live in extreme poverty and desperation.

References

Acheson, J.M. (1979). Traditional inshore fishing rights in Maine lobstering communities. In *North Atlantic Maritime Cultures*, ed. R. Andersen, 253–76. The Hague: Mouton.

Acheson, J.M. (1985a). The Maine lobster market: between market and hierarchy. *Journal of Law, Economics and Organization*, **1**, 385–92.

Acheson, J.M. (1985b). The social organization of the Maine lobster market. In *Markets and Marketing*, ed. S. Plattner, 105–30. Monographs in Economic Anthropology No. 4. Lanham: University Press of America.

Acheson, J.M. (1988a). Patterns of gear changes in the Maine fishing industry: Some implications for management. *MAST*, **1**, 49–65.

Acheson, J.M. (1988b). *The Lobster Gangs of Maine*. Hanover: The New England University Press.

Acheson, J.M., Bort, J.R. & Acheson, A.W. (1978). *Bristol, Maine*. Interim report to the National Science Foundation. Project Title, "University of Rhode Island/University of Maine Study of Social and Cultural Factors Influencing Extended Jurisdiction in the New England Fishing Industry."

Acheson, J.M. & Reidman, R.L. (1982). Biological and economic effects of increasing the minimum legal size of American lobsters in Maine. *Transactions of the American Fisheries Society*, **111**, 1–12.

Ackerman, E.A. (1941). *New England's Fishing Industry*. Chicago: University of Chicago Press.

Clifford, H.B. (1961). *The Boothbay Region: 1906–1960*. Freeport: The Cumberland Press.

Dow, R.L. (1967). Temperatures, fishing effort point up growing problem. In *The Influence of Temperature on Maine Lobster Supply*. Augusta: Sea & Shore Fisheries Research Bulletin No. 30.

Dow, R.L. (1969). Cyclic and geographic trends in seawater temperature and abundance of American lobster. *Science*, **164**, 1060–3.

Judd, R. (1988). Saving the fisherman as well as the fish: Conservation and commercial rivalry in Maine's lobster industry, 1872–1933. *Business History Review*, **62**, 596–625.

Krouse, J.S. (1972). Size at first sexual maturity for male and female lobsters found along the Maine coast. Lobster Information Leaflet No. 2. Augusta: Maine Department of Sea & Shore Fisheries.

Krouse, J.S. (1973). Maturity, sex ratio, and size competition of the natural population of American lobster, *Homarus americanus*, along the Maine coast. *Fishery Bulletin*, **71**, 165–73.

Martin, K.R. & Lipfert, N.R. (1985). *Lobstering and the Maine Coast*. Bath: Maine Maritime Museum.

McConnell, C.R. (1963). *Economics*. New York: McGraw-Hill.

McFarland, R. (1911). *A History of the New England Fisheries*. Philadelphia: University of Pennsylvania Press.

Rowe, W.H. (n.d.). *The Maritime History of Maine*. Freeport: The Bond Wheelwright Company.

Russell, H.S. (1976). *A Long Deep Furrow: Three Centuries of Farming in New England*. Hanover, NH: University Press of New England.

Stahl, J.J. (1956). *The History of Old Broad Bay and Waldoboro*, Vol. 1: The Colonial and Federal Periods. Vol. 2: The Nineteenth and Twentieth Centuries. Freeport: The Bond Wheelwright Company.

Townsend, R. & Briggs, H. III (1982). *Maine's Marine Fisheries: Annual Data 1947–1981*. Orono: New Hampshire–University of Maine Sea Grant Publication.

8

Human responses to weather-induced catastrophes in a west Mexican fishery*

JAMES R. McGOODWIN

Department of Anthropology
University of Colorado
Boulder, CO 80309, USA

The coastal plain in the south of the modern-day Mexican state of Sinaloa, known locally as *Sur de Sinaloa* (South Sinaloa), is the site of Mexico's most productive inshore marine fisheries, producing shrimp, fish and, until fairly recent times, oysters and similar mollusks. The region encompasses a strand plain and inshore marine region which is roughly 17 km wide by 120 km long (Fig. 8.1). The main features of the terrain are sandy beach ridges, most of which are heavily vegetated, well-developed mangrove marshes, open briny lagoons, and estuarine channels leading to the sea. Politically and administratively the region falls within the *municipios* of Rosario and Escuinapa, which are situated south of the urban port of Mazatlán.[†] Moreover, this semiarid coastal plain is crossed by the Baluarte, Cañas, and Acaponeta Rivers. The Rio Baluarte, the region's largest river, is the principal drainage with a mean annual discharge of 1.6 billion m³ (Schmidt, 1976, p. 34).

Fisheries prior to modern times in South Sinaloa

Human exploitation of marine resources on Mexico's Pacific coast began as early as 8,000 years ago (Hubbs & Gunnar, 1964, p. 145), and large aboriginal populations inhabited South Sinaloa's

---

* Initial field work for this study took place from fall 1971 through fall 1972, and was sponsored by the Latin American Studies Institute at The University of Texas at Austin. Brief field studies were also conducted from 1974 through 1978. Communications with key informants have been ongoing through the present.

[†] A Mexican *municipio* is analogous to a county in the US.

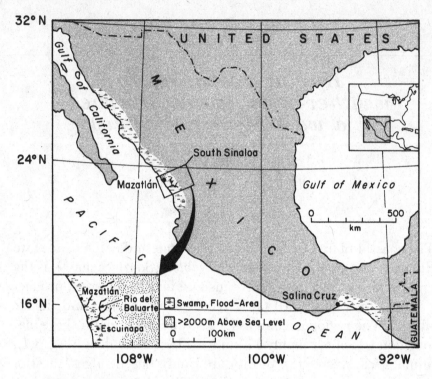

Fig. 8.1   South Sinaloa, Mexico, with swamp and flood area, and relief greater
than 2,000 m above sea level.

coastal plain for more than a millennium prior to European colonization (Scott, 1967–1972). Edward Spicer (1969, p. 788), the
region's principal ethnohistorian, has said of this coastal region
that it

> *provided fish and shellfish in abundance. The lowland
> was an area with potentialities for large concentrations
> of population, even without irrigation techniques, poten-
> tialities which were beginning to be realized at the time of
> the entry of the Spaniards.*

Today, more than 575 large shell accumulations, deriving from
the aboriginal population's long-term exploitation of oysters and
other marine mollusks, have been found scattered through the
region. According to Shenkel (1969, p. 25), an archaeologist who
studied them for several years, these shell mounds (Fig. 8.2) "range
in size from small surface scatters ... to impressive dumps up to
two miles [3.22 km] long, 600 feet [183 m] wide with peaks reaching 35 feet [10 m] in height," with some originating as early as
AD 300–400 (p. 42). These mounds, as well as the primitive fish

hooks, harpoon points, and fish bones which are sometimes found buried within them, provide mute testimony to the aboriginal population's long-term dependence upon mollusks along with other marine species.

Fig. 8.2 A shell mound constructed over several years just prior to the great flood in circa AD 1200. The mound contains pre-Columbian pottery and other artifacts, suggesting it may have served as a house platform. Hundreds of shell mounds similar to this one have been found scattered throughout South Sinaloa's coastal zone. (Photo courtesy of Stuart D. Scott.)

In the southern part of the region, in the middle of a densely vegetated and nearly impenetrable mangrove swamp, a particularly large and enigmatic shell mound has been found. Measuring approximately 79 by 89 m at its base, and 23 m in height, it is roughly quadrilateral in shape, as were most Mesoamerican pyramids. Archaeologists estimate that it was constructed sometime between AD 700 and 1300, and think it was possibly a ceremonial structure on the order of a temple mound which had been built by people living in a powerful chiefdom or perhaps a nascent state. Particularly interesting is that much of this mound is comprised of the shells of edible mollusks which were never opened. This suggests that the aboriginal people who built it enjoyed such an abundance of these resources that they could waste them as food, utilizing them instead as building materials – perhaps also in a display of conspicuous consumption to flaunt their power to their enemies (Shenkel, 1969, 1971; Snedaker, 1971).

Archaeological studies also indicate that a severe flood occurred sometime between AD 1100 and 1300, decimating South Sinaloa's shellfish populations and forcing the human population to abandon the estuarine region completely (Shenkel, 1969, pp. 43 & 46). What happened to them immediately thereafter is not known. They may have merely resettled further inland, turning their attention to agricultural pursuits, or they may have suffered cultural extinction by being killed off or enslaved by other already well-established and hostile people inhabiting the higher ground along the foothills at the edge of the coastal plain. Regardless of whatever actually occurred, the coastal plain was gradually resettled as its marine resources recovered. By the time of the Spanish conquest in the early sixteenth century, it was again densely populated with people who harvested its abundance of marine resources, particularly shellfish.

The accounts of the first Spanish explorers to visit this region describe prosperous people who were members of various centrally organized chiefdoms or nascent states (Spicer, 1969, pp. 784–5).* Of particular note was the indigenous people's active harvesting of oysters in the estuarine zone, their production of salt by evap-

---

* These mutually antagonistic chiefdoms (which were perhaps nascent states) were referred to by the first Spanish explorers traveling South Sinaloa's coastal plain as *señorios*, that is, regions under the control of a powerful headman. The part of South Sinaloa under consideration in this study fell within the *señorios* of Chiametla and Aztatlán. Chiametla, located on the coastal plain along the banks of the Baluarte River, may have politically controlled over 22 smaller, satellite villages. Less is known about Aztatlán, which had its center on the coastal plain along the banks of the Acaponeta River, although it is fairly certain that it similarly held sway over satellite villages in what is now the extreme south of the modern-day state of Sinaloa as well as the northern extreme of the modern-day state of Nayarit. Most likely the aboriginal people the Spanish first encountered along South Sinaloa's coastal plain were members of the highly cultured Tahue and Totoramé ethnic groups, both of which spoke a Uto-Aztecan dialect and were part of the Mesoamerican cultural tradition (Spicer, 1969; Segmen, 1972, p. 34).

oration at various sites on the fringes of the briny lagoons, and their active trading of smoked oysters and salt with their neighbors. Salt was extremely important to the pre-Columbian people as a preservative of fish, shrimp, and other foodstuffs.

The region's conqueror, Nuño Beltrán de Guzmán, well known among historians for his unusually ruthless and genocidal treatment of the region's Indians, made his first foray into the region in 1530, establishing a camp along the flood plain of the Acaponeta River. He did not remain there for long, however. Soon after his arrival he was abruptly forced to leave, as the sudden onset of a particularly heavy rainy season led to the loss of "a great number of his Indian allies as well as his own men and provisions during the inundating flood of the season" (Segmen, 1972, p. 41). From the Acaponeta River flood plain Guzmán and his army then traveled about 60 km further northward, eventually arriving at the head town of a chiefdom or nascent state which the Indians called Chiametla. Realizing this town was a center of some importance, Guzmán established a mission fortress there in the heart of the coastal zone, near the present-day town of Rosario, and took over the native people's thriving oyster and salt-producing industries.

Widespread decimation of the aboriginal population ensued for the remainder of the sixteenth century, not only because of ongoing warfare between the Indians and the Spanish conquerors, but also because of the diseases the Spaniards introduced for which the native people had no immunity – especially smallpox and typhus. As a result, the region's early colonial era was characterized by slow growth, and life and travel remained extremely precarious for the Spanish colonists for a long period of time. Notably, during the early 1540s warring Indians drove nearly all Spaniards from the region, including from Chiametla, and prevented their re-entry until the early 1560s.

However, with the discovery of valuable deposits of gold and silver in the nearby mountains, Spanish colonization and pacification efforts were intensified, increasing the rate of Indian exterminations, as Indians were either killed or enslaved and forced to work in the nearby mines. Chiametla was again briefly reoccupied by the Spaniards in the early 1560s, and their exploitation of the fisheries and salt works resumed there.

The region, however, was not easily pacified. Various colonial records describe the region as a sanctuary for rebelling Indians and note that trouble with warring Indians continued well into the 1700s (Segmen, 1972, p. 46).

Once the region was finally "pacified" – and indeed nearly depopulated of Indians in the process – new settlers began to move into the area, during the late eighteenth and early nineteenth centuries. They established cattle-raising haciendas and other business enterprises, particularly fisheries. By the mid-nineteenth century various historical documents describe the region as once again having a thriving trade in smoked oysters and salt. This trade reached to such distant urban centers in Mexico as Durango, roughly 160 km to the east, on the other side of the formidable western mountain range, as well as to Guadalajara, about 300 km to the southeast.

Then, in the late nineteenth century, the region's oysters and shrimp were discovered by Asian traders. Practically overnight a lively export trade in the region's seafoods developed, which saw South Sinaloa's oysters and salted shrimp being transported to distant markets in China and Japan. Correspondingly, inshore production levels of marine foods soared.

The fisheries in more recent and modern times

Whereas in pre-Columbian and early colonial times the region's mollusks were the mainstay of its local economy, today the region has a more diversified economy which is dominated by cattle raising and agriculture. Nevertheless, it is still the site of Mexico's most productive inshore fisheries, and is particularly well known for its production of shrimp. Oyster production, however, is now insignificant.

During the 1930s, following the end of the Mexican Revolution (1910–24), fishing cooperatives were established here by the federal government in order to exploit the potentialities for export trade in the region's oyster and shrimp resources. Thus, these exportable resources were placed off limits to members of the rural population who were not members of the cooperative organizations. This was no problem in the 1930s, since the new organizations were able to enroll practically all fishermen wishing to join. However, in over half a century since these organizations were first founded, their production levels have declined, whereas the rural population has soared, such that now only around two percent of the present-day

rural inhabitants are members of the government-sanctioned organizations. The remainder of the rural population which specializes
in fishing now does so illegally whenever it harvests shrimp and
oysters (McGoodwin, 1987).

South Sinaloa's weather patterns

Aboriginal myths from this region allude to its bi-seasonal
weather pattern and periodic catastrophes, emphasizing " ... con-
ceptions of stages in the creation and a universal flood, a view of
life as involved in an opposition between supernaturals controlling
wet and dry seasons" (Spicer, 1969, p. 798). The region's pre-
Columbian farmers relied greatly upon the flooding of the river
courses to provide nutrients and vitally needed water for their
crops. Their counterparts, however, who specialized in fishing,
counted on the rainy season to bring great influxes of fresh water into the inshore marine ecosystems, driving the mobile marine
species seaward, especially the inshore shrimp, thereby facilitating
their capture at weirs erected across estuary channels. However,
as already mentioned, the inundating floods were sometimes also
problematic for this region's pre-Columbian people, as well as their
Spanish conquerors.

The coastal plain is in the climatological zone characterized as
Aw, or tropical wet-and-dry (Vivo Escoto, 1964, p. 6). The annual mean temperature is 26°C, with July and August being the
warmest months (Schmidt, 1976, pp. 16 & 23). Annual rainfall
averages 1,194 mm with an annual variation of 40 percent (Scott,
1968, p. 5). Ocean temperatures immediately offshore range between a rather warm 30°C in August and September to a nearly
cold 22°C from January through March (Hubbs & Roden, 1964).
September is the month of greatest rainfall, a time during which
the rains are so copious that some flooding along South Sinaloa's
coastal plain occurs nearly every year. This period of intense rain
coincides with the region's peak hurricane season.

Tropical storms which sometimes achieve the intensity of hurricanes or tropical cyclones (locally called *chubascos*) bring about
the heaviest and most widespread rains in this region. They are
also the reason for the large year-to-year variation in annual precipitation. Deluge rainfall levels are sometimes associated with the
cyclonic activity; the severe storm which struck the south coast of

Sinaloa on September 12, 1968, for instance, precipitated 320 mm of rain in 24 hours at Siqueros (just east of Mazatlán). Such storms generally occur between late May and November, with September having the greatest number of events. However, because the storm tracks are usually oriented from southeast to northwest in this area, they usually parallel rather than directly strike the coastline (Schmidt, 1976, p. 22).

Yet, when such storms do come ashore the results are often disastrous. Over 2,000 people died when west Mexico was struck by a hurricane and floods in 1959; 21 people were killed and 5,000 were left homeless when in 1982 a hurricane, floods, and mud slides struck Manzanillo (roughly 450 km south of South Sinaloa); and, in 1984 torrential rains resulted in four dead and thousands homeless along Mexico's west coast (US AID, 1989).

South Sinaloa's bi-seasonal climate is such that for nearly eight months (December through July) almost no rain falls. During this time the landscape becomes so dry that it almost looks as if it had been burned, while rivers are often reduced to mere trickles. Then, during the four-month rainy season, there are frequent tropical thunderstorms which are often prompted by passing cyclonic storms, and which account for practically the entire yearly rainfall.

During the rainy season some flooding always occurs. Up to a point this is regarded as beneficial as it replenishes water supplies and soils. It also prompts the seaward migration of inshore marine resources, which greatly facilitates their capture. However, should a hurricane-strength storm pass too close to the coast or veer into it, catastrophic flooding often results with serious consequences for the rural population, works of infrastructure, agriculture, and the fisheries.

Weather catastrophes in the modern era

In September 1968, cyclonic storms prompted unusually heavy rains in South Sinaloa. For several days practically all lands in the coastal zone were completely covered with water. All crops in progress were lost, nearly all livestock drowned, and practically all mobile marine life was driven out of the inshore marine ecosystem. Moreover, the region's oyster beds were entirely lost as a result of several days' exposure to fresh water and because of their burial under a thick layer of silt which had washed into the marine ecosystem.

This flood was still much talked about by the local people three years later, when the author started his first field studies. Moreover, in the larger towns, particularly Escuinapa and Rosario, in public buildings and in business offices such as banks, various photographs documenting aspects of the 1968 flood were displayed. They were remarkable photos: one showing the Baluarte River flooding over its banks at Rosario, some 60 feet (18 m) above its normal level for that time of year; others showing settlements in the rural countryside out along the flat coastal plain, water halfway up the sides of the houses, people standing in waist-deep water, no ground showing anywhere. Local residents noted that several people had drowned or had otherwise been lost, and that following the withdrawal of the high waters there were widespread food shortages which had lasted for more than a year. Furthermore, they said, much of the limited aid which the federal government attempted to provide the region never reached the people who needed it most, having been sold instead by corrupt government officials in charge of distributing it.

In 1968 the low-lying part of the coastal plain had approximately 38,000 inhabitants, almost none of whom had many options for relocation, not even temporarily, following the great flood. This is because practically all of South Sinaloa, since the close of the Revolution in the mid-1920s, had experienced several decades of land reform under Mexico's ongoing *ejido* program, which had brought about the immigration of many impoverished rural people to the region from elsewhere in Mexico.* Hence, whereas there had always been ample vacant land on higher ground to retreat to in the past (at least over the past two centuries), by 1968 this was no longer the case. By then the previously vacant lands on higher ground further inland had become crowded with immigrants who

* Mexico's *ejido* program, launched in the era of post-revolutionary reform, was actively promoted until fairly recent times. Briefly, it entails the federal government's seizure of large land holdings and then redistributing the land to communities of impoverished rural people who either already live in the area or who are resettled there. The lands are then owned in common by the *ejidal* communities, while use rights to particular plots are ascribed to certain individuals for their use during their lifetimes.

had moved there as part of the post-revolutionary land reform movement. Thus, one response option for peasants living along the coastal plain who sought to flee tropical storm impacts was no longer available.

Impact on the region's fisheries

A few years prior to the great flood of 1968, the high demand posed by the lucrative export trade in oysters had already led to overexploitation. Realizing this, regional fishery managers imposed a general moratorium on oyster harvesting and began a program of dumping old shells and ceramic tiles into the estuaries in order to provide oyster bedding sites. A program to produce and release seedling oysters was also begun. Subsequently, these programs proved to be so successful that limited harvesting of oysters was again permitted by the beginning of 1968 (Shenkel, 1969, p. 46).

When the 1968 flood struck, the coastal zone remained under water for several days, and when the waters receded dead fish and shrimp (as well as livestock) were seen rotting in the fields everywhere. Less immediately apparent was the damage to the region's oyster beds. These had been flooded for several days with fresh water while also being buried under tons of silt, sand, and muck. In essence, the calcareous oyster beds had become underwater ghost towns in just a few days, bringing about a total cessation of all oyster production.

Overall, the local economy was considerably more diversified in 1968 and less dependent upon the oyster resource than it had been prior to modern times. Nevertheless, the local population was more severely impacted than its predecessors by the catastrophic flood – particularly the region's fishing people. Immediately following the flood there were few marine-food resources left in the inshore system, while the oyster resources were completely lost. Thus, some fishermen reluctantly turned to day-wage work when agricultural activities resumed, while many others moved away to various parts of western Mexico.

Again, after the 1968 flood, the federal government imposed a general moratorium on oyster harvesting in an attempt to give the stocks time to recover. However, by that time the region had become so crowded that over the ensuing years during which the moratorium was imposed many of the region's impoverished rural habitants were compelled to harvest oysters illegally. As a result, these resources have never fully recovered.

Other factors also inhibited the oysters' recovery. The increasing influx of pollutants from the region's developing agricultural sector, for instance, adversely impacted not only the oyster resources but also other living resources in the marine ecosystem. Even now, more than 20 years since the 1968 flood – *for the first time in this region's history* – its oyster resources have not fully recovered following a catastrophic flood.

Possible impacts of more frequent floods in the future

Neither the Indian population that was forced to abandon the region sometime between AD 1100 and 1300, nor the region's Spanish conqueror, Guzmán, who lost a considerable number of men and supplies in 1530, nor the rural–coastal dwellers who experienced the 1968 flood, seem to have anticipated this region's recurrent inundations. As a result, all suffered grave consequences. Ironically, the residents of South Sinaloa today seem similarly unconcerned with and unprepared for the next such event.

Cross-cultural studies of human responses to weather-induced catastrophes such as hurricanes and accompanying severe flooding have commented upon the surprising passivity of people who know they may be impacted by such events. Thus, Baumann and Sims (1974, p. 26) note, summarizing a cross-cultural study of human responses to weather-induced catastrophes done by Burton et al. (1969), that "simply bearing the loss from hurricanes is probably the most common type of individual response." Further, they state, "it is clear that neither awareness of the existence of the hurricane hazard, nor indeed past experience with it, are sufficient to produce precautionary actions" (p. 27). Such a situation seems to be the case among South Sinaloa's rural residents.

Correspondingly, a commonplace finding in studies of human responses to weather-induced catastrophes is that people seem to hope for the best in situations where there is little information about the probability of future catastrophes, and where few perceived alternatives exist concerning what they should do if disasters occur. In coastal Bangladesh, for example, where catastrophic tropical cyclones strike with far greater frequency than they do in

South Sinaloa, the local population manifests a surprising passivity regarding such events. Aminul Islam (1974), who studied intensively the Bangladesh coastal people's responses to cyclonic storms and accompanying severe flooding, notes that while "the existence of a substantial flood hazard is known to all" (p. 21), "people are reluctant to leave their homesteads and property despite grave danger" (p. 19), and, furthermore, that "when a cyclone of severe intensity is expected at longer intervals, the public mind tends to underestimate the severity of damage. Between two disasters there is a carefree period of inactivity" (p. 24). Again, these observations seem equally applicable to South Sinaloa's coastal people.

Cross-cultural differences also seem to play a role in determining how people respond to weather-induced catastrophes. Thus, Baumann and Sims (1974, p. 28) note that people with cultural orientations stressing individual autonomy tend to adopt "active-rational" means of coping with such natural hazards, whereas people with cultural orientations stressing the power of outside forces such as "God, fortune, luck," tend to adopt more "passive-fatalistic" means of coping. "Those who see themselves as less autonomous," they say, "or who acknowledge the power over their lives of outside forces, would be more fatalistic, more accepting, and yet more frightened of the hurricane threat and more undone by its consequences" (p. 30).

Because fatalism and passivity are cardinal attributes in the cultural orientations of most of South Sinaloa's impoverished, politically powerless, and wage-dependent rural population, it is no surprise that most rural dwellers speak of severe cyclonic storms and accompanying severe flooding as "acts of God," as things over which they have no control and about which they can do little. Thus, in view of their scarce resources, their few options for relocating even temporarily, and their low expectations that the government would provide assistance should a catastrophe occur, it is perhaps rational on their part to think that about all they can do in the face of a natural catastrophe is to bear it.

However, the current lack of contingency planning for flood disasters in South Sinaloa can also be accounted for by other more concrete factors: the inability of meteorological researchers to forecast with some degree of reliability when such disasters will strike in the future; the greater demands that a now record-level rural population makes on local resources; the reduced flexibility of the region's economic organization; and the general lack of resources at the national level which could be committed to disaster relief.

Today's population in South Sinaloa's coastal zone, which is now about 65,000 (as compared with approximately 38,000 in 1968), makes record-level demands on the region's natural resources. Moreover, as mentioned above, these rural people are considerably constrained in their abilities to move to higher ground – even temporarily – should flooding occur, now that formerly available lands which provided alternative living sites in the past are filled with recent immigrants to the region.

The economic organization of the region is also considerably less flexible than it was only a few decades ago. There has been a steady shift throughout the twentieth century from economic reliance upon diverse, individualistic subsistence and small-scale commercial activities, to a predominant and still increasing reliance upon wage labor. Whereas back at the turn of the century practically the entire rural–coastal population had independent and diverse means of subsistence, by the time of the great flood just 20 years ago around 63 percent of the population was wage dependent, while today more than 75 percent is wage dependent (McGoodwin, 1973, 1987).

Furthermore, there has been a steady decline in the proportion of the rural population devoted to fishing as compared with agriculture. Whereas at the beginning of the twentieth century nearly all local rural people had a variety of means available to them for providing their subsistence needs, surviving mainly by engaging in small-scale subsistence fishing and agricultural pursuits, today's wage-dependent people rely to a great degree upon incomes earned from working on corporately run farms, plantations, and cattle-raising enterprises. Unfortunately, these organizations are far less able to respond to sudden ecological change than were the subsistence-oriented fishermen and farmers of former years. And, again, as a consequence of Mexico's serious internal economic problems, significant relief aid is lacking, leaving this region even more vulnerable to calamity than it was in 1968, should another great flood occur.

The key to maintaining the productivity of marine resources in estuarine systems such as those in South Sinaloa is the maintenance of optimal mixing of fresh and sea waters. Should this balance be disturbed more often in the future by weather-induced anomalies – such as may be prompted by global warming – the region's fishery production can be expected to decline even more markedly.

Global warming in the near future could have the following consequences in this region: (1) increased cloud cover and precipitation causing an increased frequency and intensity of periodic inundations, bringing even more unfavorable influxes of fresh water, silt, and pollutants; (2) a change in the cyclonic pattern, causing more often hurricanes to directly strike the coastal area; (3) a rise in sea level which might reduce the amount of protected inshore marine habitat along the coastal plain while also converting the brackish ecosystem into one which is more purely marine; (4) an increase in sea surface temperatures, prompting a reduction of the currently most dependable inshore resources, particularly the less mobile ones such as mollusks, while also prompting a northward migration of the more mobile ones, corresponding to a greater diversity of warmer-water tropical and subtropical species which are not as valuable in terms of human subsistence or fishing industry; (5) similarly, marine-species transitions prompting the need for greater flexibility in the fisheries, increased capital expenditures for new types of gear, and other changes in fishing techniques; (6) a decrease in agricultural production resulting from the inundation of arable lands; and (7) a decrease in the region's population as some people abandon the coastal zone and seek more secure places to live and work. In essence, any of the foregoing problems could lead to the long-term decline of what is currently a very productive region of west Mexico.

In the past there seems to have been an undue emphasis on the part of South Sinaloa's fishery managers to ascribe too much importance regarding the variability of marine resources to the effects of human fishing effort, while overlooking the decisive role that extreme meteorological events often play. This is undoubtedly because local fishermen are easier to identify and control than the environment *per se* – much less the weather. Thus, future management of this region's fisheries must take into greater consideration the role which meteorological events may play in the availability of local marine resources.

Lessons regarding climate change and fisheries

The case of South Sinaloa's fisheries suggests three general lessons concerning how similar fisheries might respond to similar climatic anomalies, or to an exacerbation of environmental or societal problems such as might be prompted by global warming:

- *A change in climatic patterns prompted by global warming will lead to less predictable and perhaps chaotic transitions in patterns of marine-resource availability in a region's fisheries, requiring higher degrees of flexibility and adaptability on the part of fishers themselves.*

In South Sinaloa, the regional impacts of a global climate change may bring about a new and stable pattern of marine-resource availability within a fairly short time, or it may instead prompt the development of patterns which could take several human generations before they reach stable and predictable states, or it may bring about chaotic and long-term instability in local patterns of marine-resource availability. In any of the foregoing scenarios, local fishers would be wise to retain cultural attributes which stress diversified and flexible approaches to fishing, as well as diversity in their local economies.

- *The severity of adverse impacts on the human population stemming from climate-induced catastrophes will be inversely proportional to the availability of economic resources and the degree of flexibility in the socioeconomic and political structure.*

Thus, if South Sinaloa's rural coastal population continues to increase while the region's economic strength and socioeconomic and political flexibility continue to decline, the impacts of future catastrophic floods in this region will be even more severe than they have been in the recent past.

- *The degree of government commitment to contingency planning for climate-induced catastrophes will be proportional to the frequency with which such events occur.*

As noted earlier in the case of Bangladesh, high frequencies of weather-induced catastrophes did not prompt decisive precautionary or mitigating responses on the part of the rural–coastal population. However, government officials in this region did manifest high degrees of concern and made active efforts to plan for such catastrophes, their scarce resources notwithstanding. In South Sinaloa, where one human generation spans approximately 20 years, more than a generation has passed since the last catastrophic flood in 1968. As a result, most local people and government officials seem unconcerned about the possibility that such an event could adversely impact their lives in the near future. But, if such catastrophic events begin to occur with greater frequency in the future, then there will likely be an increased concern on the part of local governmental officials for mitigating their impacts on the rural–coastal population.*

As with forecasting extreme meteorological events (timing, magnitude, location, etc.), predicting future marine stocks has proven similarly recalcitrant to the most sophisticated scientific endeavors. While one can say with a fair degree of certainty what the "weather" or the state of a stock of a certain species in a fishery will be tomorrow, our powers of prediction drop quickly as we extend the future time frame. In addition, there are major difficulties in coupling the interactive effects of weather, fishing effort, and countless other variables influencing the dynamics of marine populations, making accurate forecasting of marine stocks at any future time (beyond the immediate) still an unsolved problem. About all we know with any degree of certainty is that fishery resources, like the "weather," have heretofore seemed to fluctuate within certain predictable upper and lower bounds over the long run.

Global warming, however, would change all of this. It seems certain, for instance, that global warming would alter patterns of living marine resource availability in ways we cannot currently forecast with any degree of confidence. Some presently abundant marine resources may even become extinct, while others may rise to a prominence until now never seen. In other words, the overall species composition of the oceans may be radically altered.

---

* Paradoxically, Aminul Islam (1974) notes that in coastal Bangladesh "One major factor encouraging people to stay where they are is availability of aid and relief" (p. 22), which "has further caused an increase in flood-loss potential" (p. 24).

In general, global warming may not prove to be as problematic for small-scale fishers as for their more specialized, larger-scale, and more industrialized counterparts. Generally speaking, small-scale fishers already manifest higher degrees of flexibility in their approaches to fishing, as well as greater degrees of occupational pluralism and economic diversity in their local cultures.

References

Aminul Islam, M. (1974). Tropical cyclones: Coastal Bangladesh. In *Natural Hazards: Local, National, Global*, ed. G.F. White, pp. 19–25. New York: Oxford University Press.

Baumann, D.D. & Sims, J.H. (1974). Human response to the hurricane. In *Natural Hazards: Local, National, Global*, ed. G.F. White, pp. 25–30. New York: Oxford University Press.

Burton, I., Kates, R. & Snead, R. (1969). *The Human Ecology of Coastal Flood Hazard in Megalopolis.* Chicago: University of Chicago, Department of Geography, Research Paper No. 115.

Hubbs, C.L. & Gunnar, G.I. (1964). Oceanography and marine life along the Pacific coast of Middle America. In *Handbook of Middle American Indians*, vol. 1, vol. ed. R.C. West, gen. ed. R. Wauchope, pp. 143–86. Austin: The University of Texas Press.

McGoodwin, J.R. (1973). *Economy and Work on the Northwest Mexican Littoral: An Analysis of Labor Recruitment among the Shark Fishermen of Teacapan, Sinaloa.* Ph.D. Dissertation, University of Texas at Austin.

McGoodwin, J.R. (1987). Mexico's conflictual inshore Pacific fisheries: Problem analysis and policy recommendations. *Human Organization*, **46**, 221–32.

Schmidt, R.H., Jr. (1976). *A Geographical Survey of Sinaloa.* Southwestern Studies, monograph no. 50. El Paso: Texas Western Press.

Scott, S.D. (Ed.) (1967–1972). *Archaeological Reconnaissance and Excavations in the Marismas Nacionales, Sinaloa and Nayarit, Mexico.* Preliminary Reports, Summers. West Mexican Prehistory, Parts 1 through 6. Buffalo: Department of Anthropology, State University of New York at Buffalo.

Segmen, P.K. (1972). Historical notes on the 16th and 17th centuries, related to studies of the human paleoecology of the Marismas Nacionales in Sinaloa-Nayarit, Mexico. In *Archaeological Reconnaissance and Excavations in the Marismas Nacionales, Sinaloa and Nayarit, Mexico.* Preliminary Reports, Summers. West Mexican Prehistory, Part 6, ed. S.D. Scott, pp. 30–63. Buffalo: Department of Anthropology, State University of New York at Buffalo.

Shenkel, J.R. (1969). Shell mound archaeology in the Marismas Nacionales. In *Archaeological Reconnaissance and Excavations in the Marismas Nacionales, Sinaloa and Nayarit, Mexico*, Preliminary Reports, Summers. West Mexican Prehistory, Part 2, ed. S.D. Scott, pp. 24–46. Buffalo: Department of Anthropology, State University of New York at Buffalo.

Shenkel, J.R. (1971). El Calón: Revisited. In *Archaeological Reconnaissance and Excavations in the Marismas Nacionales, Sinaloa and Nayarit, Mexico,*

Preliminary Reports, Summers. West Mexican Prehistory, Part 5, ed. S.D. Scott, pp. 19–24. Buffalo: Department of Anthropology, State University of New York at Buffalo.

Snedaker, S.C. (1971). The Calón shell mound: An ecological anachronism. In *Archaeological Reconnaissance and Excavations in the Marismas Nacionales, Sinaloa and Nayarit, Mexico*, Preliminary Reports, Summers. West Mexican Prehistory, Part 5, ed. S.D. Scott, pp. 15–18. Buffalo: Department of Anthropology, State University of New York at Buffalo.

Spicer, E.H. (1969). Northwest Mexico: Introduction, and The Yaqui and the Mayo. In *Handbook of Middle American Indians*, vol. 8, ed. E.Z. Vogt, pp. 830–45. Austin: The University of Texas Press.

US AID (Agency for International Development) (1989). Report of the Office of US Foreign Disaster Assistance, Agency for International Development, July. Washington, DC: US Department of State.

Vivo Escoto, J.A. (1964). Weather and climate of Mexico and Central America. In *Handbook of Middle American Indians*, vol. 1, ed. R.C. West, pp. 187–215. Austin: The University of Texas Press.

9

Irruption of sea lamprey in the upper Great Lakes: analogous events to those that may follow climate warming

HENRY A. REGIER and JOHN L. GOODIER

Institute for Environmental Studies
University of Toronto
Toronto, Ontario M5S 1A1, Canada

Introduction

Ecosystemic science as applied in the Great Lakes

The irruption of the sea lamprey into the upper Great Lakes – Huron, Michigan, and Superior (see Fig. 9.1) – occurred at a time when both cultural and natural aspects of the Basin ecosystem were under increasing stress by factors other than the invading sea lamprey. At the time there was intense disagreement among some experts about the causes of particular fishery effects in the Great Lakes. Thus, J. Van Oosten inferred that overfishing was mostly to blame for decreases in catches of preferred species; R. Hile argued that the sea lamprey was the main culprit, at least in the three upper lakes; and T.H. Langlois invoked pollution, based on his experiences in Lake Erie (Egerton, 1985). About four decades later we note that not all the disagreements have been resolved, but that all the strong protagonists for only one of the possible explanations have passed on. With hindsight we opine that each of these "one-cause experts" had strong evidence for his views from some locales within the Basin, but insufficient evidence to generalize that inference far beyond those locales.

The 1971 Symposium on Salmonid Communities in Oligotrophic Lakes (Loftus & Regier, 1972) was an attempt to transcend the polarizations and biases generated by "one-stress experts." This SCOL Symposium sought to build on the more ecosystemic initiatives of F.E.J. Fry, R.A. Vollenweider and others. Other ecosystemic symposia have followed since then, though one can still en-

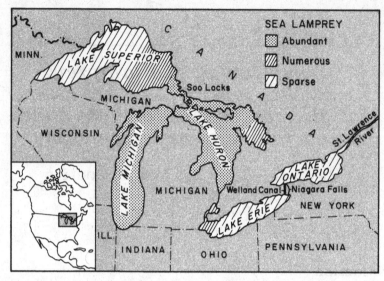

Fig. 9.1 Distribution and abundance of sea lamprey invasion into the Great
Lakes. (Adapted from Applegate & Moffett, 1955.)

counter "one-stress experts," usually of a rather dated reduction-
istic mindset.

Ecosystemic effects of some 18 human-induced stresses acting
on the Great Lakes ecosystem were characterized by Francis et al.
(1979). These included overharvesting, irruption of exotic species,
nutrient and sediment loading, shore and bottom restructuring,
water entrainment and diversion, and introduction of toxics, con-
taminants, and heat.

Separately, jointly, at times synergistically but seldom antago-
nistically, many cultural stresses have acted on Great Lakes Basin
ecosystems during the past 200 years. They have strained, overrid-
den or crippled the self-organizing processes of these ecosystems.
The result has been a debilitated condition, a general impairment
of ecological integrity expressed throughout the whole ecosystem.
Features of this syndrome include: a weakening and breakdown of
isolating mechanisms between subsystems; a blurring or disappearance of biotic spatial boundaries between and among subsystems; a
suppression of long-lived, larger, slow-reproducing (i.e., "K-type")
species and expansion of smaller, short-lived ("r-type") species;
and an increase in the frequency of fluctuations in ecosystem-level
phenomena (Rapport et al., 1985; Harris et al., 1988; Regier et al.,
1989; Regier et al., 1990).

Climatic changes (in future decades) will likely contribute on balance to this "general distress syndrome" and may impede progress currently underway toward "restoring and maintaining the chemical, physical and biological integrity of the waters of the Great Lakes Basin ecosystem," the stated purpose of the 1978 Great Lakes Water Quality Agreement (IJC, 1987). Research efforts are underway to assess effects of climate change on Great Lakes fish and fisheries (see, e.g., TAFS, 1990). Here we offer the irruption of the sea lamprey in the upper Great Lakes as a possible analogue of some unpredictable phenomena that will accompany climate warming. We have reexamined data of decades ago related to sea lamprey and relevant studies, both from the ecosystem perspective and current understanding as sketched above. The definitive history of the sea lamprey's irruption and subsequent fortunes in the Great Lakes remains to be written.

The sea lamprey irruptions as an analogue

The sea lamprey (Fig. 9.2) is a parasitic snake-like fish with a suction-cup mouth, horny teeth and rasping tongue; it attaches to the fish, penetrates the skin, and feeds on the blood of its victim. It prefers soft-scaled, large fish that live in cold waters during summer. In the Great Lakes such fish include lake trout, lake whitefish, suckers, and introduced salmon. Except for suckers, it shares some of our human preferences for fish! Native of the North Atlantic Ocean, the sea lamprey gathers in bays and estuaries in early spring, and migrates up streams to lay its eggs and then die.

Historical ecologists are not certain when the sea lamprey first invaded the lower Great Lakes Basin, that is, Lake Ontario and the finger lakes. If this occurred before about 1850, then the effects on lake trout and lake whitefish could not have been severe, because no historical accounts of serious harm have been found. But it is clear that the successful invasion of the four upper lakes – Erie, Huron, Michigan and Superior – occurred after 1920. Niagara Falls, as a barrier, is too big for the sea lamprey. They presumably entered the upper lakes via the Welland Canal or the Erie Canal or both. Perhaps they hitch-hiked on vessels with the aid of their suction-cup mouths; they have been observed to do so. In early years, unusual occurrences in locales far beyond their normal migratory range may be explained by hitch-hiking. Some

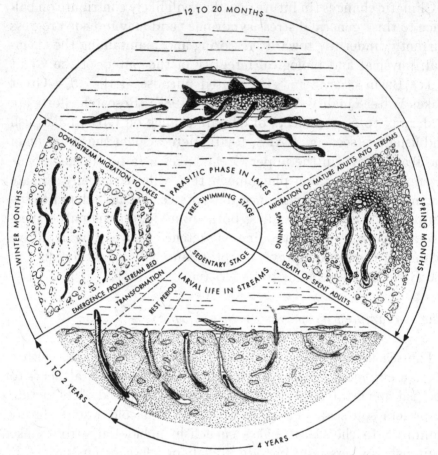

Fig. 9.2 Life cycle of the sea lamprey (Applegate & Moffett, 1955). The lamprey lives 6.5 to 7.5 years. The size of each of the segments in the chart does not correspond to the length of time the lamprey spends in each stage. The lamprey spends most of its life as a larva (Applegate, 1950).

sea lamprey larvae may have been introduced by anglers who used the larvae as bait; there was an organized trade in such bait.

Limited damages caused by the native silver lamprey had been reported for Lake Erie. "Great runs" of silver lamprey were known in the Cuyahoga River of Lake Erie around 1850 (Great Lakes Sea Lamprey Committee, 1946). Problems with sea lamprey in Lake Ontario had been reported by Cobb (1900) and Dymond et al. (1929). However, neither the fishing industry nor government management agencies were prepared to deal effectively with the spread of sea lamprey and the devastating mortality it inflicted upon the resident salmonid populations of Lakes Huron, Michigan, and Superior. Here "salmonids" include both the salmonines (like lake trout) and coregonines (like lake whitefish).

The sea lamprey case study provides glimpses of an industry and policymakers attempting to cope with a rapidly destabilizing fishery. At a time when it was necessary to adapt quickly and revamp fishing methods, certain conditions made this difficult. These included the following:

- the rapidity and unprecedented nature of the sea lamprey invasion (there is evidence that they hitch-hiked on the hulls of ships);
- the division of authority among a number of Great Lakes agencies, each wary of having its rights to manage its waters and fish usurped;
- the previous depletion of certain fisheries as a result of some other cultural stresses;
- a fishing industry committed to traditional techniques with insufficient capital, knowledge, and inclination to adapt; and
- the outbreak of World War II, which increased demand for fish, and also for young fishermen for the navies.

These factors are examined in more detail below.

This case study also relates to the emergence of fisheries policy at an international level and to the coordination of the actions of state, provincial, and federal agencies. These actions led to the establishment of the Great Lakes Fishery Commission. Thus, the sea lamprey problem may be seen as being directly responsible for the establishment of the first permanent international body concerned with the fishing interests of the Great Lakes. Many attempts, during the preceding 60 years, to create such a commission had failed. We do not recount the story of the recovery of many stocks of Great Lakes fish after the low point of Great Lakes fishery fortunes in the early 1960s. Thus, this is a case study of a catastrophe in which the sea lamprey played a key role.

The spread of the sea lamprey in the Great Lakes

Completion of the Welland and Erie Canals in the 1820s provided passage to Lake Erie (Fig. 9.1) but those routes were apparently not used by sea lamprey for a century. From Lake Erie the sea lamprey moved progressively through the upper Great Lakes over three decades. For Lake Huron, trout production declined first in the main lake and then in Georgian Bay and Saginaw Bay. Lake Michigan's decline was from north to south. Lake Superior losses proceeded from east to west along both the north and south shores. Shetter (1949) traced the lamprey's progress through Lakes Erie, Huron, and Michigan. The first sea lamprey positively identified in Lake Superior was captured off Isle Royale in 1946, although there was an earlier newspaper account from 1938. Questionnaires to fishermen in that year suggested that lamprey were already established in eastern Lake Superior but that losses to the fishery were not severe (MacKay & MacGillivray, 1949). In 1949 it was present in five American streams of Lake Superior but abundant in only one (US House of Representatives, 1949). The first large year class of lamprey near Marquette (Michigan) on the south shore occurred in 1952 (Halverson, 1955).

Certain regions were affected more severely than others. In Lake Michigan, the more northerly waters were more strongly "infested" than southerly waters (Shetter, 1949; US House of Representatives, 1949). Replies from Canadian fishermen of Lake Huron in 1946 described high levels of sea lamprey predation and little regional variation, with the exception that wounding may have been more severe in the southwesterly part of Georgian Bay than along its eastern shore (MacKay & MacGillivray, 1949).

Some island and offshore stocks of lake trout managed to escape severe predation entirely and may remain extant today. These included certain reef trout in Lake Superior and small populations in McGregor Bay and Parry Sound of southern Georgian Bay (Goodier, 1981; Berst & Spangler, 1973). In this study we primarily use data on the sea lamprey–lake trout interaction as indicative of the overall impact of sea lamprey. Other valued species, notably lake whitefish, were also harmed or killed by sea lamprey, but to a somewhat lesser extent.

Prior to the irruption of sea lamprey in the upper lakes, lake trout populations did not fluctuate markedly in abundance. However, this was not the case with lake whitefish. Regier and Loftus (1972) contrasted the population dynamics of the salmonines and coregonines of the Great Lakes as relatively stable and unstable, respectively, under moderate levels of natural and cultural stresses. It is difficult to sort out cause-and-effect relationships in naturally unstable populations. Perhaps for this reason most

studies of the impact of sea lamprey have used lake trout as a
surrogate of ecosystemic impact.

Among the marketed species, lake trout and lake whitefish were
harmed the most by sea lamprey. These particular salmonids prefer cold deep water, except for spawning runs into clear rivers,
reef, and shoal areas in the fall. Thus, they were not as vulnerable
to the pollution that occurred mostly in the shallower inshore and
bay areas in the decades before about 1950. Emphasis in this chapter is on the lake trout and lake whitefish populations that were
strongly affected by sea lamprey before pollution was sufficiently
intense to cause damage.

Upper Great Lakes ecosystems:
background and role of the sea lamprey

In considering the sea lamprey as a case study, one must assess its particular role in the decline of the fisheries on the different Great Lakes. Production figures are summarized in Tables 9.1a,b,c.

Lake Huron

The fisheries of Lake Huron are shared by Ontario and Michigan. The International Board of Inquiry (1943) noted that on
the American side, production (i.e., landed catch of all marketed
species) from Saginaw Bay (Michigan) at the end of the 1930s was
nearly double that of Lake Huron proper. Other centers of the
industry were in the St. Ignace–Cheboygan area, and the Alpena
and Port Huron districts. In Canadian waters, Georgian Bay production was approximately equal to that of the main lake and
about three times that of the North Channel. These differences
reflect the degree of shallowness and fertility of the various areas
(Egerton, 1989). Salmonids seldom occur in the most shallow and
most fertile waters.

The years from 1895–1939 were identified by Hile (1949) as the
"modern normal" period for American Lake Huron, that is, years
showing no long-term trends in lake trout production. Average
lake trout production was 842 tons (763 metric tons). By 1945, 13
years after the first recorded lamprey in Lake Huron, production
had dropped by 90 percent of the average to 86 tons (78 mt).

Table 9.1a
Average annual production of certain species
in the combined commercial fisheries of the
United States and Canadian waters of Lake Huron,
1879–1964, in short tons*

Period	Lake trout	Lake whitefish	Lake herring[b]	Chubs[b]	Others	Total
1879–1909[a]	3072	1792	c	c	7022[d]	11886
1910–1919[a]	2911	1184	2291	431	3430	10249
1920–1929	2584	1426	1945	566	2838	9360
1930–1934	2492	2543	2124	709	3336	11205
1935–1939	2660	1324	2567	393	3169	10114
1940–1944	1388	468	1355	256	3034	6502
1945–1949	294	1076	836	180	2698	5085
1950–1954	207	2306	820	212	2510	6054
1955–1959	15	444	83	791	2202	3537
1960–1964	1	506	29	2116	2298	4950

[a] Average for years of record
[b] Catches of lake herring and chubs combined through 1908
[c] Canadian data not available until 1902
[d] Nearly half of this was herring and chubs
*1 short ton = 0.907 metric tons

Table 9.1b
Average annual production of certain species
in the commercial fisheries of Lake Michigan,
1879–1963, in short tons*

Period	Lake trout	Lake whitefish	Lake herring	Chubs	Others	Total
1879–1909[a]	3612	1358	8488	1030	3574[b]	18062
1910–1919[a]	3382	835	4321	1762	3448[b]	13748
1920–1929[a]	3500	1056	2008	1451	2214[b]	10229
1930–1934	2671	1865	2492	2183	2558	11770
1935–1939	2519	600	2310	2647	4032	12110
1940–1944	3289	675	915	985	4971	10837
1945–1949	1337	1878	3022	2717	3248	12203
1950–1954	7	718	4000	5241	4625	14591
1955–1959	–	52	1819	4974	6818	13664
1960–1963	3	134	71	5421	6167	11796

[a] Average for years of record
[b] Incomplete data
*1 short ton = 0.907 metric tons

Table 9.1c
Average annual production of certain species
in the combined commercial fisheries of the
United States and Canadian waters of Lake Superior,
1879–1967, in short tons*

Period	Lake trout	Lake whitefish	Lake herring	Chubs	Others	Total
1879–1909[a]	2662	1415	566[b]	–	336	4978
1910–1919[a]	2110	413	4785[b]	–	566	7873
1920–1929	2132	334	4270[b]	–	858	7593
1930–1934	2078	358	5582[b]	–	208	8226
1935–1939	2350	385	7194[b]	–	260	10188
1940–1944	2208	532	8288	252	408	11686
1945–1949	2280	653	7268	130	200	10530
1950–1954	1942	510	5679	80	332	8541
1955–1959	940	438	6150	390	406	8324
1960–1964	162	256	5684	538	1046	7686
1965–1967	114	261	2982	1042	912	5311

[a]Average for years of record
[b]Catches of lake herring and chubs combined through 1940
*1 short ton = 0.907 metric tons

In 1946 production declined to 19 tons (17 mt) and by 1950 the fishery had disappeared.

R. Hile (US House of Representatives, 1952) also defined the period 1923–39 as "normal" for Canadian waters (average 1,590 tons or 1,442 mt). By 1954 lake trout production had dropped over 95 percent to 84 tons (76 mt). Berst and Spangler (1973) concluded that "the sea lamprey was the single critical factor in the final decline of the lake trout in Lake Huron." Nevertheless, a majority of the Canadian Lake Huron and Georgian Bay fishermen interviewed by Loftus (1980) felt that the smelt had also been a major factor in the decline, although the proposed causal mechanism was unconvincing.

With the introduction of the deep trap net into Lake Huron in 1928, production of lake whitefish rose sharply from 4,400 tons (3,990 mt) to over 8,000 tons (7,260 mt) in 1931. Although use of these nets was limited by regulation in 1935, and never permitted in Canada, American whitefish populations may already have fallen into decline when the lamprey arrived to prey on them. In

Saginaw Bay, whitefish, except for the period 1946 to 1948, had been scarce since the mid-1930s, following previous years of heavy production. Smith (1968) pointed to a progressive fishing-up of local stocks in different areas of the lake prior to the trap-net ban. Spangler (1970) later inferred that the lamprey contributed to the mortality of whitefish in northern Lake Huron, and related seasonal mortality of whitefish to scarring by sea lamprey.

Lake Michigan

Hile et al. (1951a) identified the period 1927–44 as "modern normal" for Lake Michigan as a whole: average annual lake trout production was 2,825 tons (2,562 mt). The last year in which the average was exceeded was 1944; five years later (in 1949), production had dropped by 95 percent to 171 tons (155 mt). Production in 1950 was 27 tons (24.5 mt) and by 1953 the fishery had disappeared.

Hile assigned major blame to the sea lamprey for the decline of lake trout stocks in both Lake Huron and Lake Michigan. Smith (1968) presented a somewhat more complicated picture; he noted the relatively low lamprey abundances during the period of trout decline and suggested that the trout were sensitized to the lamprey by an increase in exploitation in the early 1940s, particularly in Illinois waters. He attributed this increase to the close proximity of the Chicago markets, greater demand during World War II, and declining harvests. Wells and McLain (1973) contended, however, that an increase in production confined to such a small area would not have affected the whole lake. Eschmeyer (1957) demonstrated that lake trout continued to decline, even after the collapse of the Lake Michigan fishery. Fry (1953) showed the effects of sea lamprey predation on lake trout in South Bay, Lake Huron, an area closed to commercial fishing. We infer that the dominant cause for the rapid loss of trout stocks in Lake Huron and Lake Michigan was the sea lamprey, although certain stocks were undoubtedly already stressed by intensive fishing and/or pollution. Van Oosten (1949) noted, for example, the gradual decline in production between 1893–1938 in the face of increased fishing effort.

Wells and McLain (1973) have traced the history of the whitefish fishery in Lake Michigan, relating accounts of substantial decline in the 1800s due to overfishing and sawmill refuse. Increased fishing intensity was also implicated in production losses following peak years of 1928–32 and 1942–43. They concluded that the sea lamprey did not significantly attack whitefish until the loss of its preferred prey, the lake trout.

Lake Superior

In the late 1800s there were complaints of fish stock failures from both fishermen and officials. Kumlien (1887) and Smiley (1882) reported losses at Chequamegon Bay. Declines of Apostle Islands and Isle Royale trout were cited by the US House of Representatives (1897), and the Canadian Department of Marine and Fisheries annual report for 1893 noted a great scarcity of whitefish along the entire American north shore as far as the Canadian border. Low production from US waters prompted many Americans to extend operations to Canada. By 1910 there was damage to Canadian lake trout stocks of the Slate, the Lizard, and the Michipicoten Islands.

Lawrie and Rahrer (1972) and Pycha and King (1975) inferred that an increase in fishing effort during the 1930s and 1940s, coupled with the growth of a large angling industry, had stressed lake trout stocks by the time of the sea lamprey's arrival. New innovations such as automatic net lifters, fathometers, special nets and, more importantly, the change from cotton to the long-lasting, more effective nylon gill nets in 1950–53 led to immediate increases in fishing intensity. In the Michigan waters of Lake Superior it was widely thought, even among the fishermen themselves, that overfishing was a major cause of the decline in the harvest of whitefish and lake trout (Gallagher & Van Oosten, 1945). As discussed below, increased pressure arose in part from a shifting of vessels and gear from the lower lakes where trout stocks had failed. Pycha and King (1975) suggested that lake trout mortality, due to lamprey predation, probably did not become serious until the early to mid-1950s, after stock decline was already apparent.

Hile (US House of Representatives, 1952) defined as "normal" for lake trout production, the periods 1926–49 for US waters (1,524 tons or 1,382 mt) and 1930–49 for Canadian waters (762 tons – 691 mt). US production had fallen 90 percent of the average by 1961 (162 tons – 147 mt), and Canadian harvests declined 92 percent to 61 tons (55 mt) in 1960.

The responses to the sea lamprey

Governmental and institutional response

Background

Prior to the mid-1940s, the various agencies of the Great Lakes tended to regard the resources of these waters as self-sustaining. The commercial fishery was the favored user-group; access to the fishery was open to fishermen who understood the politics involved.

Frick (1965, p. 89) noted:

> *Regulation of the Great Lakes fisheries has been carried out through a patchwork accumulation of laws and directives rather than by well-organized uniform legislation. Agreement on policy and uniformity in regulations [was] made difficult or impossible by divided jurisdiction: eight states, one province and the two federal governments have all claimed some degree of jurisdiction over the Great Lakes fisheries.*

The influx of American fishermen into Canada in the late 1800s, along with fishermen of both countries poaching illegally in each others waters, and the monopsonistic domination by large US firms, highlighted the international character of the fishing industry. Biologists recognized a decline in certain fish populations and called for widespread action, which led to the establishment of an *ad hoc* joint international commission in 1893. This was followed by a treaty between the US and Great Britain in 1908 proposing a fisheries commission empowered to determine nets, boats and other fishing methods by international regulation. This treaty – a parallel to the 1909 Boundary Waters Treaty – failed to receive US Congressional ratification and ultimately died about 1917. In all, 27 formal and informal Great Lakes conferences were held between 1883 and 1943. Only after the sea lamprey was acknowledged to be a serious pest was coordinated international action undertaken. But the resulting Great Lakes Fishery Commission had its executive responsibilities limited to sea lamprey control.

From the beginning, fishermen played an active and vocal role in stimulating agency actions against the lamprey. In the years following 1936, Michigan fishermen in the Rogers City area of Lake

Huron complained of lost production. Nevertheless, approximately 300 American Lake Huron and Lake Michigan fishermen, surveyed by the International Board of Inquiry in 1940, failed to identify the sea lamprey as involved in production losses – overfishing and pollution were regarded as most detrimental. J.W. Moffett, Chief of US Great Lakes Fishery Investigations, reported before the US House of Representatives in 1952 that complaints from fishermen, especially those on Lake Huron, began to increase in the early 1940s; because of wartime shortages neither the Province of Ontario nor the State of Michigan investigated or engaged in serious preventative measures. He noted that it was not until production of trout in Lake Michigan began to fall off in about 1945 "that any widespread alarm was voiced."

An editorial in *The Fisherman* (Anon., 1948) gave the lack of a statewide fishermen organization as one reason for the poor participation in lamprey programs and weak lobbying strength. The lamprey problem appears to have fostered more interaction. In 1940 the Michigan Fish Producers Association was organized, and several municipalities, counties, and associations lobbied state legislators and congressional delegations.

Small-scale lamprey surveys were initiated by Michigan and the US Fish and Wildlife Service as early as 1938. The International Board of Inquiry (1943) conducted 29 public hearings in the Great Lakes region, attended by some 1,500 commercial fishermen, sportsmen, and officials. Again it was recommended that a joint agency be established. In 1946 a fisheries treaty was signed between the US and Canada, but failed to be ratified because it required that the states relinquish some authority to a federally created regulatory commission.

Also in 1946, a Great Lakes Sea Lamprey Committee was formed and included technical representatives from the US Fish and Wildlife Service, the Province of Ontario, and the states bordering the Great Lakes. The trapping of lamprey in spawning streams was begun – over 11,000 were captured in the Little Thessalon River of the North Channel in Ontario in 1946.

The Great Lakes Committee, established in 1948, acted as an early forum for the exchange of ideas and for some coordination of programs. Nevertheless, the actions of the various agencies were largely independent and annual funding was not assured. Such funding first occurred for the US program in 1950 (Egerton, 1989).

Thus, early delay in responding to the sea lamprey threat arose partly from a failure to gain support for coordinated action from the agencies involved and from the fishing industry. Spaulding and McPhee (1989, p. 73) explained:

> *American fishermen did not want any regulation that would even temporarily reduce their catch and earnings. Also they did not want to be put in a situation of dependence upon a geographically remote international agency to deal with problems that traditionally had been taken care of by their individual governments.*

Representatives from the State of Ohio were most strongly opposed to the convention, partly on the grounds that equal representation for Canada was unfair for American fishermen who harvested the major portion of the catch. Canada continued its long, if cautious, support for the creation of a Commission.

It was not until December 1952 that continuing problems of overfishing, pollution, and the sea lamprey were perceived widely enough to demand renegotiation of a modified treaty. Meetings of the delegations followed and ratification of a treaty finally came in October 1955. The Great Lakes Fishery Commission held its first meeting in November 1956. It gained responsibility for sea lamprey control and a mandate to foster rehabilitation of the valued fish stocks but, unlike the agency envisaged in 1946, it was not empowered to regulate and enforce fisheries regulations.

Legislation and control programs in the 1950s and 1960s

The initial sea lamprey control program utilized mechanical and electromechanical barriers erected in the lamprey spawning streams. The first US weir was erected in 1951 and Ontario began construction the following year. These barriers were superseded by the use of lampricides selected from among some 6,000 chemicals tested at the sea lamprey research laboratory at Hammond Bay, Michigan. Chemical treatment was initiated in Lake Superior in 1958, because this lake's trout stocks were still relatively healthy and so held the greatest promise of being saved. Treatment was extended to Lake Michigan in 1960, to Lake Huron in 1966, to Lake Ontario in 1971, and to Lake Erie in 1986. Between 1958 and 1960, all Lake Superior tributaries known to contain larval sea lamprey were chemically treated.

In the early decades of this century, most states and Canada introduced legislation for net mesh sizes, closed seasons and catch limits based on fish weight or length. As sea lamprey control measures were instituted and lake trout rehabilitation was begun, increasingly strict regulation was implemented to control the commercial industry. Michigan legislated a closed season for lake trout in 1953 (State of Michigan, 1953). Lake Superior was closed to lake trout fishing in 1962, except to the extent necessary to provide researchers with data. Policies of limited entry to the fishery were instituted by stages.

Ecosystem responses

Prior to 1950, Lake Michigan contained seven species of chub (small coregonines) of varying sizes and growth rates, occupying different depth niches. The sea lamprey heavily attacked the two largest of these (along with the larger salmonids and burbot), and the two large chub species were extinct by 1960. With their demise, the lamprey preyed on the four intermediate-sized species of chub which occupied the lamprey's preferred deep water range and, by 1963, they constituted only one percent of chubs taken in test trawls. On the other hand, bloater chub, now relieved of predation by lake trout and not large enough to be of interest to commercial fishermen or lamprey, gained a competitive advantage and increased to such great numbers during the 1950s that they finally constituted 99 percent of the chub population. They also invaded northern Lake Michigan where chubs had not previously been abundant, and in 1961 were 2.5 times more numerous than the combined chub species present before the lamprey irruption (Hile & Buettner, 1955).

Economic costs and market responses

During this period of destabilization, certain market changes had strong implications for the Great Lakes industry. Reduced costs of meat and poultry, due to improvements in rearing livestock and fowl, together with a rise in the standard of living, led to an increase in the consumption of meat and a decrease in the consumption of fish. Frick (1965) noted the necessity for the fishing industry to improve quality and reduce costs in order to remain competitive. Also, consumer tastes in the lucrative Jewish market were changing, with a greater demand for packaged ready-to-cook or pre-cooked foods at the expense of fresh or whole-smoked fish, as Great Lakes whitefish, pike, and trout had traditionally been sold. In general, the quantity of fish sold as packaged fillets, breaded fish sticks and portions increased.

With these changing consumer tastes and the severe decline in supply of Great Lakes prime quality fish, markets opened for processed fish from western Canadian lakes. These fish were lower in quality, produced with lower labor costs and lower in price to the consumer. Also, processed marine fish became more prominent in the markets. Thus, the sea lamprey problems may have contributed to a shift of some North American marketing patterns, especially in Chicago and New York.

Economic losses to the sports fisheries from sea lamprey predation are difficult to assess. The early history for the sport fisheries has not been documented well. For all three of the upper Great Lakes, this history is a long one, although especially significant was the growth in popularity of lake trout trolling in the 1920s to 1940s. The sport was especially popular in Grand Traverse Bay and southern Georgian Bay. Considerable funds flowed into these regions.

The costs for the sea lamprey control program, as conducted by the Great Lakes Fishery Commission, were calculated on the basis of the historic value of the lake trout fishery in the two countries; for sea lamprey research and control Canada paid 31 percent and the United States 69 percent. Administrative and general research costs were shared equally.

By 1950 the Ontario government was spending about US$50,000 a year for Great Lakes investigations; US expenditures were US$286,500 (US House of Representatives, 1952). In 1958 the total budget allocations were US$914,300 by the US and US$424,500 by the Canadian governments. In 1987 they were US$4,662,000 and US$2,279,300, respectively (Spaulding & McPhee, 1989).

Industry responses

As policymakers in the US and Canada sought common ground for dealing with the lamprey problem, the industry struggled to adapt. Uncertainty about the future of the fish stocks and a worsening economic situation discouraged long-term planning; many individuals simply opted to "muddle through." Fishermen tended to respond by either abandoning their trade, shifting their efforts to new species, or "gypsy" fishing, that is, searching for new grounds and ignoring any territorial rights by extant fisheries.

Resignation

The rising costs of labor, gill nets, and other gear, after World War II, pushed the industry to a marginal level in some areas. The increased efficiency of nylon gill nets was offset in part by their tendency to entangle great numbers of the small fish species (chubs, alewives, and smelt that burgeoned in the absence of the predatory lake trout). Labor costs of removing these fish from the nets exceeded their value and, often, gill nets could not be fished in certain seasons (Anon., 1960).

The loss of traditional high-value species forced many fishermen out of the industry. Between 1946 and 1961, the number of Ontario fishermen declined progressively from 3,000 to about 1,700. The numbers of fishermen on Lake Huron and Georgian Bay in the late 1930s had fallen to less than one-third by the late 1950s. The numbers of US fishermen dropped steadily from 5,019 in 1940 to 3,734 in 1960. As observed by J.W. Moffett (US House of Representatives, 1952), "Everyone appeared willing to do something about the sea lamprey but no-one knew just what should be done."

Many enterprises were one-boat operations run by a single family or small partnership. By the mid-1960s, fishing on American Lakes Huron and Michigan had become largely a part-time or seasonal enterprise, with formerly full-time fishermen spending the major part of their time employed in other occupations. Many of the remaining full-time fishermen were "old-timers" who could not easily find a new trade. Such people were unwilling or unable to invest the time and money in their operations to make improvements to deteriorating equipment.

In Lake Superior in the late 1950s and early 1960s some fishermen could not cover their costs and foreclosures occurred. The number of processors or "first handlers" on American Lake Superior declined from 21 in 1959 to only 10 in 1964. The peak number of persons employed by these processors fell from 527 in 1959 to 289 in 1964 (Anon., 1969).

Hile and Buettner (1959, p. 30) described the grim situation on Saginaw Bay:

> *To describe the commercial fishery of Saginaw Bay as threatened by failure would be unduly optimistic, for it is now collapsing and the process is well advanced. The greatly decreased fishing activities would be much lower except that many fishermen resist abandonment of a lifetime occupation or are protecting capital investment in the hope of better days. This "hanging-on" cannot continue much longer.... Saginaw Bay fisheries like other segments of the industry have been caught in the "squeeze" created by a rise in cost of production much greater than the increase in the price of fish. Limitations of capital, the small size of producing firms, the division of the fishery among numerous ports of landing, legal restrictions on grounds, methods, and time of fishing, and the resistance to change that these conditions encourage, have made most difficult the adjustment that might have eased the stringency. The Saginaw Bay fishermen, in the main, are taking and handling fish in the same way as for several decades past.*

Increasing industrialization in the Saginaw Bay area during the early 1960s drew people away from fishing on a full-time basis (Anon., 1968, pp. 45–6):

> *The industry is on the verge of total collapse. Firms who have "consumed" their existing gear and boats and have borrowed to the limit of their credit are nearing the end of the tether. The small independent and the part-time operators who accounted for considerable production are disappearing through superannuation or abandonment of work that pays too little. Young men are not attracted to an industry where rewards are scant for independent operations and pay, as an employee, is the Wage Board minimum. Unless the trend is halted, Saginaw Bay will have no in-*

dustry to take advantage of that hoped-for day when sea lamprey control and other measures may bring the return of plentiful high-value fish.

Diversification

The fishermen who survived with least difficulty were those located at places having a multi-species fishery, such as Grand Haven on Lake Michigan, rather than those depending on one or two high-value species. Also, "vertically integrated fishermen" with their own retail outlets were able to purchase fish from other areas, when their own catches failed.

Some fishermen managed to adapt to the loss of high-value lake trout and whitefish by turning their attentions to lower-valued species. During the period 1960–66, seven species – lake herring (also coregonines), chubs, alewives, carp, yellow perch, smelt, and whitefish – contributed 500 tons (453 mt) or more to Michigan's production (Crowe, 1968). The Great Lakes Fishery Commission (1956, p. 23) considered the implications:

> *Undoubtedly, most fish populations in the upper Great Lakes are affected by the sea lamprey directly and very definitely affected indirectly. Shifts in the distribution of the fishery placed pressures on fish populations that normally were not heavily exploited. These pressures resulted in some temporary maladjustments in the shoal and bay fisheries and certainly had their effect on the relationships among the fish species.*

In Lakes Michigan and Superior the deep-water fisheries concentrated on chubs, and increased production after 1948 and 1957, respectively. However, the pattern was different on American Lake Huron, where chub production remained relatively low during the period of lake trout decline (Anon., 1968). The revival of the chub fishery in Lake Huron did not occur until 1958 (following publication of reports of abundance by the US Bureau of Commercial Fisheries), temporarily increasing both numbers of fishermen employed and gross earnings, until the slump in production after 1961.

New markets for less-valued fish such as carp and yellow perch were also opened. Markets for species such as carp, yellow perch, and suckers were highly competitive and characterized by fluctuating prices; fishermen were forced to raise their efficiency and adopt new fishing and processing methods. The growing abundance of small species, such as smelt, alewives, and bloaters, and the loss of the large predators were making the traditional production gear – gill and pound nets – obsolete. These new target species demanded a shift from selective harvesting of larger fish to volume production. Fishermen had traditionally sold their catch of whitefish and trout either in the round or gutted. Now, mechanization of packing facilities was necessary. Also, improved processing techniques such as reduction plants, as well as new and improved methods of preservation (due to the tendency of the lower-value frozen fish to spoil more quickly), were essential.

Thus, the small size and labor-intensive character of fishing enterprises on the upper Great Lakes began to change. On Lake Superior, for example, as individual fishermen left the industry, their licenses were purchased by firms which could afford extensive investment and which possessed capital sufficient to permit them to weather the vagaries of changing markets and unpredictable resources. In the 1960s, the fishing enterprises of Jackfish and Port Coldwell on the Canadian North Shore failed along with the lake trout on which they depended; their people moved and these villages ceased to exist. During these years, licenses frequently changed hands until, finally, the eastern and western Lake Superior Canadian fisheries came to be dominated by larger firms, such as Ferroclad Fisheries (north of Sault Ste. Marie, Ontario) and Kemp Fisheries (based in Duluth, Minnesota).

Emigration and intensification

Did some fishermen, confronted with the apparent inexorable loss of high-value species and without confidence in future benefit to be derived from conservation, opt to increase fishing intensity in order to harvest fully all those fish which remained? This does not appear to have been the case on American parts of Lakes Huron and Michigan. Hile (1949) reported that fishing intensity for lake trout generally declined after 1939. Hile et al. (1951a) also showed that intensity indices for the Michigan waters of Lake Michigan generally declined after World War II (with the exception of Green Bay, Wisconsin), a phenomenon the authors attributed to the increased targeting of whitefish, a coregonine.

Declining trout abundances, furthermore, coincided with World War II. In the early years of the war, many younger fishermen were drafted as valued sailors for the navy. The fishermen's associations lobbied strongly for exemption, noting that exemptions were already given to farmers as food producers. They eventually won their case, preventing further destabilization of the industry through the loss of manpower.

It is possible, however, that wartime demands for food led to the escalation of fishing effort in some waters. Conservation was accorded a minor place in resource development programs during World War II (Highsmith et al., 1962). The increase in lake trout production from the Illinois waters of Lake Michigan in the early 1940s, for example, has already been noted. According to Frick (1965), price inflation during World War I stimulated heavy fishing in the Canadian waters of Lake Erie, a phenomenon repeated to a lesser degree during World War II.

In the face of failing resources during the early lamprey years, certain fishermen sought new grounds. Some Lake Huron gill net fishermen, during the lean years after the late 1930s, moved their operations to Lake Michigan where fish stocks were still productive, remaining there until the loss of that lake's trout or in some cases entering the chub fishery (Anon., 1968). In their turn, fishermen from northern Lake Michigan ports such as Charlevoix moved south to Grand Haven, etc. Many of these men already had previous ties to the area, having traditionally fished these southern open waters during winter months.

As the lake trout fisheries of Lakes Huron and Michigan were failing, those of Lake Superior still remained lucrative. In Canada there did not appear to be any major transfer of men or equipment from Lake Huron to Lake Superior. For the American shore, however, J.W. Moffett (US House of Representatives, 1952, p. 55) noted:

> *Many fishermen moved into other fisheries, greatly increased the pressure on certain fish stocks and seriously overcrowded several areas where fishing for species other than lake trout was good. Lake Superior trout fisheries were overexploited because of heavy market demands until they too are now in a precarious condition. The Lake Superior lake trout fishery would have ceased or certainly become drastically reduced had market prices remained as*

> *they were before the loss of Lake Michigan and Lake Huron fisheries. Increasing sea lamprey populations in Lake Superior are threatening seriously to repeat the history of their predations in the other lakes. . . . In the State of Michigan waters the output has been maintained by increasing fishing pressure more and more as the abundance of fish has declined rapidly toward a critical level.*

In 1949 trout production from American Lake Superior had increased 6 percent from the 1929–43 mean, but fishing intensity was up 62 percent (Hile et al., 1951b). Moffett noted:

> *Species not subject to extensive sea lamprey attacks are threatened by transfers of fishing pressure by fishermen forced to seek new grounds. In the highly productive waters of northern Green Bay [Lake Michigan], for example, the fishing pressure on some of the major species has increased five- to tenfold or more within the last 5 or 6 years* (US House of Representatives, 1952, p. 74).

Concluding comments

The fortunes of most Great Lakes fisheries ebbed to their lowest points in the early 1960s. In the offshore waters of the three upper Great Lakes, the main cause (direct and indirect) of the collapse of various salmonid stocks (i.e., salmonines plus coregonines) was the sea lamprey. In shallow bays of the upper lakes, as well as throughout the two lower lakes, eutrophication and toxic pollution played predominant roles (Egerton, 1989). But everywhere an already intensive fishery did not practice restraint and was not constrained to compensate for these other intensifying stresses.

In the upper lakes the 25-year period from the late 1930s to the early 1960s was difficult for the commercial fishermen, sport-fishermen service enterprises and the small port communities. The 25-year period since then has had difficulties of another kind. Effectively, commercial fishermen have lost their favored position as harvesters of these resources. They have been relegated to second or third place behind anglers and Indian quasi-artisanal fisheries. Many port communities are again benefitting from the fisheries interests although somewhat nervously. Some major difficulties remain unsolved. For example, should expensive efforts to restore lake trout continue? Should the states and provinces pay for the lamprey control program? Should the fisheries' interests learn to live indefinitely with sea lamprey populations that are under only partial control? Should fisheries, including sport fisheries, targeted on seriously contaminated stocks be banned entirely?

Our case study effectively covers the 25-year period from the late 1930s to the early 1960s. Today, fifty years after their initial irruption, the harm done by the sea lamprey has been only partially reversed.

Summary

The sea lamprey irrupted sequentially in the upper Great Lakes in the period 1935–55. The invasion by sea lamprey was facilitated by ship canals and by ships themselves. Shipping interests were never held directly accountable for this consequence, perhaps because the effect followed long after the cause.

The preferred prey of lamprey included salmonids, both salmonines and coregonines, which are also preferred by fisheries, whether commercial, recreational, or artisanal. Salmonids are ecologically sensitive species and were increasingly being stressed by many adverse cultural influences, including a fishery practiced with an intensity that eventually bordered on biological overfishing.

When the sea lamprey began competing with fishermen for their preferred fish, fishermen did not temper the intensity of their fishing activities to compensate for the new stress exerted by sea lamprey.

Many relatively discrete stocks of salmonids were extinguished altogether, as a result of the combination of several stresses in which sea lamprey predation played a predominant role. Generally, the low point of fisheries fortunes in the Great Lakes occurred in the early 1960s.

During the decades that the sea lamprey irrupted sequentially in the upper lakes, fishery politics were in turmoil with respect to a number of difficult issues, one of which was the perennial struggle in the United States about states versus federal rights. The irruption of sea lamprey as an issue did not become politicized in the sense that any jurisdiction or group of stakeholders was blamed for it, not even shipping interests. Control of sea lamprey provided an appropriate focus for joint interjurisdictional action and became the beachhead for expanding interjurisdictional governance of fishing issues in the Great Lakes.

Many small fisheries and fishing communities succumbed as a result of the new sea lamprey stress imposed on an ecosystem that was already experiencing moderate cultural stresses throughout the ecosystem and some intense stresses locally. Many small fisheries ceased operations in ways reminiscent of how farming ceased in marginal areas during hard times; the young left for the cities and towns and the old used up their "capital" in part-time fishing in a life of relative poverty.

Some larger fisheries consolidated their vertically integrated operations, diversified fishing practices, diversified their businesses to include nonfishing enterprises and survived for the two decades until some preferred species returned as a result of lamprey control and other rehabilitative reforms.

As of 1989, the sea lamprey was under partial control in most parts of the Great Lakes; in large rivers and their mouths, the sea lamprey is not under control because available control techniques are not cost-effective in those waters. Where sea lamprey have been partially controlled, and because other rehabilitive reforms are also underway in these ecosystems, fisheries for some valued species are again thriving. But the recovery of lake trout has been very slow and is far from complete.

The sea lamprey invasion of the Great Lakes occurred at the time that various other effects of cultural malpractices in the Great Lakes Basin were combining synergistically to create a major crisis for the fisheries and for other valued aspects of the Basin ecosystem. For the fisheries interests, themselves culpable for some of the ecosystem abuses, the sea lamprey became a scapegoat. No fishery interest was held responsible for this common enemy. Fisheries factions that could not collaborate previously then joined forces to combat the sea lamprey. Limited collaboration against the repulsive lamprey became a basis for expansion of collaboration to other issues, such as overfishing. This may be the main lesson that this case study brings by analogy to a consideration of possible consequences of climate change.

Acknowledgments

We were assisted by the late F.E.J. Fry and by the late C. Verduin. Our draft report was reviewed critically by D.W. Coble, R.L. Eshenroder, W.M. Spaulding Jr., and W.M. Sprules. Nevertheless we know that our interpretations must contain errors for which we bear responsibility. The Donner Canadian Foundation helped with grant funds.

References

Anon. (1948). Too little and too late: An editorial on the critical lamprey problem on the Great Lakes. *The Fisherman*, **5**, 11, 14.

Anon. (1960). Problems related to the establishment of a trawl fishery on Lake Michigan. Public Archives of Canada, Ottawa RG23 v. 1610 (784-8-7(9)). Ann Arbor: US Fish & Wildlife Service.

Anon. (1968). Report on commercial fisheries resources of Lake Huron Basin. US Fish & Wildlife Service, Bureau of Commercial Fisheries.

Anon. (1969). Report on commercial fisheries resources of the Lake Superior Basin. US Fish & Wildlife Service, Bureau of Commercial Fisheries.

Applegate, V.C. (1950). Natural history of the sea lamprey, *Petromyzon marinus*, in Michigan. US Fish & Wildlife Service, Special Science Report No. 55.

Applegate, V.C. & Moffett, J.W. (1955). The sea lamprey. *Scientific American*, **192**, 26–41.

Berst, A.H. & Spangler, G.R. (1973). Lake Huron: The ecology of the fish community and man's effects on it. Great Lakes Fishery Commission, Technical Report No. 21.

Cobb, J.N. (1900). The commercial fisheries of Lake Erie, Lake Ontario, and the Niagrara and St. Lawrence Rivers. *Annual Report*. Albany: Commissioner of Fisheries, Game and Forests of the State of New York.

Crowe, W.R. (1968). Profile of an industry. *Michigan Conservation*, March–April, 19–23.

Dymond, J.R., Hart, J.L. & Pritchard, A.L. (1929). The fishes of the Canadian waters of Lake Ontario. *University of Toronto Studies, Biological Series*, **37**, 335.

Egerton, F.N. (1985). Overfishing or pollution? Case history of a controversy on the Great Lakes. Great Lakes Fishery Commission, Technical Report No. 40/41.

Egerton, F.N. (1989). Missed opportunities: US fishery biologists and productivity of fish in Green Bay, Saginaw Bay and Western Lake Erie. *Environmental Review*, **13**, 33–64.

Eschmeyer, P.H. (1957). The near extinction of lake trout in Lake Michigan. *Transactions of the American Fisheries Society*, **85**, 102–9.

Francis, G.R., Magnuson, J.J., Regier, H.A. & Talhelm, D.R. (1979). Rehabilitating Great Lakes ecosystems. Great Lakes Fishery Commission, Technical Report No. 37.

Frick, H.C. (1965). Economic aspects of the Great Lakes fisheries of Ontario. Fisheries Research Board of Canada. Bulletin No. 149.

Fry, F.E.J. (1953). The 1944 year class of lake trout in South Bay, Lake Huron. *Transactions of the American Fisheries Society*, **82**, 178–92.

Gallagher, H.R. & Van Oosten, J. (1945). Supplemental report. *International Board of Inquiry for the Great Lakes Fisheries, Report and Supplement*, 90–3. Washington, DC: International Board of Inquiry for the Great Lakes Fisheries.

Goodier, J.L. (1981). Native lake trout (*Salvelinus namaycush*) stocks in the Canadian waters of Lake Superior prior to 1955. *Canadian Journal of Fisheries and Aquatic Sciences*, **38**, 1724–37.

Great Lakes Fishery Commission (1956). The sea lamprey: Its effects on Great Lakes fisheries and the status of efforts to control it. Ann Arbor: Great Lakes Fishery Commission.

Great Lakes Sea Lamprey Committee (1946). Condensed transcript of Proceedings of the Great Lakes Sea Lamprey Conference, 14–15 November 1946. Ann Arbor: Great Lakes Sea Lamprey Committee.

Harris, H.J., Harris, V.A., Regier, H.A. & Rapport, D.J. (1988). Importance of the nearshore area for sustainable redevelopment: The Baltic Sea and Great Lakes. *Ambio*, **17**, 112–20.

Halverson, L.H. (1955). The commercial fisheries of the Michigan waters of Lake Superior. *Michigan History*, **39**, 1–17.

Highsmith, R.M., Jensen, J.G. & Rudd, R.D. (1962). *Conservation in the United States*. Chicago: Rand McNally and Co.

Hile, R. (1949). Trends in the lake trout fishery of Lake Huron through 1946. *Transactions of the American Fisheries Society*, **76**, 121–47.

Hile, R. & Buettner, H.J. (1955). Commercial fishery for chubs (ciscoes) in Lake Michigan through 1953. US Fish & Wildlife Service, Special Scientific Report No. 163.

Hile, R. & Buettner, H.J. (1959). Fluctuations in the commercial fisheries of Saginaw Bay, 1885–1956. US Fish and Wildlife Service, US Department of Interior, Research Report No. 51.

Hile, R., Eschmeyer, P.H. & Lunger, G.F. (1951a). Decline of the lake trout fishery in Lake Michigan. *Fishery Bulletin*, **60**, 77–95.

Hile, R., Eschmeyer, P.H. & Lunger, G.F. (1951b). Status of the lake trout fishery in Lake Superior. *Transactions of the American Fisheries Society*, **80**, 278–312.

International Board of Inquiry for the Great Lakes Fisheries (1943). Report and Supplement. Washington, DC: International Board of Inquiry for the Great Lakes Fisheries.

IJC (International Joint Commission) (1987). *Revised Great Lakes Water Quality Agreement of 1978*. Detroit: IJC.

Kumlien, L. (1887). The fisheries of the Great Lakes. In *The Fisheries and Fishery Industries of the United States*, Section 4, Part 14, ed. G.B. Goode. Washington, DC: US Commission on Fish and Fisheries.

Lawrie, A.H. & Rahrer, J.F. (1972). Lake Superior: effects of exploitation and introductions on the salmonid community. *Journal of the Fisheries Research Board of Canada*, **29**, 765–76.

Loftus, D.H. (1980). Interviews with Lake Huron commercial fishermen. Owen Sound, Ontario: Lake Huron Fisheries Assessment Unit, Ontario Ministry of Natural Resources.

Loftus, D.H. & Regier, H.A. (1972). Introduction to the Proceedings of the 1971 Symposium on Salmonid Communities in Oligotrophic Lakes. *Journal of the Fisheries Research Board of Canada*, **29**, 613–6.

MacKay, H. & MacGillivray, E. (1949). Recent investigations on the sea lamprey, *Petromyzon marinus*, in Ontario. *Transactions of the American Fisheries Society*, **77**, 148–59.

Pycha, R.A. & King, G.K. (1975). Changes in the lake trout population of southern Lake Superior in relation to the fishery, the sea lamprey, and stocking, 1950–70. Great Lakes Fishery Commission, Technical Report No. 28.

Rapport, D.J., Regier, H.A. & Hutchinson, T.C. (1985). Ecosystem behavior under stress. *American Naturalist*, **125**, 617–40.

Regier, H.A., Holmes, J.A. & Pauly, D. (1990). Influence of temperature changes in aquatic ecosystems: An interpretation of empirical data. *Transactions of the American Fisheries Society*, **119**, 374–89.

Regier, H.A. & Loftus, D.H. (1972). Effects of fisheries exploitation on salmonid communities in oligrotrophic lakes. *Journal of the Fisheries Research Board of Canada*, **29**, 959–68.

Regier, H.A., Welcomme, R.L., Steedman, R.J. & Henderson, H.F. (1989). Rehabilitation of degraded river ecosystems. In *Proceedings of the International Large River Symposium*, ed. D.P. Dodge, pp. 86–97. *Canadian Journal of Fisheries and Aquatic Sciences, Special Publication No. 106*.

Shetter, D.S. (1949). A brief history of the sea lamprey problem in Michigan waters. *Transactions of the American Fisheries Society*, **77**, 160–76.

Smiley, C. (1882). Changes in the fisheries of the Great Lakes during the decade 1870–1880. *Transactions of the American Fisheries Society*, **11**, 28–37.

Smith, S.H. (1968). Species succession and fishery exploitation in the Great Lakes. *Journal of the Fisheries Research Board of Canada*, **25**, 667–93.

Spangler, G.R. (1970). Factors of mortality in an exploited population of whitefish, *Coregonus clupeaformis*, in northern Lake Huron. In *Biology of Coregonid Fishes*, ed. C.C. Lindsey & C.S. Woods, pp. 515–29. Winnipeg: University of Manitoba Press.

Spaulding, W.M. & McPhee, R.J. (1989). Preliminary report of the evaluation of the Great Lakes Fishery Commission by the bi-national evaluation team. Ann Arbor: Great Lakes Fishery Commission.

State of Michigan (1953). Public and local acts of the legislature of the State of Michigan passed at the regular session of 1953, pp. 259–65. Lansing: State of Michigan.

TAFS (Transactions of the American Fisheries Society) (1990). *Proceedings of the Symposium on Effects of Climate Change on Fish*, Toronto, Ontario, 14–15 September 1988. *Transactions of the American Fisheries Society*, **119**, 173–389.

US House of Representatives (1897). Preservation of the fisheries in waters contiguous to the United States and Canada. 54th Congress, 2nd Session, 24 February 1897. Doc. 315. Washington, DC: US House of Representatives.

US House of Representatives (1949). Commercial fishing in the Great Lakes area. Hearings before the Subcommittee of the Fisheries and Wildlife Conservation of the Committee on Merchant Marine and Fisheries. 81st Congress. Washington, DC: US House of Representatives.

US House of Representatives (1952). Further research and control of sea lampreys of the Great Lakes area. Hearings before the Subcommittee of the Fisheries and Wildlife Conservation of the Committee on Merchant Marine and Fisheries. 82nd Congress. Washington, DC: US House of Representatives.

Van Oosten, J. (1949). A definition of depletion of fish stocks. *Transactions of the American Fisheries Society*, **77**, 283–9.

Wells, L. & McLain, A.L. (1973). Lake Michigan: Man's effects on native fish stocks and other biota. Great Lakes Fishery Commission, Technical Report No. 20.

10

North Sea herring fluctuations*

R.S. BAILEY

Scottish Office,
Agriculture and Fisheries Department
Marine Laboratory
Aberdeen AB9 8DB, Scotland

and

J.H. STEELE

Woods Hole Oceanographic Institution
Woods Hole, MA 02543, USA

Introduction

In historical terms, the North Sea herring is one of the most important marine fish resources in the world. It supported major fisheries in many countries of northwest Europe for hundreds of years (Cushing, 1988); yet in 1977 the directed fisheries were closed following a collapse of the stocks to a small fraction of their earlier levels. There is voluminous literature documenting the sequence of events that led up to the closure, the interpretations that were placed upon them, and the arguments that ensued. Only in hindsight, however, can a reasonably convincing account of the causes be assembled and, even now, there are aspects of the problem that defy explanation. The simple fact is that the demography of fish populations in the sea can only be studied by indirect means – from information on what fishermen remove and from the results of sampling on research vessel surveys.

In this contribution, the history of the collapse and subsequent recovery is briefly summarized. In particular, the question addressed is the relevance of environmental changes and whether action could have been taken to prevent or mitigate stock collapse. The impact of the collapse on the fishing industry is discussed and an evaluation is given of what might be required to avoid the consequences of stock collapse in similar instances in the future.

The biological problem

The landings of herring from the North Sea from 1920 to 1988 are shown in Fig. 10.1. These landings derive, however, from a sequence of more or less isolated fisheries in different parts of

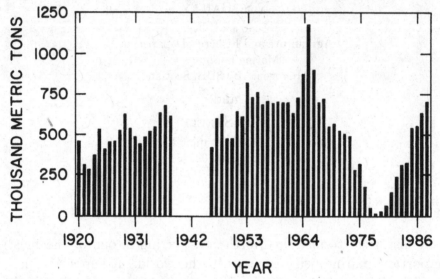

Fig. 10.1 Total international landings of herring from the North Sea, 1920–88.

the North Sea at different times of the year. The decline in landings first became noticed in the southern North Sea, while only in the late 1960s was there a noticeable decrease in the northern North Sea (Fig. 10.2). These differences in timing are partly attributable to the fact that the North Sea herring is not a homogeneous, randomly-mixing population, but an assemblage of discrete spawning groups that mix on feeding grounds prior to spawning (Fig. 10.3). The first uncertainty was, therefore, that relating to the stock identity of the fish caught: this received a massive input of research without any convincing general conclusion.

Research biologists first addressed the cause of the fishery decline seriously in the mid-1950s, following the collapse of the important East Anglian fishery. It was not accepted by all scientists that the problem was a global one (in the sense of the total North Sea) until after 1965, in which year a large fleet of purse seiners from Norway entered the North Sea following the collapse of the stock of Norwegian spring spawning herring.

Rather than considering restriction of fishing effort, much of the scientific discussion at this time hinged around the need for con-

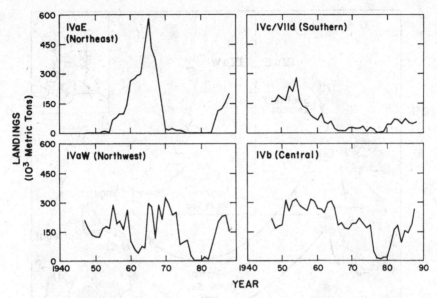

Fig. 10.2 North Sea herring landings in the four subareas shown in Fig. 10.3, 1947–88.

trols on industrial fisheries for juvenile herring, and on fishing for herring on the spawning grounds. Calculations of equilibrium yield per recruit (that is, the potential yield over its life span from each cohort of herring produced by each year's spawning) gave no indication that there would be any benefit from limiting fishing effort, so long as the number of recruits entering the fishery each year was maintained. Evidence on changes in recruitment was equivocal. There was some indication of a decrease in year class strength for the North Sea as a whole, but it was uncertain whether this was simply part of the natural variability in recruitment or related to the egg production of the stock. Uncertainty in interpretation was transmitted into ambivalence in the advice with quite serious disagreement among the scientists involved.

As evidence for a sustained decrease in recruitment accumulated in the 1970s, the key question, as in many other fish stocks, was what controlled recruitment. A plausible relationship between recruitment and the spawning stock size that produced it can be caused either by a real relationship between these two variables, or purely by environmental control. If recruitment is poor over an extended period because of an environmental change, then the spawning stock decreases regardless of whether there is a causal relationship between recruitment and stock. The environmental

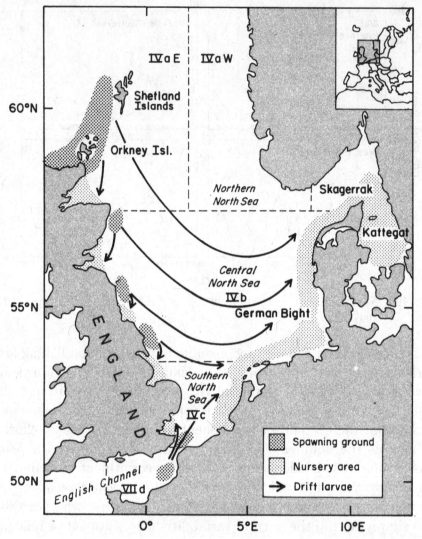

Fig. 10.3 Herring spawning grounds, nursery areas and drift routes of larvae,
with subareas used in Fig. 10.2. (Adapted from Corten, 1986, by
permission of the General Secretary, ICES.)

effect is exacerbated by heavy fishing pressure, because the adult
stock decreases that much more rapidly. The problem is summarized pictorially in Fig. 10.4.

   In the family of curves in Fig. 10.4a, there is strong dependence
of recruitment on parent stock modified by environmental variation. In Fig. 10.4b, variations in egg production, for which spawning stock is a proxy, are fully compensated by density-dependent
factors acting in the early life history; and variation in recruitment

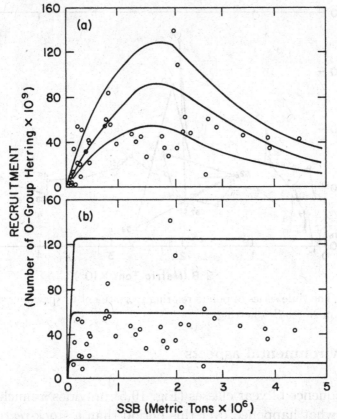

Fig. 10.4 Hypothetical stock–recruitment relationships for North Sea herring: (a) strong dependence of recruitment on spawning stock size and low levels of spawning stock modified by environmental factors; (b) recruitment entirely controlled by environmental factors except at very low levels of stock size. For further explanation, see text.

is entirely driven by environmental noise. There is no information in the time series plot of recruitment and stock to distinguish between these two extreme possibilities. The problem of distinguishing cause and effect is clearly shown by the time series of stock and recruitment shown in Fig. 10.5.

During the period of low recruitment in the 1970s, a relationship between recruitment and stock was postulated for North Sea herring. But it is clear that a subsequent sequence of improved year classes from 1980 onwards has been produced from a low spawning stock size. This fails to resolve the issue completely, because the closure of the fishery from 1977 allowed the spawning stock to recover from its very low levels of 1974–77. However, it changes one's perception of the relative importance of stock and environment.

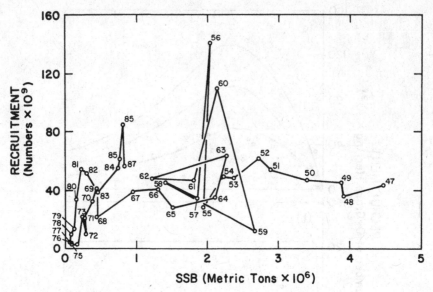

Fig. 10.5 The time series of points relating recruitment to spawning stock size in North Sea herring.

The environmental aspects

The sequence of year classes (Fig. 10.5) provides a much better sense of what happened over this period than a stock-recruit plot without temporal changes. It is clear that there was a period of low relative survival in the mid-1970s followed by much enhanced survival from 1980 onwards. This enhanced survival cannot be attributed to increased stocks. The opposite appears to be true (Fig. 10.6). One possibility is that the main causative factor was variation in the transport of larvae from the northwestern spawning grounds to the southeast nursery areas. This hypothesis, proposed by Corten (1986), is based on larval survey data for the years 1976–83.

An alternative hypothesis, recently proposed by Munk and Christensen (1990), suggests that the variation in recruitment was due to changes in the survival of the larvae and juveniles produced in the central regions of the North Sea rather than to a change in the pattern of drift of larvae produced in the northern region. There are no adequate hydrographic data to substantiate the first of these hypotheses directly (Corten, 1990), although major variations in North Sea circulation have been observed. An important question is whether these variations are internally generated within

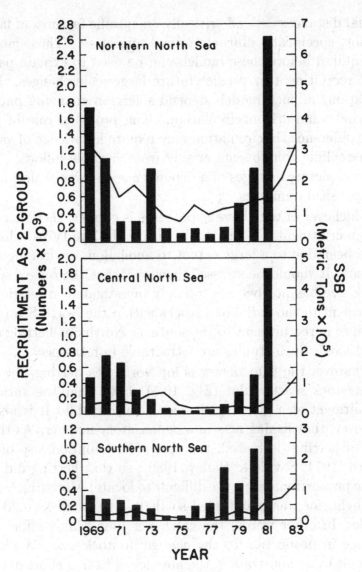

Fig. 10.6 Stock size (drawn lines) and recruitment (histograms) for separate components of North Sea herring. (Reproduced from Corten, 1986, by permission of the General Secretary, ICES.)

the North Sea, or whether exchanges with the Northeast Atlantic and the Norwegian Sea play a role (Dooley, 1983), thereby linking the North Sea changes with larger-scale variability in the North Atlantic (Dooley et al., 1984).

Recently, Bartsch et al. (1989) have modeled the winter transport of larvae and demonstrated that advection is an adequate mechanism to take larvae across the North Sea. However, the actual distances depend critically on specific features of larval behavior, specifically diurnal vertical migration. Thus, more work is required before these models can be used to explain particular past recruitment or predict future large-scale changes. Further, variations in such models depend solely on changing patterns of internal wind-driven circulation. The probable role of year-to-year differences in circulation may require knowledge of variations in baroclinic components arising from variable inflows. Yet, the general picture emerges of a dependence on large-scale features in the physical dynamics.

Whichever of the above hypotheses is correct, there is no doubt that recruitment was anomalous during the late 1970s, which must have been due to a large extent to anomalous environmental conditions. It should be stressed, however, that the role of spawning stock size has not been adequately investigated in relation to recruitment to the individual stocks within the North Sea (e.g., the collapse in recruitment to the southern North Sea herring stock in the 1950s). Such studies are intractable in retrospect.

Whatever the true answer is for North Sea herring, the recruitment/stock scatter plot (Fig. 10.5) shows that the variation of recruitment at a given stock size is considerable. It is also clear, however, that fishing acts in a depensatory manner. As the stock size of herring decreased, the area of distribution also decreased (Burd, 1974; Saville & Bailey, 1980), so that herring did not become proportionately more difficult to locate. Moreover, the shoaling behavior was maintained so that large catches could still be made. In other words, the catch per unit fishing effort did not change in proportion to the change in stock size. While this is difficult to demonstrate in the absence of fishing effort data, there is a clear inverse relationship between fishing mortality rate (a measure of the proportion of the stock taken) and the stock size (Fig. 10.7).

To summarize, therefore, it appears that recruitment to the total North Sea herring stocks declined during the 1970s. While the existence of a recruitment stock relationship cannot be ruled out as a contributory cause, the decline was probably mainly due to environmental changes, which may include competition, predation, etc., as well as purely physical factors. With increased fishing mortality, the stock collapsed rapidly (according to Cooke, 1984, a collapse is when the stock is reduced to a negligible proportion

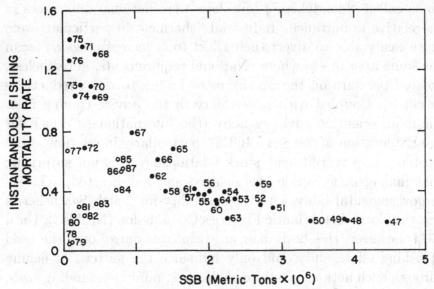

Fig. 10.7 The instantaneous fishing mortality rate on North Sea herring plotted against spawning stock biomass. Figures indicate years. Filled circles show the depensatory nature of fishing mortality in the years up to the fishery closure in 1977. Open circles show the rapid increase in fishing mortality following the reopening of the fishery in 1983.

of its former level) and remained at a low level during the period of closure until conditions for survival of larvae improved around 1982–83. Recent year classes have been at or above the long-term average level (Anon., 1989).

The management problem

The North Sea herring was exploited directly by about 10 countries. The national fisheries differed in character and objectives, the main fisheries being drift net and trawl fisheries for adults and recruiting age classes for human consumption (either fresh or cured); small mesh trawl fisheries for juvenile herring for reduction to meal and oil; purse seine fisheries for both human consumption and reduction. The relative importance of each of these methods and of each country's catch has changed; the major developments being the introduction of the small mesh industrial fishery in Denmark in the 1950s and of the Norwegian purse seine fishery around 1965. The North Sea herring stock was of considerable importance to many countries (including some that did not fish directly but transported herring from the fishing vessels on so-called "klondykers") but there were national differences in its relative importance. Industrial fishermen, in particular, were more easily able to divert their effort to other resources either in the same area or elsewhere. National requirements, nevertheless, created pressure on the management bodies to resist reductions in catch. Coupled with uncertainty in the advice coming from the main scientific advisory body (the International Council for the Exploration of the Sea – ICES), particularly in relation to the existence of a recruitment–stock relationship, it is not surprising that management action was delayed and half-hearted. The intergovernmental body with responsibility for North Sea fisheries was the Northeast Atlantic Fisheries Commission (NEAFC). Until 1974, however, this body had no statutory control over the level of fishing effort and could only influence the pattern of fishing through such actions as seasonal closures, minimum landing sizes, etc. Even when total allowable catches (TACs) were accepted in 1974, they were subject to powers of veto. In this situation, the TACs agreed upon were much larger than recommended and did not restrict catches. The result of this inaction, combined with failing recruitment, was that ICES advised a closure of the fishery in April 1976 (Anon., 1977). Only on the declaration by the UK of enlarged fishery limits in December 1977, however, closely followed by a wider decision by the European Community and Norway, was it possible to close the fishery.

In retrospect it is still uncertain what the results would have been if there had been stricter controls at an earlier stage. With decreased recruitment, there is little doubt that the stock size would have decreased, but a larger buffer stock could have been maintained to sustain a fishery over the poor years. The existence of a larger spawning stock might have improved recruitment, and allowed a quicker recovery. Even so, it is not clear whether the fishing industry could have been altogether protected from a collapse of the fishery during the years of abnormally low recruitment. Although not discussed in this chapter, it should also be noted that the fisheries for herring in some adjacent areas (e.g., the west coast of Scotland, the Celtic Sea off southern Ireland) were also closed at this time as a result of stock collapse, while others were at low levels (e.g., northwest Ireland, the Irish Sea, the Firth of Clyde). The societal impacts of the closure of the North Sea herring fisheries are addressed in the following section.

Societal impacts

In the last century and earlier parts of this century whole communities around the coasts of the North Sea depended on herring (e.g., Coull, 1986). This reliance has decreased but herring fishing is still important to some coastal communities. The effects of the collapse of the stock, therefore, varied in different ports bordering the North Sea. There is also some information available on imports of herring over the period of closure and the extent of supplies from other areas. Thus, imports from Canada maintained supplies of herring in the UK while catches in the Baltic were available to Denmark and Germany. As a generalization, however, worldwide supplies of herring diminished to low levels. Whereas the total world catch reached a peak of 4.6 million metric tons in 1966, it had decreased to 1.1 million metric tons by 1979 (UN FAO, 1989).

The effect on the market of the collapse is well demonstrated by the quantities and percentages of herring that were sold to the home market in Scotland before and after 1978 (Table 10.1). In the period 1971–77 never more than 20 to 25 percent were "klondyked," while most of the landings were absorbed by the home market, which accounted for 72,000 to 94,000 metric tons from 1971 to 1974. During the 1977–82 closure, much of what was available went to the home market and the unit price in both Scotland and other countries in the European Community responded accordingly (Table 10.2). In Scotland the unit price rose by a factor of over 20 between 1969 and 1979. Since the reopening of the North Sea and west of Scotland fisheries, the home market has been satiated at a level of under 20,000 metric tons and the unit price decreased and has stabilized at around £120 per metric ton from 1982–88. This lack of growth of the European market has given rise to considerable disquiet in the European Community (Commission of the European Communities, 1987). Clearly, the nonavailability of herring for a number of years has had a major and lasting effect.

The effect of the collapse in the small mesh trawl industrial fisheries was less noticeable, because catches of alternative species (mainly sprat, Norway pout, and sandeels) increased during this period (Fig. 10.8). Similarly, effort in the purse seine industrial fishery was transferred progressively to the North Sea stocks of

Table 10.1

Landings and disposal of herring by UK vessels in Scottish ports
1971–88

| Year | Total landings ($t \times 10^3$) | Landings to | | | |
| | | Home Market | | Klondyked | |
		Quantity	%	Quantity	%
1971	132.9	72.2	54.3	27.1	20.4
1972	137.9	78.4	56.9	25.1	18.2
1973	145.3	88.3	60.8	37.4	25.7
1974	129.4	94.1	72.7	23.2	17.9
1975	98.5	78.5	79.7	11.4	11.6
1976	73.1	62.9	86.0	7.7	10.5
1977	38.3	32.9	85.9	5.2	13.6
1978	13.8	13.3	96.4	0.4	2.9
1979	2.0	2.0	100.0	0.0	0.0
1980	2.2	2.1	95.5	0.0	0.0
1981	32.6	8.5	26.1	13.0	39.9
1982	41.5	7.9	19.0	16.7	41.2
1983	51.3	10.4	20.3	26.2	51.9
1984	67.9	12.2	18.0	39.3	61.9
1985	87.0	12.0	13.8	54.8	64.6
1986	100.7	16.4	16.3	77.1	77.3
1987	96.4	18.1	18.8	65.6	68.0
1988	87.4	17.4	19.9	62.4	71.4

Source: Scottish Sea Fisheries Statistical Tables

Table 10.2

The unit price (£ per metric ton) of herring in Scotland, 1965–88

Year	Unit Price	Year	Unit Price	Year	Unit Price
1965	27.5	1973	59.4	1981	130.1
1966	23.6	1974	91.3	1982	122.3
1967	24.8	1975	97.3	1983	126.3
1968	25.6	1976	128.5	1984	125.7
1969	26.0	1977	292.6	1985	121.2
1970	31.7	1978	392.5	1986	111.2
1971	33.3	1979	569.7	1987	119.6
1972	36.8	1980	416.5	1988	119.8

Source: Scottish Sea Fisheries Statistical Tables

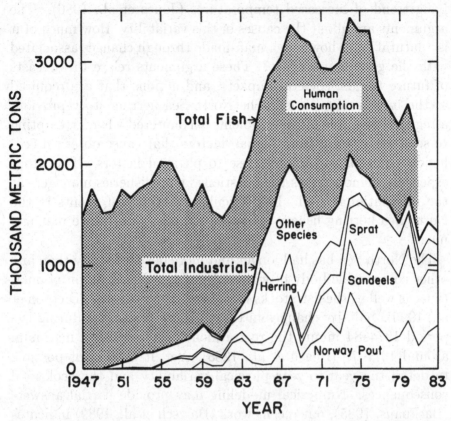

Fig. 10.8 Landings of fish used for reduction and landings of all fish from the North Sea shown as cumulative amounts within each year. (From Bailey, 1987.)

mackerel and sprat and, more recently, to the migratory western stock of mackerel.

Since the reopening of the North Sea herring fisheries in 1983, the European Community and Norway have agreed to a basis for management of the North Sea herring stock in which a set quota allocation procedure is applied to the TAC agreed on each year. In practice, however, the herring fisheries have developed rather unevenly. While major landings are made in Norway, Denmark, The Netherlands, and Scotland, there has been almost no recovery of markets in England, and landings in Germany are relatively low.

Discussion

Over the last century we have experienced significant climatic fluctuations which, at the global level, appear to have an underlying trend of increased temperatures (Jones et al., 1986). The arguments are about the causes of this variability. How much of it is "natural" and how much man-made through changes associated with the greenhouse effect? These arguments relate to forecasts of future trends, societal impacts, and actions that are required at the international level. The North Sea herring story provides a regional example of the problems encountered when attempting to separate natural from human factors that cause observed ecological changes and relate those to physical factors. This story especially demonstrates the questions facing fisheries managers at the international level. In particular, do the past events in the North Sea herring fishery prescribe strategies for future management?

The North Sea has had longer and more detailed study than any other comparable body of water. Trends in the plankton populations as well as the fish stocks have been demonstrated (Dickson et al., 1990). Salinity changes in the northern North Sea during the period 1974–84 mimic the recruitment pattern with a minimum around 1977–79 (Heath et al., 1985). Yet, it is still not possible to detail the year-to-year physical variability and the ecological consequences. Numerical modeling may provide partial answers (Backhaus, 1985), but recent work (Bartsch et al., 1989) indicates that details of behavior of plankton and fish larvae are needed to convert this information into ecological forecasts.

Evidence is now mounting that natural environmental factors played some part in the collapse of the North Sea herring stocks. What is not clear is whether, with different management, the "collapse" would only have been a short-lived "decline." Technically, this requires the recruitment–stock relationship to be expressed in terms of causal explanations for the mean and the variance. In practice this is not possible and so predictions of recruitment are limited to one or two years. This is much less than the time scale of the period of anomalous recruitment. Without a separation of these factors, explicit incorporation of environmental trends is not possible. But would the appreciation of such environmental factors alter the precepts on which management is based?

The countries around the North Sea have different methods of fishing for herring and different uses for the product – human consumption, fish meal, export. Some countries have alternative stocks available. Thus, the social and economic responses to the

collapse and return of herring varied significantly. Given the uncertainties about causes and also about the results of management
alternatives, consistent policies are not easily achieved. General
consensus on concepts of conservation and environmental change
do not necessarily lead to agreement on particular policies, so long
as there is a high level of uncertainty.

The decade of the 1970s is regarded as anomalous in terms of fish
stock fluctuation (Corten, 1986) and North Atlantic hydrography
(Dooley et al., 1984). This latter anomaly may be connected with
changes in atmospheric pressure distributions at high latitudes
in winter. Any causal relation with fish stock recruitment could
be directly through the water circulation patterns (Corten, 1986)
or indirectly through changes in wind-driven currents (Bartsch et
al., 1989). Thus there may be some correspondence between regional pelagic fish fluctuations and "global" climatic variations and
trends. It may be questioned whether the 1970s were "anomalous"
or whether this kind of pattern should be expected in the future
– and with increasing frequency, if the pressure from fishing is
maintained and if climatic changes, for example, global warming,
increase.

The possible effects of global climate change on the hydrography
and ecosystem of the North Sea are not clear. This uncertainty is
likely to persist and one therefore has to accept that forecasting
the herring stocks more than one or two years ahead will not be
possible. In this situation, how should one advise the decision-
makers?

The characteristics of the North Sea herring stock germane to
this discussion are as follows:

- The historic stability of North Sea herring as a resource.

- The severity of the final stages of the collapse caused by the
  combination of depensatory fishing pressure and recruitment
  failure.

- The reversibility of the collapse and the reasonably speedy
  recovery of the stock when the fishery was closed and recruitment returned to normal. In this respect, North Sea herring
  appears to differ markedly from some other pelagic fish stocks
  which have remained at low levels for several decades (e.g.,
  Sharp & Csirke, 1983).

- The fact that the anomalous hydrographic influences on recruitment persisted for a number of years.

- The slow response by the scientific community and, ultimately, the managers.

- The lasting effects of the closure on the markets for herring.

The implications of these characteristics are that the North Sea herring stock responds to hydrographic variability just as do many other stocks of pelagic fish. Within a constant regime, however, the stability and resilience of the stock appears to be reasonably high, so that it is appropriate to manage the stock as though it has persistence, that is, not to exploit year classes as though they are adventitious bounty.

The duration of the hydrographic anomaly, if repeated in the future, suggests that it would be possible to maintain a spawning stock at a reasonable level over a few years of poor recruitment, so long as catches were successfully restrained and so long as the management response was swift and resolute. This would, in turn, protect the fisheries from the need for total closure, and the markets from collapse.

In the past, action has been retarded by uncertainty on the grounds that a restriction on fishing is unwarranted unless the need for conservation is unequivocal. This uncertainty is likely to persist because, by their nature, short-term anomalies of the sort experienced in the North Sea are unpredictable and difficult to interpret at the time. If the worst effects of stock collapse are to be mitigated in the future, the implication of this is that fisheries managers would have to react to strong inference as opposed to scientific proof. This is likely to be contentious, because there is more than one way of responding to uncertainty.

If climatic change were to result in more persistent changes in the ecosystem, then any changes in the herring and other stocks are likely to persist over much longer periods. The scenario presented in this chapter does not address this eventuality, but it is relevant to the situation in which anomalous events recur with a greater frequency.

References

Anon. (1977). Reports of the Liaison Committee of ICES to the Northeast Atlantic Fisheries Commission and to the International Baltic Sea Fishery Commission, November 1975 to September 1976. Cooperative Research Report. *International Council for the Exploration of the Sea*, **56**, 117.

Anon. (1989). Report of the Herring Assessment Working Group for the Area South of 62°N, 4–14 April 1989, Copenhagen. ICES Council Meeting. 1985/Assess:15 (mimeo). Copenhagen: International Council for the Exploration of the Sea.

Backhaus, J.O. (1985). A three-dimensional model for the simulation of shelf sea dynamics. *Deutsche Hydrographische Zeitschrift*, **38**, 165–87.

Bailey, R.S. (1987). Industrial and underutilised fish resources. In *Developments in Fisheries Research in Scotland*, ed. R.S. Mailey & B.B. Parrish, pp. 75–87. Farnham: Fishing News Books Ltd.

Bartsch, J., Brander, K., Heath, M., Munk, P., Richardson, K. & Svendsen, E. (1989). Modelling the advection of herring larvae in the North Sea. *Nature*, **340**, 632–6.

Burd, A.C. (1974). The northeast Alantic herring and the failure of an industry. In *Sea Fisheries Research*, ed. F.R. Harden-Jones, pp. 167–91. London: Elek Science Publishers.

Commission of the European Communities (1987). *Report to the Council on the Situation of the Herring Market*. COM-87. Brussels: Commission of the European Communities.

Cooke, J.G. (1984). Glossary of technical terms. In *Exploitation of Marine Communities*, ed. R.M. May, pp. 341–8. Berlin: Springer-Verlag.

Corten, A. (1986). On the causes of the recruitment failure of herring in the central and northern North Sea in the years 1972–1978. *Journal du Conseil. Conseil International pour l'Exploration de la Mer*, **42**, 281–94.

Corten, A. (1990). Long-term trends in pelagic fish stocks of the North Sea and adjacent waters and their possible connection to hydrographic changes. *Netherlands Journal of Sea Research*, **25**, 227–35.

Coull, J.R. (1986). The herring fishery in Peterhead at the turn of the century: revolution through steam power. *Aberdeen University Review*, **174**, 323–32.

Cushing, D.H. (1988). *The Provident Sea*. Cambridge: Cambridge University Press.

Dickson, R.R., Kelly, P.M., Colebrook, J.M., Wooster, W.S. & Cushing, D.H. (1990). North wind and production in the eastern North Atlantic. *Journal of Plankton Research*, **10**, 151–69.

Dooley, H.D. (1983). Seasonal variability in the position and strength of the Fair Isle current. In *North Sea Dynamics*, ed. J. Sündermann & W. Lenz, pp. 108–19. Berlin: Springer-Verlag.

Dooley, H.D., Martin, J.H.A. & Ellett, D.J. (1984). Abnormal hydrographic conditions in the Northeast Atlantic during the 1970s. *Rapports et Procès-*

Verbaux des Réunions. Conseil International pour l'Exploration de la Mer, **185**, 179–87.

Heath, M.R., Henderson, E.W., Hopkins, P.J., Martin, J.H.A. & Rankine, P.W. (1985). Hydrographic influences on the distribution of herring larvae in the Orkney–Shetland area. ICES Council Meeting. 1985/H:29 (mimeo). Copenhagen: International Council for the Exploration of the Sea.

Jones, P.D., Wigley, T.M.L. & Wright, P.B. (1986). Global temperature variations between 1861 and 1984. *Nature*, **322**, 430–4.

Munk, P. & Christensen, V. (1990). Larval growth and drift pattern and the separation of herring spawining groups in the North Sea. *Journal of Fish Biology*, **37**, 135–48.

Saville, A. & Bailey, R.S. (1980). The assessment and management of the herring stocks in the North Sea and to the west of Scotland. *Rapports et Procès-Verbaux des Réunions. Conseil International pour l'Exploration de la Mer*, **177**, 112–42.

Sharp, G.D. & Csirke, J. (Eds.) (1983). *Proceedings of the Expert Consultation to Examine Changes in Abundance and Species Composition of Neritic Fish Resources, 18–29 April 1983, San Jose, Costa Rica. FAO Fisheries Report No. 291*, Vols. 2–3. Rome: FAO.

UN FAO (1989). *Yearbook of Fishery Statistics: Catches and Landings No. 64.* Rome: Food and Agriculture Organization.

11

Atlanto-Scandian herring:
a case study

ANDREI S. KROVNIN and SERGEI N. RODIONOV*

VNIRO (Fisheries Research Institute)
Moscow 107140, USSR

Introduction

Not long ago, Atlanto-Scandian herring was the most common among the herring inhabiting the world's oceans. Nowadays, despite the fact that its stock has been considerably depleted, it remains an important species from the point of view of commercial potential.

The term "Atlanto-Scandian herring" is commonly used to refer to three stocks: Norwegian spring-spawning herring, Icelandic spring-spawning herring, and Icelandic summer-spawning herring. Since the Norwegian spring-spawning herring is the largest of these three stocks, the term "Atlanto-Scandian" is often used as a synonym for this stock.

This chapter deals primarily with Norwegian spring-spawning herring, although some attention is paid to both of the Icelandic stocks as well. All three stocks have been subjected to similar changes in abundance during the last several decades. They were overexploited in the 1960s and 1970s but now two of them – Norwegian spring-spawning herring and Icelandic summer-spawning herring – have begun to recover.

The main countries dependent on the exploitation of Atlanto-Scandian herring are Norway, Iceland, and the USSR. Norway has jurisdiction over the main spawning grounds, feeding, and nursery areas, and also over one of the wintering areas; Iceland has jurisdiction over the southern wintering area, spawning, and feeding grounds. These two countries are the major biomass owners. Within the Soviet economic zone in the Barents Sea, there are nursery grounds of some year classes spawned off the Lofoten Islands (Norway). At times the Soviet zone can be of great importance. For example, about 70 percent of the recruits to the 1983

* Presently a postdoctoral fellow in the Advanced Study Program, National Center for Atmospheric Research, Boulder, Colorado.

year class were distributed in the areas which are under Soviet jurisdiction. This year class, on which the recovering fisheries are based, was the first relatively strong one since 1959.

From an economic point of view, the herring fishery has been one of the most important for Iceland. Recovery of the stock up to the level of the 1960s and restoration of traditional migration routes, which drastically changed when the stock decreased, are extremely vital for this country. In fact, about 40 percent of Iceland's total national export in the 1960s consisted of herring products (Jakobsson, 1985a).

A sharp decline in the Atlanto-Scandian herring stock in the late 1960s and early 1970s was one of the most striking events in the history of European fisheries. Its analysis is very important in order to assess societal abilities to cope with drastic changes in resource abundance, changes which are associated to a great extent with climatic variability. The history of the expansion and collapse of this fishery provides an appropriate model in our attempt to better understand the interactions between climate, fish populations and human activities.

This study is divided into seven sections, the first of which describes the major features of distribution, migration pattern, and environmental conditions of the Atlanto-Scandian herring. The second section deals with the historical review of the fishery and its exploitation. The effects of climate change on stocks and their availability are elaborated in the third section. Problems of management are reviewed in the fourth. In the fifth section we discuss how countries dependent on the Atlanto-Scandian herring reacted to the collapse. Some lessons from the history of this fishery that might be useful to better understand the possible future impact of climatic change on society are presented in the sixth section. In the final section key points of this study are summarized.

General biology and environmental conditions

Although the distribution and migration patterns of Atlanto-Scandian herring have changed during the last decades, it may be fruitful to study their common characteristics based on situations observed in the 1950s when the stock was at a high level.

Spawning of Norwegian spring-spawning herring occurs in February–March along the Norwegian coast in bottom layers affected by warm Atlantic waters with temperatures of 4–7.5°C and salinity 32–35 ‰ (Fig. 11.1). The surface layers, as a rule, are occupied by colder, less saline waters. Soon after spawning, larvae

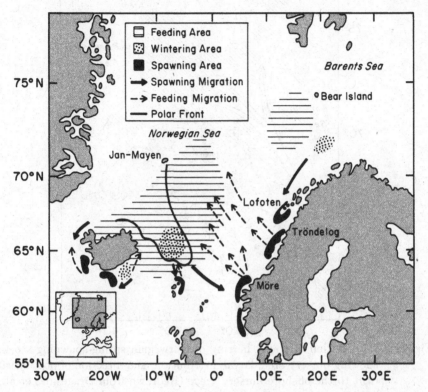

Fig. 11.1 Migration pattern of the Atlanto-Scandian herring during a period of high level stock, showing feeding area, wintering area, spawning area, spawning migration, feeding migration, and Polar Front. (Modified from Anon., 1972; and Hamre, 1991.)

rise into the upper layers and drift northward (Fig. 11.2). Under favorable recruitment conditions, most of the larvae are found in the Barents Sea. Spent herring (that is, post-spawning) migrate northwestward into the Norwegian Sea where they feed on zooplankton. Larger individuals reach the Polar Front in June–July and some of them cross into cold waters. The feeding area extends from Spitsbergen to Jan Mayen in the north to the western borders of the East Icelandic Current to the south. During the fall, the feeding area shrinks and the herring concentrate in the southwestern part of the Norwegian Sea along the southern and southwestern borders of the East Icelandic Current.

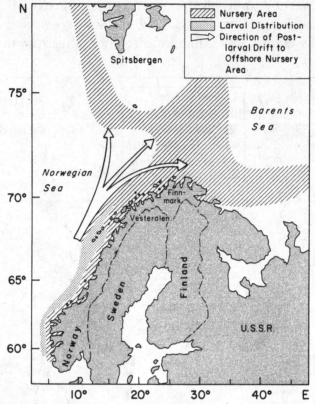

Fig. 11.2 Distribution of young herring of Norwegian spring-spawning stock, showing nursery area, larval distribution, and direction of post-larval drift to the offshore nursery area. (Redrawn from Dragesund et al., 1980.)

Norwegian spring-spawning herring spend the winter off the east Icelandic coast. In December and January prespawning concentrations migrate toward the Norwegian coasts. Course and speed of spawning migrations are regulated by optimal surface temperatures in the range of 6–8°C. Devold (1963), describing these migrations in detail, found that herring gather in cold water pockets before crossing the warm Norwegian Current into colder coastal waters. Usually, the herring approach the Norwegian coast (off Möre) and then spread to the south and north.

The described migration pattern is normal, when the stock of Norwegian spring-spawning herring is at a high level. The decline in stock, which occurred during 1950–70, resulted in changes in migration, distribution and location of spawning grounds.

From 1950 to 1962 the spawning grounds of the Norwegian spring-spawning herring gradually moved northward; since 1959, spawning south of Bergen (Norway) had been negligible (Drage-sund, 1970). In the second half of 1960s a separate component of stock was found in addition to the main one. It had a distinct feed-ing and wintering area south of Bear Island. Between 1965 and 1969, the herring fed mainly in the Spitsbergen to Jan Mayen area (Anon., 1978). In late autumn the herring gathered, as before, in the traditional wintering grounds off eastern Iceland.

Sharp changes in migration patterns took place in the 1970s, when the stock was depleted. In 1972 spawning was negligible. Two components of immature herring revived after the overex-ploitation of the 1960s: the first one in the Barents Sea and the second one off the western Norwegian coast. Both of them spawned for the first time in 1973; the Barents Sea (northern) component at Lofoten, and the Norwegian (southern) component off Möre. Un-like the previous years, the spent herring did not leave the coasts but migrated into interior waters where they fed during summer and fall. After 1983, when relatively rich year classes added to the spawning stock, separation into two components was not observed (Hamre, 1991).

In the summer of 1988 Soviet vessels found herring schools to the north of Iceland, southeast of Jan Mayen. Sometime later, Norwe-gian purse seiners, returning from fishing near Iceland, reported sighting herring shoaling on the way to the Norwegian coast. To validate this, a Norwegian research ship collected data which in-dicated that at that time there were more herring along the coast than there had been at any time since the mid-1960s (Anon., 1989). It should be noted that in 1989 the small component of Norwegian spring-spawning herring spawned at the southern grounds for the first time since 1959 and that spawning began rather late (in the first half of March).

During the 1970s, the individual rates of growth increased. For example, the herring of the 1969 year class reached 36.5 cm in length compared to 34 cm at the same age in the 1950s and 1960s. At the same time the average age of maturation declined from six to seven years in the 1950s to four years in the 1970s (Bakken, 1983).

Icelandic spring-spawning herring should probably be consid-ered as a component of Atlanto-Scandian herring living at the outer limits of its distribution (Jakobsson, 1985b). Usually this component is relatively small, but it can reach high levels of abundance under favorable environmental conditions in Icelandic waters.

This herring stock spawns in open waters off the southern coast of Iceland in March–April. By June, after spawning, herring move to the north coast of Iceland for feeding. There, the herring mix with the feeding Norwegian spring-spawning herring, which migrate to the same area. In late fall the mature fish begin their spawning migrations to the south coast.

Icelandic summer-spawning herring spawn in July–August, mostly in the same location as the Iceland spring spawners. In fall the spent herring migrate to the north. At this time their area of distribution overlaps with that of the spring-spawning stock.

History of Atlanto-Scandian herring fishery

The Atlanto-Scandian herring fishery began about one thousand years ago (Seliverstov & Seliverstova, 1991). Traditionally, the main fishery of adult herring had been a winter fishery along the Norwegian west coast prior to and during the spawning season. Available data show that periods of rather sharp fluctuations of catches occurred in the nineteenth century (Fig. 11.3). Gradual increases of catches occurred during the period 1810–30; from several hundred metric tons in 1810 to 70,000 metric tons in 1830. In the 1860s annual catches exceeded 100,000 metric tons followed by a sharp decline in the 1870s. In the 1880s and 1890s the annual catches totalled only 13,000 metric tons.

With decreases in winter adult herring catches, the fishery of immature (fat)* herring of two to five years and of young herring of zero to two years began to develop in the late 1890s and in the early 1900s. This fishery took place along the Norwegian coast in the fjords of northern Norway and in the Barents Sea along the Russian coast. At the beginning of the twentieth century the winter catches began to increase. An extremely strong year class appeared in 1904 which formed the base for 10 years of high catches.

* Fat herring means immature herring of two to five years of age in earlier decades and two to three years nowadays, caught in the fjords of northern Norway. Fat and immature are used here synonymously.

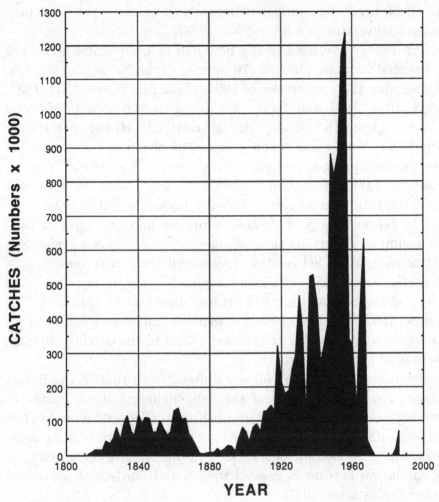

Fig. 11.3 Catches of winter herring off western Norway. (Modified from Devold, 1963.)

Other important fisheries were in summer and fall on the feeding grounds along the Polar Front to the north and northeast of Iceland. Usually, the fishery season extended from June to September and, until 1950, the stock was mainly exploited by Icelandic and Norwegian vessels.

Development of mainly the winter herring fishery in the early 1900s was associated not only with increased strength of year classes at that time but also with advances in the utilization of herring for reduction to fish meal. Until World War I, all of the herring catch was used for human consumption but after the war, fish-meal processing plants began to develop. In fact, by the 1940s, about 60 percent of adult herring catches was used for those purposes (Seliverstov & Seliverstova, 1988).

Ten rich year classes in the first half of the twentieth century accounted for approximately 70 percent of the Norwegian catches. Those were the year classes of 1904, 1918, 1923, 1930, 1934, 1937, 1943, 1944, 1947 and 1950. According to Marty and Fyodorov (1959), during the period 1923–50 relatively strong year classes appeared about every seven years. The sharp rise in catches of Norwegian spring-spawning herring began after World War II, reaching 1,000,000 metric tons in the early 1950s. In 1954 and 1956 the total annual catch exceeded 1,500,000 metric tons.

The Soviet fishery of Atlanto–Scandian herring began in 1950 as a summer fishery on the feeding grounds in the Norwegian Sea (Ponomarenko & Seliverstov, 1989). All the Soviet vessels were drifters. Since 1952, a fall and winter herring fishery developed along the migration routes of prespawning herring (Marty & Yudanov, 1962). Soviet vessels also exploited immature fish in inlets and bays along the Murman Coast. Most of the herring landings were used for food (curing).

Herring processing in Norway differed from that in the Soviet Union. In Norway, most of the fish, including small immature herring, were reduced to oil and fish meal (Seliverstov and Seliverstova, 1991). Only a small part of the immature herring landings was processed for delicatessen products. Norwegian catches of immature herring in coastal waters and fjords have increased considerably since 1951.

In the first half of the 1950s the Norwegian spring-spawning herring stock was at a high level of 8 million metric tons. In 1957 it reached a peak value of 10 million metric tons and then sharply decreased to about 3 million metric tons in 1963 (Anon., 1978). With this decline, the main source of its exploitation was lost to the Norwegian fishery, resulting in a serious economic crisis. The number of vessels significantly declined and fishermen were forced to find new ways to increase efficiency and reduce the costs of fishing. Norway carried out a technical reconstruction of its fishing fleet, as Iceland had done some years earlier. In the early 1960s, the two-dory system was replaced by the ring-net technique using power blocks. This allowed for a reduction in crew size by half. In addition, the new equipment was less affected by weather, allowing the fleet to operate much farther offshore. The development of hydroacoustic equipment was also of prime importance for the herring fishery (Dragesund et al., 1980).

These factors resulted in a sharp increase of catches, mainly through Icelandic efforts in the 1960s. In 1966, when the total catch of adult herring reached its maximum, the Icelandic share was 40 percent, whereas the USSR and Norway accounted for 29 and 26 percent, respectively (Table 11.1). The Faroes and West Germany caught the remaining 5 percent.

Table 11.1
Total catch (in thousands of mt) of adult and pre-recruit
Norwegian spring-spawning herring, 1950–71.

Year	Iceland	Norway	USSR	Faroes	Fed. Rep. Germany	Total
1950	30.7	781.4	14.0	–	–	826.1
1951	48.9	902.3	43.0	–	–	994.2
1952	9.2	840.1	69.9	–	–	919.2
1953	31.5	692.2	110.0	16.2	–	849.9
1954	15.2	1103.6	160.0	27.6	–	1306.4
1955	18.1	979.3	207.0	13.1	–	1217.5
1956	41.2	1160.7	235.0	23.7	–	1460.6
1957	18.2	813.1	300.0	17.0	–	1148.3
1958	22.6	356.7	388.0	17.7	–	785.0
1959	34.5	426.9	408.0	13.7	–	883.1
1960	26.7	318.4	465.0	11.0	–	821.1
1961	85.0	111.0	285.0	16.9	–	497.9
1962	176.2	156.2	209.0	9.8	–	551.2
1963	177.5	130.4	350.0	12.9	–	670.8
1964	367.4	366.4	365.8	18.3	–	1117.9
1965	540.0	259.5	489.2	31.5	5.6	1325.8
1966	691.4	497.9	447.4	60.7	26.1	1723.5
1967	359.3	423.7	303.9	34.9	9.7	1131.5
1968	75.2	55.7	124.3	16.1	1.8	273.1
1969	0.6	15.6	3.2	4.4	0.3	24.1
1970	–	20.3	–	0.6	–	20.9
1971	–	6.9	–	–	–	6.9

From Anon. (1978).

It should be noted that changes in migration routes during the 1960s greatly affected the location of summer and fall fisheries. They practically ceased off Iceland and relocated in shelf waters south and west of Bear Island and Spitsbergen.

A sharp decline in the spawning stock occurred after 1966, due to negligible recruitment and to progressively increasing fishing

mortality (Fig. 11.4). By 1970 it was less than 50,000 metric tons
(Bakken, 1983). Total catch of three main herring-fishing countries
– Norway, Iceland and USSR – declined by a factor of six in only
two years (from 1966 to 1968), but because of increased prices
of herring, Norway continued to exploit the depleted stock until
1971. However, Norwegian fishing of immature herring continued
after this date (Seliverstov & Seliverstova, 1991).

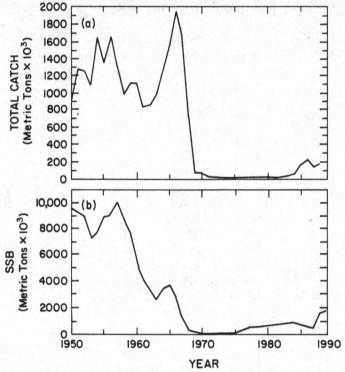

Fig. 11.4 Norwegian spring-spawning herring. (a) Total catches, 1950–1988
(from Hamre, 1991 and Anon., 1990a), and (b) estimated spawning
stock biomass (from Hamre, 1991).

During the 1970s and the 1980s fishing was restricted, and the
spawning stock of Norwegian spring-spawning herring slowly increased, reaching about 650,000 metric tons by 1984. When the
strong 1983 year class was recruited to the spawning stock, its
biomass rose to 1,360,000 metric tons in 1988 and 1,500,000 metric tons in 1989 (Hamre, 1991).

Variations in the catches and stock size of Icelandic spring-
spawning herring are shown in Fig. 11.5. The Icelandic fleet
exploited herring off both northern and southern Iceland. The
main fishing effort took place on the feeding grounds during sum-

mer. Sometimes Icelandic spring spawners were caught together with Icelandic summer spawners and Norwegian winter spawners. Mixed catches were divided among three herring stocks by means of biological criteria – mainly by fish-scale type.

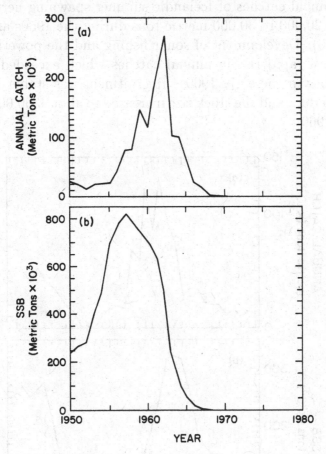

Fig. 11.5 Icelandic spring-spawning herring. (a) Annual catch, 1950–1971, and (b) spawning biomass. (From Jakobsson, 1985b.)

In the early 1950s catches were at a very low level but, with the introduction of sonar and powerblock purse seining in the late 1950s, catches rose quite sharply (see Fig. 11.5) to a peak of about 280,000 metric tons in 1962 (Jakobsson, 1980a). A rapid decline in catches then followed until an almost total depletion of the stock by 1968.

During the 1950s, fishing mortality was very low, allowing the spawning stock biomass to increase rapidly due to good recruitment. Since that time, increases in fish mortality as well as a complete change in recruitment were observed. This change in the state of the stock occurred before any regulatory measures came into force. The sharp decline in recruitment in the late 1960s seems to have been prompted by the deterioration of environmental conditions (discussed in the next section).

The annual catches of Icelandic summer-spawning herring varied from 20,000 to 30,000 metric tons during the 1950s and 1960s (Fig. 11.6). Development of sonar fishing and the powerblock resulted in a rapid rise in annual catches which exceeded 100,000 metric tons in the early 1960s. Recruitment was stable and high in that period, and the stock size increased to about 300,000 metric tons in 1960.

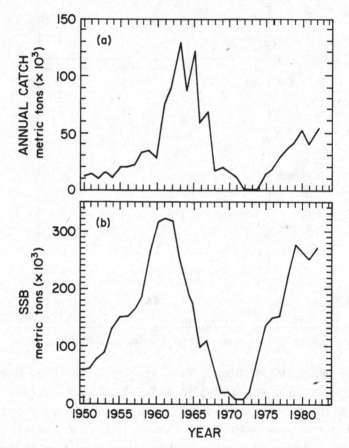

Fig. 11.6 Icelandic summer-spawning herring. (a) Catches for 1950–1971, and (b) spawning stock biomass. (From Jakobsson, 1985b.)

During 1964–70, recruitment was at a very low level but in 1971 and 1974, when the stock size was small, relatively high year classes appeared. These year classes which were protected by regulatory measures contributed considerably to the stock. Later, continued good recruitment resulted in the rapid recovery of the Icelandic summer-spawning herring stock which reached the level of the middle 1960s. In the early 1980s annual catches totaled 45,000 to 50,000 metric tons.

The impact of climate

As shown by Hjort (1914), the early stages of the life cycles are the crucial ones for Atlanto-Scandian herring. Abiotic factors and food supply play an important, maybe decisive, role during this period.

The spawning season begins earlier in warm years than in cold years. According to Dr. Axel Boeck (referenced in Devold, 1963), one of the symptoms of decline in the Norwegian herring fishery toward the end of the previous century was that herring arrived later each year for spawning at the Norwegian coast.

Also, in warm years the development of phytoplankton and zoo-plankton starts earlier, the process is more intensive and its duration is shorter. The herring already begin to feed in spring; they attain maximum fattiness in June–July. On the other hand, when the water temperature is low, plankton development is slower and takes more time. The feeding season lasts longer and the herring become fattiest in August–September.

When the advection of warm Atlantic waters by the Norwegian Current increases, herring larvae tend to be distributed in the northern and northeastern parts of the Norwegian Sea and in the Barents Sea. In colder years, when the inflow of warm waters from the south is relatively weak, the larvae concentrate in coastal waters along southern and central Norway (Seliverstov & Seliverstova, 1991).

Fluctuations in the intensity of the Norwegian Current determine not only the distribution pattern of larvae and juveniles but also influence growth rate, maturation age and, hence, variations in recruitment to the spawning stock. Normal and abundant year classes, appearing as a rule in warm years, recruit to the spawning stock at the age of four to five years. About 70 percent of individuals in the strongest year classes reach the mature stage at five to six years and the remaining 30 percent at seven to eight years.

The weak year classes, most often associated with cold years, enter the spawning stock at an earlier age – more than 60 percent of individuals at three to four years, and the maturing process is completed by the age of five years.

Water temperature is also of great importance during the first wintering. There is evidence that the relationship between the appearance of strong year classes and environmental conditions is likely to be more pronounced for annual temperature and ice than for temperature during the spawning season (Gershanovich, 1986). Bogdanov et al. (1969, 1976) established a good correlation between variations of relative abundance of herring stocks and water temperature in the 0–200 m layer at Kola Section; this was considered a good indicator of the temperature regime in the whole Northeast Atlantic (Izhevsky, 1961).

Changes in water temperature in the Northeast Atlantic are determined to a great extent by atmospheric circulation patterns. In their study of winter and spring circulation, Rodionov and Krovnin (1991) showed that strong year classes appeared when the deep Icelandic low became coupled with the developed Azore high. In this case, the distribution pattern of sea level atmospheric pressure anomalies indicates increased heat transport to the Northeast Atlantic and the Norwegian Sea.

On the other hand, the principal feature of atmospheric circulation in the years when poor year classes occur is the formation of the positive anomaly cell over Greenland and, as a result, an increased frequency of northerly winds over the region.

Linkages between catches of Atlanto-Scandian herring and secular variations in ice conditions were studied by Yudanov (1964) and by Beverton and Lee (1965). As shown in Fig. 11.7, periods of high catches of Norwegian spring-spawning herring start when there is a decrease in the length of time of the ice cover, and all of them end as the length of time of the ice cover increases. It was also shown that rich year classes of herring occurred less frequently in cold periods than in warm periods. In addition, Bogdanov et al. (1969) noted that recruitment of the spawning stock fluctuated substantially from year to year in the cold periods, whereas in the warm periods its interannual fluctuations were much reduced.

Increases in catches and the more frequent appearances of rich year classes of Atlanto-Scandian herring in the first half of the twentieth century coincided with climatic warming, which became

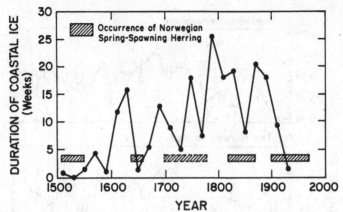

Fig. 11.7 Association of Norwegian spring-spawning herring with the number of weeks of coastal ice on the north coast of Iceland. (From Beverton & Lee, 1965.)

most pronounced after the 1920s. Air temperature both in the Northeast Atlantic and averaged over the Northern Hemisphere reached its maximum value by the late 1930s, after which a cooling started, however, the ocean temperature continued to remain at a high level into the 1950s (Fig. 11.8).

The subsequent climatic cooling was most marked in the 1960s. The sharp decline of ocean temperatures occurred in the Norwegian and Greenland Seas. At Jan Mayen six of the seven severest winters since 1922 occurred between 1965 and 1970 (Rodewald, 1972).

As mentioned earlier, the formation of a positive anomaly cell of atmospheric pressure over Greenland is important for the cold climatic regime in the Northeast Atlantic region. For the period 1956–65 mean winter atmospheric pressure over Greenland was 7 mb higher than during the period 1900–39; by the end of the 1960s it had increased by another 5 mb. Frequent outbreaks of cold air east of Greenland promoted the increase in ice cover in this region. Expansion of the ice cover along the eastern coast of Iceland by 1968 was at its southernmost limit in this century (Malmberg, 1984).

The East Icelandic Current had changed from being an ice-free arctic current from 1948 to 1963 to a polar current transporting drift ice between 1964 and 1971. There is evidence that the adverse environmental conditions which prevailed north and east of Iceland played an important role in changing the migratory pattern of the Atlanto-Scandian herring (Dickson & Lamb, 1972; Jakobsson,

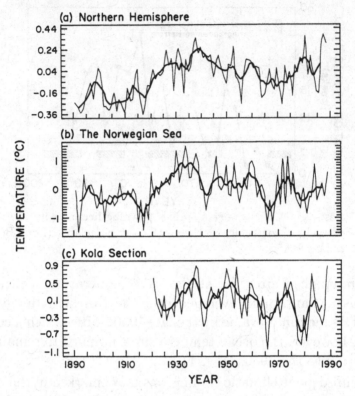

Fig. 11.8 Annual surface air temperature for (a) the Northern Hemisphere (2.5-
87.5° N), (b) for the Norwegian Sea, and (c) annual water temperature integrated over the 0–200 m layer on the Kola Section (70°30'N,
33°30'E - 72°30'N, 33°30'E). Departures from the 1951–80 reference
period. Heavy lines represent 5-year running means. Descriptions of
surface air temperature and sea temperature data files are given in
Gruza et al. (1990) and Bochkov (1982), respectively.

1980b). One could say that climatic changes, coupled with heavy
exploitation, were the main reasons for the collapse of herring
stocks in the late 1960s.

It seems that climatic variability has increased in recent decades.
Environmental conditions may change abruptly both from year to
year and from one set of years to another. For instance, since
1971, the character of atmospheric circulation of the North Atlantic changed radically (Dickson et al., 1975). The increase of
atmospheric pressure over Norway, Greenland, and the Irminger
Sea, which had been observed for a long period of time, ceased suddenly. The positive pressure anomaly cell over Greenland almost
vanished. The mean atmospheric pressure in that area dropped by
9.5 mb between the periods 1966–70 and 1971–74. A reduction of

northern outbreaks over the Norwegian Sea resulted in a warming of these waters. It is important to note that the positive water temperature anomaly was at a maximum in higher latitudes and decreased toward the south. In the Faroe-Shetland Channel, during 1971–75, negative anomalies predominated (Sobchenko, 1979). As a whole, the climatic regime in that period may be considered to have been favorable for the Atlanto-Scandian herring but, because of the state of depletion of the stock, strong year classes did not appear.

In the second half of the 1970s there was a reversal in regional climate conditions once again. According to Rodewald (1978), during the winter of 1977–78 a shift from the regime of strong zonal circulation over the North Atlantic to the regime of weakened circulation occurred. The frequent northern outbreaks over the Norwegian and Barents Seas resulted in a cold climatic regime in that area, which generally resembled the climate of the 1960s. However, Bochkov et al. (1987) pointed out that the main reason for water temperature decrease was the weak heat advection by the North Atlantic, and the Norwegian and Nordcap Currents. From 1977 to 1981, the area in the Barents Sea occupied by warm Atlantic waters (with temperatures of more than 2°C) contracted by a factor of 4, compared to the 1972–76 conditions.

The subsequent warming, which peaked in 1983, undoubtedly promoted the appearance of the strong 1983 year class. Preliminary results show that, because of the favorable thermal conditions in 1989, the recruitment of that year class would be relatively high.

Management

The fishery based on the Norwegian spring-spawning herring stocks collapsed completely before any restrictive management measures were accepted. The question of fishery regulation was raised as early as the 1950s, when the stock size and catches were at high levels. In 1956 a Soviet delegation visited Norway to discuss the problem of fishing in northern European waters. It was noted that, despite the high abundance of the Norwegian herring, this stock would be unable to withstand for any length of time the increased fishing pressure.

The question of a more rational exploitation of the Norwegian stock was raised once again at the 45th International Council for the Exploration of the Sea (ICES) Session in November 1957, but it did not get a favorable response. Norwegian scientists assessed the stock biomass at 40 million metric tons with a fishing mortality of two to three percent. Hence, they concluded that there were no reasons to worry about its future. Unfortunately, it became known later that there had been an error in the method of stock assessment used by the Norwegians and, as a result, the stock biomass was overestimated by at least a factor of 3 or 4 (Marty & Martinsen, 1969).

Three years later it became clear that the anxiety about the fate of the Norwegian herring had not been unfounded. At the end of 1960, the Norwegian Minister of Fisheries noted at a meeting of the Norwegian Herring Society that systematic escalation in the fishing of juveniles by Norwegian fishermen had been the main reason for low catches in recent years.

In the 1960s and the early 1970s the effect of fishing young herring was heavily debated within the Atlanto-Scandian Herring Working Group (Anon., 1970, 1972, 1978). The main reason that the Working Group could not reach a conclusion about the imposition of a ban on a young herring fishery was that the available data indicated that the major nursery areas for the 0 and 1 age groups were far offshore and outside the area of this fishery (Jakobsson, 1985b).

After the drastic decline of catches in the late 1960s, the Working Group considered the state of Norwegian herring stock and concluded that, since 1966, recruitment to the adult stock was negligible because of the overfishing of young herring of the 1963 and 1964 year classes (Anon., 1978). The Working Group recommended to maintain the exploitation of adult and juvenile herring at much lower levels than in previous years. A minimum landing size of 20 cm was accepted in the spring of 1970 for the Norwegian herring fishery north of 62°N. The use of herring for reduction was prohibited in Norway, beginning in January 1971. Those measures prevented the complete extinction of the 1969 year class which at that time was subjected to heavy fishing of the 0 and 1 age groups.

The total ban on the winter fishery for adults was accepted by Norway only in 1972. Iceland and the USSR had ceased their fishing of adult and fat herring from 1969 onward, and Soviet fishing of small (young) herring had ceased as early as 1963 (Alexeev & Seliverstova, 1982).

From 1972 to 1975 the herring fishery was limited by trilateral agreements between Norway, the USSR, and Iceland. In 1973 and 1974 the minimum landing size of 20 cm was replaced by catch quotas on small and fat herring which were 6,800 and 6,300 metric tons, respectively (Hamre, 1991). In 1975 ICES advised a complete end to the commercial fishing; in fact it continued. In 1976 at the North-East Atlantic Fisheries Commission (NEAFC) session Norway, the owner of the remaining stock, was finally convinced to accept a total ban on the fishing of immature herring. In 1978, however, the fishery was reopened with a national quota for Norway set at 7,500 metric tons. Reported catches were 9,800 tons, but the total catch, including an estimate of unreported figures, was estimated to be about 20,000 metric tons (Jakobsson, 1985b; Anon., 1989). In 1979 the fishery was closed once again, except for catches for scientific purposes. Between 1980 and 1983, the national quota for Norway was gradually increased from 930 to 20,000 metric tons. Those quotas were shared between 450–560 ringnetters and about 2,500 vessels with gill nets (Jakobsson, 1985b). It is clear that dividing such a small quota among more than 3,000 vessels was a difficult task. In addition, it was difficult to enforce fishery regulations, given that there were several thousand fishermen. To prevent catches that were too large, with 3,000 vessels in operation, areas of densest herring concentrations were closed to fishing.

The ban on the adult herring fishery which lasted until 1983 (the catches of spawning herring did not exceed 1,000 metric tons in the period 1977 to 1982) and restrictions on immature herring by-catches, for example in the sprat fishery, gave positive results, as the spawning stock began slowly to recover. In 1984 The Atlanto-Scandian Herring Group of ICES accepted a decision to reopen the fishery on the Norwegian stock.

The future fate of Norwegian spring-spawning herring stock depends to a great extent on Soviet–Norwegian cooperation. Presently, relations between these two countries are good and in the past both parties have made concessions that have not been regretted. When Soviet fishermen were accused of fishing herring in 1985, it was pointed out that an abundance of the species in the Barents Sea had driven part of the stock into traditional capelin grounds where the Russians were fishing an agreed-upon quota. Once it was discovered that they had captured a large proportion of herring in the capelin hauls, the Russians stopped fishing and left the area. They did not resume until the capelin shoals had moved closer to the coast (Anon., 1985a).

As mentioned above, a large part of the 1983 year class was distributed in the Barents Sea (presumably within the Soviet economic zone) where special measures to protect young herring were introduced. Due to the total ban on young herring fishery in the Barents Sea, the Soviet Union received a quota of 15,000–20,000 metric tons in Norwegian waters in subsequent years. The quota is set every year at meetings of the Joint Soviet–Norwegian Fishery Commission.

ICES considers the Norwegian spring-spawning stock as depleted because of heavy overexploitation and, therefore, maximum catches must not exceed the established quotas. The minimum preferred level of 2.5 million metric tons for spawning stock biomass is set but has not yet been reached (Hamre, 1991; Ponomarenko & Seliverstov, 1989). Also, since 1983 all but the 1989 year classes have been poor. ICES experts estimate that the rate of Norwegian spring-spawning herring stock recovery will be low up to 1993–94 (Anon., 1990b).

Regulation of fishing the Icelandic stocks came into operation before they were totally depleted. In 1966, Iceland accepted regulations on the minimum allowable landing size. Scientists also gave their recommendations on maximum catch quotas and established a closed fishing season (Bakken, 1983). Those recommendations were gradually accepted but it was too late to stop the process of stock decline. By the time effective regulatory measures came into force, Icelandic spring-spawning herring had already collapsed. The future of Icelandic spring-spawning herring depends upon the type of management. The recovery of this herring stock seems to be associated with the state of the Norwegian spring-spawning herring stock. For example, Jakobsson (1985b) supposed that Icelandic spring-spawning stock would not recover without immigration.

From 1965 onwards there was a change in the exploitation pattern of Icelandic spring-spawners, when a fishery based on young herring developed (Jakobsson, 1985b). To prevent overexploitation of juveniles it was suggested in 1966 to set the minimum landing size at 25 cm. Nevertheless, the Icelandic government accepted a minimum size of 23 cm which in fact was too small to stop, or even reduce, the fishing of 2-ringers but prevented any fishing of 1-ringers. In 1967 it was once again recommended that the minimum landing size be increased to 25 cm, to set quotas and to enforce a closed season from 1 March to 15 May. Except for the closed season, these recommendations were ignored.

Without any quota, the sharp increase in fishing mortality of all exploited age groups of both Icelandic stocks occurred. In 1968 the minimum landing size of 25 cm was accepted, a quota suggested by ICES in 1967 was set and the closed season was extended to 15 August. Unfortunately, these regulatory measures came too late and were not strong enough to be effective. Between 1969 and 1971, the closed season lasted from 1 February to 1 September and a quota of 25,000 metric tons was set. Even this quota proved to be too large and, by the end of 1971, the Icelandic government proposed a ban on all herring fishing except drift-netting. This ban was put into effect on 1 February 1972.

Since its reopening in 1975, the fishery was managed through quotas and individual allocations to purse seines. In addition to these measures, a minimum landing size of 27 cm existed and there was a closed season for about nine months each year. To enforce these regulations several inspectors were placed on board the herring boats and in the harbors where the catches were landed. Thus, biological control and a system of strong regulation based on it, were the main factors for the recovery of the Icelandic summer-spawning herring stock.

Discussion

Technical reconstruction of the fishery capacity (and most importantly the introduction of purse seines) exerted the primary influence on the state of Atlanto-Scandian herring stocks. As shown by Ulltang (1980), despite the beginning of a decline in stock size, from the late 1950s catchability had increased because, for such a schooling species as herring, the constant fishing effort would generate constant catch rather than constant fishing mortality. It could not last long, however, and, ultimately, increased fishing mortality coupled with unfavorable climatic conditions resulted in the collapse of the Atlanto-Scandian herring stocks.

Bakken (1983) identified the relationship between fleet and stock during the 1950–80 period in the following way: at first the stock size "managed" the quantity of vessels but later the fleet determined the state of stock. The decrease of the stock was promoted to a great extent by the development of a fishery based on exploiting juvenile and immature herring. Although the catches of these age groups did not exceed 15 percent of the total catch in weight in the 1950s and 1960s, the quantity of these fish amounted to more than 75 percent. During later years, the proportion of young fish in Norwegian catches reached 95 percent (Ponomarenko & Seliverstov, 1989).

During the 1960s, the intensification of fishing pressure coincided with a pronounced change in the regional climate. Of course, it is difficult to separate the impact of climate from that of increased exploitation in causing the collapse of the Atlanto-Scandian herring stock. Nevertheless, analyses of about a thousand years of herring fishery off the coast of Norway indicated that there had been approximately analogous situations in the past, when the fishery disappeared. Before its disappearance toward the end of the previous century, the fishery tended to shift northward. A similar situation was observed in the 1950s and 1960s. Hence, by analogy we could have expected the end of the period of abundant stock in the late 1960s or the 1970s, even if recruitment overfishing had not supervened (Cushing, 1982).

As a result of the sharp decline in herring, Norwegian fishermen were forced to change their habits and to exploit other species such as mackerel, capelin, and blue whiting. In 1987 cod landings in Norway were the highest among the commercial species in that country. However, the large processing plants which relied on cheap, plentiful supplies of high quality adult herring for curing have almost disappeared. This crisis has affected fish-meal plants to a lesser extent, since quantity of raw material, not quality, was the main requirement for them; almost any species of fish could be used. Thus, the feed requirements of broiler chickens and pigs dictated the fate of the herring-processing industries (Burd, 1974).

The collapse of the herring stocks was not as drastic for the Soviet Union as for Norway and Iceland. The reason was that Atlanto-Scandian herring formed a relatively small part of the total annual catch of the Soviet fishing fleet which operated in all of the world's oceans. Before the establishment of the 200-mile exclusive economic zone, it was rather easy to send vessels to other highly productive regions such as the Barents Sea, the North Sea, the Northwest Atlantic (Georges Bank), and the Southeast Atlantic, among others. Moreover, as a result of the renewal of the Soviet fleet, which happened to coincide with the beginning of the collapse, many old vessels were withdrawn from operation (i.e., they were destroyed).

Herring is a most important food fish. But to put it in proper perspective, if Atlanto-Scandian herring stocks recover and catches increase, most of the catch would probably be used for fish meal processing as had been before. Even when herring was much more popular and more widely consumed than now (and when there were many more plants and processes for handling it) only 300,000 to 400,000 metric tons a year were used for direct human consumption (Anon., 1985b).

Today, there is an opinion that it would be better to use fish, rather than grain, for animal feed. There is a new and rapidly growing market for special feeds for farm fish that did not exist in the 1950s and 1960s. By 1990 production of "farmed" salmon in Norway alone is expected to exceed 80,000 metric tons a year. If one adds to this the harvest of fast-expanding similar industries in Scotland, the Faroes, Ireland and Iceland, production will reach about 120,000 metric tons (Anon., 1985b).

Since the Atlanto-Scandian stock collapsed, herring has become a delicatessen product. Given the existing situation with limited markets, difficulties in marketing and expected competition, it can be assumed that a realistic estimate of food demand for Norwegian-caught herring may be around 100,000 metric tons a year (Anon., 1984).

With regard to the Soviet Union, most of its catches would probably be used for direct human consumption. Because food demand is high in the domestic market, herring products would mainly be consumed within the country.

There are no grounds for optimism nowadays, however, because the Atlanto-Scandian stock is recovering much more slowly than expected. Substantial herring by-catches (about 50%) in the sprat fishery, among other factors, have to be taken into account.

Russian and Norwegian scientists have identified the high predation pressure of cod on the relatively strong 1984 and 1985 year classes of herring. The success of both cod and herring recruitment is governed by common environmental factors, that is, the inflow of warm Atlantic waters to the area, but the production capacity of herring is lower than that of cod. In addition, at low levels of abundance the schooling mechanism of herring renders herring more vulnerable to a stock of predators, thereby resulting in a higher rate of natural mortality. Between 1983 and 1986, the rapidly growing stock of young cod, together with other predators, ate the plankton feeders such as herring and other available food in the area. The absence of sufficient food resulted, in turn, in the depletion of the cod stock in the Barents Sea. Thus, the imbalance in the predator–prey relationship may change dramatically with a shift in the region from a cold to a warm climate, explaining the paradox that a crisis in stocks and fisheries may develop when recruitment conditions of the main species are improved (Hamre, 1991).

One of the most important consequences of the herring collapse was that the production of one of the richest areas in the Northeast Atlantic became unavailable for higher trophic-level carnivores, including fishermen. Depletion of herring may have fundamentally changed the balance in the regional ecosystem and this has probably been the prime reason for the severe crisis which has developed in the Barents Sea's stocks and fisheries in recent years. The collapse of the Atlanto-Scandian herring stocks, resulting in a breakdown of the life cycle of one of the largest fish stocks in the world, is likely to be recorded as one of the most destructive encroachments on a marine ecosystem by a commercial fishery.

Conclusion

The post-World War II history of the Atlanto-Scandian herring fishery allows us to learn a number of lessons which may be helpful in the future.

It is clear that by now the capacity of fishery equipment has reached a level that threatens the survival of schooling species such as herring, even at high levels of abundance. Coupled with unfavorable climatic conditions for recruitment, this threat multiplies and may result in collapse. Disturbances in an ecosystem may be so drastic that, even under a return to favorable environmental conditions, the rate of recovery could be extremely slow.

In recent decades climatic variability has increased markedly, with sharp changes of pronounced warm and cold periods. However, the duration of periods with favorable climatic conditions might be insufficient to allow a considerable increase in stock size.

Unfortunately, the level of our knowledge about the biological features of Atlanto-Scandian herring did not allow us to avoid mistakes in the assessment of stock biomass in the 1960s (these were overestimated by a factor of 3 or 4, as noted earlier). Besides the existing notions about climate, an understanding of its variability and impacts on Atlanto-Scandian herring abundance is also far from complete.

At present there is only reliable knowledge of the qualitative relationship between recruitment to the spawning stock and the thermal regime of the sea. One can confirm for certain that, during warm epochs, the appearance of strong year classes is more frequent than during cold epochs. However, the rich year classes may appear in cold years and, conversely, the warm years do not necessarily result in rich year classes; that is, a simple strong relationship between recruitment and environmental factors is absent.

Most climatologists believe that anthropogenically induced global climate warming will occur in the near future. Such a warming will likely be favorable for Atlanto-Scandian herring stocks. Unfortunately, available general circulation models do not provide useful regional resolution. Thus, it is rather difficult to say how a global warming would be manifested in the Northeast Atlantic, although there is some evidence of a correlation between climatic changes in this region and the global climate regime (Lamb, 1972). It should be stressed that the expected global warming would be anthropogenic in origin and, therefore, the natural existing climatic relationships would break down. The nature of the water temperature changes in the Northeast Atlantic at present does not permit us to forecast with confidence the effects of an apparent increase in global temperature.

Another important lesson is the necessity to accept efficient and timely fishery regulatory measures. In the 1960s, despite the existence of symptoms of changes occurring in Atlanto-Scandian herring stocks, the importance of those symptoms was underestimated and regulatory measures were set much too late to be successful. To some extent this might be explained by the fact that at that time it was rather difficult to agree on a common management policy because of the need for consensus of all countries which had a share in the herring's exploitation.

Many questions concerning Atlanto-Scandian herring management still remain. For instance, is the aim of managing herring stock to catch the quota at the lowest possible cost in order to maximize profits? Or is the aim to provide the maximum number of people with employment by utilizing the resource? Although it is clear that the overall management objective must be a complex combination of biological and socioeconomic considerations, its practical realization is faced with the problem of balancing short-term economic losses due to a variety of constraints against long-term gains.

Obviously, improvement of Atlanto-Scandian herring stock management demands a more effective dialogue between scientists and decision-makers. The latter should react more quickly to scientific recommendations, particularly in situations of great uncertainty related to unpredictable and possibly unfavorable changes in the herring stock with the advent of a global warming.

Key points

- Climatic change may lead to changes in traditional migration routes of fish. The spawning, nursery, feeding, and wintering areas may become redistributed between the economic zones of neighboring countries. Proceeding from this observation the role of regional cooperation must be strengthened. The joint Soviet–Norwegian efforts to support the recovery of the herring stock may be considered as a good model for other regions facing similar resource management situations.

- The atmospheric warming in the first half of this century may be considered to provide a first approximation of ecological and societal responses to changing environmental conditions. This means that the global warming expected to occur in the next century will possibly be favorable for the development of the Atlanto-Scandian herring fishery. However, changes of regional environmental factors in the Norwegian Sea and Barents Sea have not yet exhibited a marked climatic tendency towards warming. Under changed temperature conditions, different regulatory measures restricting fishing activities must be developed.

- Technical development of the fishing fleet has reached a very high level and improvements are continuing. The vulnerability of herring makes it susceptible to the rapid depletion of its stock under modern fishing practices, even at high levels of abundance. Thus, management needs to be more effective and decision-makers should react more quickly to scientific recommendations.

- The depletion of stocks of commercial species may result in the serious destruction of relationships between ecosystems. Therefore, even under favorable environmental conditions, the anticipated recovery of depleted stocks may not occur or at the least may be significantly delayed. Thus, it is necessary to develop an ecosystems approach to the analysis of population dynamics, which must then be taken as the fundamental principle for a scientifically based management of fishery resources.

References

Alexeev, A.P. & Seliverstova, E.I. (1982). Atlanto-Scandian herring stocks size and perspectives of their recovering. *Rybnoye Khozyaystvo*, **11**, 32–5. (In Russian.)

Anon. (1970). Report of the Working Group on Atlanto-Scandian Herring. Cooperative Research Report, *Conseil International pour l'Exploration de la Mer*, Series A, **17**, 1–48.

Anon. (1972). Report of the Working Group on Atlanto-Scandian Herring. Cooperative Research Report, *Conseil International pour l'Exploration de la Mer*, Series A, **30**, 1–27.

Anon. (1978). Reports 1975 and 1977 of the Working Group on Atlanto-Scandian Herring. Cooperative Research Report, *Conseil International pour l'Exploration de la Mer*, **82**, 1–42.

Anon. (1984). Herring return! *Fishing News International*, **23**(9), 20–1.

Anon. (1985a). Europe's herring revival brings it problems. *Fishing News International*, **24**(11), 20.

Anon. (1985b). Herring catchers' problems of plenty. *Fishing News International*, **24**(11), 4.

Anon. (1989). Norway herring heading west. *Fishing News International*, **28**(3), 1.

Anon. (1990a). Cooperative Research Report. *Conseil International pour l'Exploration de la Mer*, **168**, part 1.

Anon. (1990b). Ressursoversikten 1990 Norsk vargytende sild. *Fiskets Gang*, **2**, 23–4.

Bakken, E. (1983). Recent history of Atlanto-Scandian herring stocks. *FAO Fisheries Report*, **291**, 521–36.

Beverton, R.J.H. & Lee, A.J. (1965). Hydrographic fluctuations in the North Atlantic Ocean and some biological consequences. In *Britain*, ed. C.G. Johnson & L.P. Smith, pp. 79–107. London: Institute of Biological Symposia.

Bochkov, Yu.A. (1982). Retrospection of water temperature within the layer 0–200 meters on Kola section in the Barents Sea (1900–1981). In *Ecologiya i Promisel Donnyh Ryb Severe-Evropeiskogo Basseina*, pp. 113–22. Murmansk: PINRO. (In Russian.)

Bochkov, Yu.A., Dvinina, E.A. & Tereschenco, V.V. (1987). Features of recent long-term changes of thermal regime in Barents Sea (1951–1985). *Mezhvuzovsky Sbornic Trudov LGMI*, **99**, 91–106. (In Russian.)

Bogdanov, M.A., Potaychuk, S.I. & Solyankin, E.V. (1969). On the long-term prediction of hydrometeorological conditions and the yield of commercial species of fish. *Rybnoe Khozyaystvo*, **9**, 7–12. (In Russian.)

Bogdanov, M.A., Elizarov, A.A., Potaychuk, S.I. & Solyankin, E.V. (1976). On the system analysis of natural phenomena in the North Atlantic and adjacent polar seas. In *Abiotic Factors of Bioproductivity in the Deep Water Areas of the World Ocean*, ed. D.E. Gershanovich, pp. 7–15. Moscow: Pishevaya Promyshlennost. (In Russian.)

Burd, A.A. (1974). The northeast Atlantic herring and the failure of an industry. In *Sea Fisheries Research*, ed. F.R.H. Jones, pp. 167–91. London: Elek Science Publishers.

Cushing, D.H. (1982). *Climate and Fisheries.* London: Academic Press.

Devold, F. (1963). The life history of the Atlanto-Scandian herring. *Rapports et Procès–Verbaux des Réunions. Conseil International pour l'Exploration de la Mer*, **154**, 98–108.

Dickson, R.R. & Lamb, H.H. (1972). A review of recent hydrometeorological events in the North Atlantic Sector. International Commission for the Northwest Atlantic Fisheries, Special Publication No. 8, 35–62.

Dickson, R.R., Lamb, H.H., S.A. Malmberg & Colebrook, J.M. (1975). Climatic reversal in northern North Atlantic. *Nature*, **256**, 479–82.

Dragesund, O. (1970). Distribution, abundance and mortality of young and adolescent Norwegian spring spawning herring (*Clupea Harengus*, Linne) in relation to subsequent year-class strength. *Fiskeridirektoratets Skrifter, Serie Havundersokelser*, **15**, 451–6.

Dragesund, O., Hamre, J. & Ulltang, O. (1980). Biology and population dynamics of the Norwegian spring-spawning herring. *Rapports et Procès–Verbaux des Réunions. Conseil International pour l'Exploration de la Mer*, **177**, 43–71.

Gershanovich, D.R. (Ed.) (1986). *Fisheries Oceanography.* Moscow: Agropromizdat. (In Russian.)

Gruza, G.V., Ranykova, E.Ya. & Rocheva, E.V. (1990). *Data on Structure and Variability of Climate, Surface Air Temperature, The Northern Hemisphere.* Obninsk: VNIIGMI-MCD.

Hamre, J. (1991). Life history and exploitation of the Norwegian spring spawning herring. *Proceedings of the 4th Soviet–Norwegian Symposium on Biology and Fishery of Blue Whiting in the Northeastern Atlantic and Norwegian Spring-Spawning Herring* (Bergen, June 1989). Bergen: Institute of Marine Research. (In press.)

Hjort, J. (1914). Fluctuations in the great fisheries of Northern Europe viewed in the light of biological research. *Rapports et Procès–Verbaux des Réunions. Conseil International pour l'Exploration de la Mer*, **20**, 1–228.

Izhevsky, G.K. (1961). *Oceanographic Basis of Formation of Fish Productivity in Seas*. Moscow: Pishchepromizdat. (In Russian.)

Jakobsson, J. (1980a). Exploitation of the Icelandic spring- and summer-spawning herring in relation to fisheries management, 1947–1977. *Rapports et Procès–Verbaux des Réunions. Conseil International pour l'Exploration de la Mer*, **177**, 23–42.

Jakobsson, J. (1980b). The North Icelandic herring fishery and environment conditions, 1960–1968. *Rapports et Procès–Verbaux des Réunions. Conseil International pour l'Exploration de la Mer*, **177**, 460–5.

Jakobsson, J. (1985a). The Atlanto-Scandian spring herring stock: Recovering slowly. In *Icelandic 1984 Fisheries Yearbook*, ed. J.J. Hamar, p. 27. Reykyavik: Icelandic Review.

Jakobsson, J. (1985b). Monitoring and management of the Northeast Atlantic herring stocks. *Canadian Journal of Fisheries & Aquatic Sciences*, **42**, 207–21.

Lamb, H. (1972). *Climate: Present, Past and Future*. Vol. 1. London: Methuen.

Malmberg, S.A. (1984). Hydrographic conditions in the East Icelandic Current and the sea ice in the North Icelandic waters, 1970–1980. *Rapports et Procès–Verbaux des Réunions. Conseil International pour l'Exploration de la Mer*, **185**, 170–8.

Marty, Y.Y. & Fyodorov, S.S. (1959). Atlanto-Scandian herring stocks and their exploitation. *Rybnoye Khozyaystvo*, **5**, 17–27. (In Russian.)

Marty, Y.Y. & Martinsen G.V. (1969). *Problems on Formation and Utilization of Biological Production of the Atlantic Ocean*. Moscow: Pishevaya Promyshlennost.

Marty, Y.Y. & Yudanov, I.G. (1962). Abundance, state of stock and perspectives of fisheries on Atlanto-Scandian herring. *Trudy PINRO*, **14**, 151–82. (In Russian.)

Ponomarenko, V.P. & Seliverstov, A.S. (1989). From the history of the herring fisheries in the Barents and the Norwegian Seas. *Rybnoye Khozyaystvo*, **1**, 60–4. (In Russian.)

Rodewald, M. (1972). Temperature conditions in the North and Northwest Atlantic during the decade 1961–1970. Special Publication No. 8, 9–34. Capenhagen: International Commission for the Northwest Atlantic Fisheries.

Rodewald, M. (1978). 1977 ein Jahr der Wende in der Nordatlantiscen Luftzircculazion? *Seewart*, **39**, 249–54.

Rodionov, S.N. & Krovnin, A.S. (1991). Northern Hemisphere climate changes and their effects on the environment of Atlanto-Scandian herring. *Proceedings of the 4th Soviet–Norwegian Symposium on Biology and Fishery of Blue Whiting in the Northeastern Atlantic and Norwegian Spring-Spawning Herring* (Bergen, June 1989). Bergen: Institute of Marine Research. (In press.)

Seliverstov, A.S. & Seliverstova, E.I. (1991). Atlanto-Scandian herring. In *Biological Resources of Neritic and Oceanic Zones of the Atlantic Ocean*, ed. D.E. Gershanovich. Moscow: Nauka. (In press. In Russian.)

Sobchenko, E.A. (1979). Thermal and dynamic characteristics of Atlantic waters in the Faroe–Shetland Channel. *Trudy GOIN*, **150**, 83–8. (In Russian.)

Ulltang, O. (1980). Factors affecting the reaction of pelagic fish stocks to exploitation and requiring a new approach to assessment and management. *Rapports et Procès–Verbaux des Réunions. Conseil International pour l'Exploration de la Mer*, **177**, 489–504.

Yudanov, I.G. (1964). Yield of Atlanto-Scandian herring with regard to secular climatic changes. *Materialy Rybokhozyaystvennykh Issledovanii Severnogo Basseyna*. Murmansk: PINRO, 9–13.

12

Global warming impacts on living marine resources: Anglo-Icelandic Cod Wars as an analogy

MICHAEL H. GLANTZ

Environmental and Societal Impacts Group
National Center for Atmospheric Research
Boulder, CO 80307, USA

In Iceland "fishing is politics." Iceland's economy has been largely dependent on fishing activities for economic development and employment as well as for the generation of foreign exchange. Therefore, any decision related to its fisheries generates high levels of interest in Iceland's political circles.

On four occasions since World War II, Iceland unilaterally extended its fishing jurisdiction. These extensions put Iceland in direct conflict with other nations with fleets fishing within these limits, principally those of Great Britain and West Germany (Fig. 12.1). While West Germany opposed these extensions, it took a less militant stand than the British and eventually acquiesced to Icelandic demands. Great Britain, however, opposed any precedent-setting extensions of coastal jurisdictions, especially those that might infringe on its activities on the designated high seas. The ensuing conflicts between Iceland and Great Britain are popularly referred to as the "Cod Wars."

Since the first Anglo-Icelandic conflict over cod in the early 1950s, the share of Iceland's foreign exchange generated by its fisheries declined from a high of about 90 percent in the early 1950s to about 80 percent at the end of the last Cod War in the mid-1970s. This still represented a sizable national dependence on the exploitation of one variable natural resource. Iceland was (and still is) the nation that is most dependent on the exploitation of fish populations, outdistancing its closest competitors by a wide margin (Fig. 12.2). Needless to say, great dangers are associated with dependence on one resource for export (i.e., export-led development; see, for example, Roemer, 1970). Disruption in the form

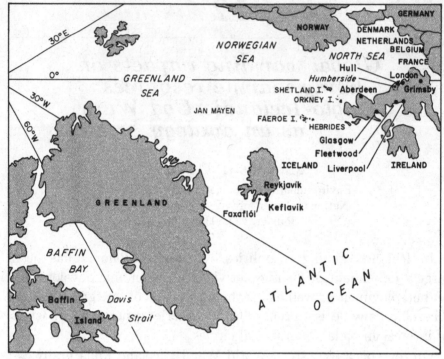

Fig. 12.1 Map of study area.

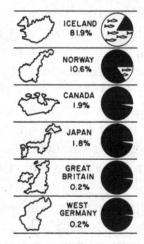

The chart clearly shows that no nation comes anywhere near Iceland in their dependence on fish and fish products for their economic existence. Fish and fish products make up less than 1% of the total export of Great Britain and West Germany, less than 2% of such great fishing nations as Canada and Japan, only 10.6% of Norway's export, but over 80% of Iceland's export is fish and fish products.

Fig. 12.2 National dependence on fish and fish products, as of 1969. (From Jónsson, 1972.)

of variability or change in any aspect of the production-marketing system (from spawning to capture to processing to pricing in the marketplace) would have adverse effects on Icelandic society.

The first section of this chapter discusses the analogy between the Anglo-Icelandic Cod Wars and the possible societal responses

to changes in the marine environment that might be associated
with a global warming of the atmosphere. The second section
discusses the history of the Anglo-Icelandic conflicts over unilateral
extensions by Iceland of its fishing jurisdiction. A summary of key
issues that appear in each of these conflicts is presented in the third
section. The fourth section provides lessons to societies on factors
that can affect the effectiveness of future responses to changes
in the availability or abundance of living marine resources. The
concluding section provides comments on the limitations as well as
strengths of using the "forecasting by analogy" approach to gain
a glimpse of possible future societal responses to global warming
impacts on the marine environment.

Introduction

Between 1952 and 1976 the UK and Iceland were engaged in
conflict about fishing activities in and around the Icelandic continental shelf area. Through a series of unilateral declarations,
Iceland expanded its fishing jurisdiction from three to four, four
to 12, 12 to 50, and finally from 50 to 200 nautical miles. These
extensions, in effect, locked UK trawlers out of what Iceland considered to be its national fishing grounds, thereby creating crises
in different parts of the UK fishing industry.

In contrast to case studies like those of the impacts of El Niño
events on the Peruvian anchoveta fishery (e.g., Glantz & Thompson, 1981), or the impacts of a warming in the 1920s and 1930s
of sea surface temperatures in the North Atlantic on Atlanto-
Scandian herring (see Krovnin & Rodionov, this volume), which
show a relatively clear link between environmental variability and
biological productivity, the Cod Wars case study provides a polit-
ical/legal situation analogous to climate impacts on a fishery.

The analogy applies to the responses of fishing nations to the
impacts of possible changes in abundance and, more specifically,
changes in the availability of favored, valued living marine resources. While each of these unilateral extensions has unique features and provides a set of lessons about the ability of societies to
cope with change, I have chosen to discuss the various extensions
as a set rather than individually. Several existing publications
provide detailed descriptions about each of these Anglo-Icelandic
conflicts. As a set, the conflicts provide a rich variety of examples

of how societies cope with drastic adverse changes in the availability of living marine resources which they have traditionally exploited.

What is the analogy?

With each extension of the Icelandic fishing jurisdiction, UK distant-water trawlers lost access to areas that they had exploited for cod since the turn of the century. Each extension was met with immediate opposition by the UK, and eventual acceptance after several years of verbal and physical conflict. From a UK viewpoint, these situations can be viewed as losses in availability of fish populations rather than as extensions of fishing jurisdictions. What were the motivations on the part of the Icelandic decision-makers to challenge UK activities and existing international law? What were the responses in the UK to these challenges? How was Iceland able to achieve all of its goals in each of these conflicts? Addressing these questions can possibly provide insights into societies' reactions to regional changes in the availability or abundance of living marine resources in the event of global climate change.

Anglo-Icelandic Cod Wars

Iceland, an island nation, is poorly endowed with land-based natural resources. With no minerals to speak of, its main potential land-based resource lies in energy production from hydrological and geothermal sources. Its natural energy supply might be used to attract foreign industrial development to the country. However, the valuable resources that Iceland does not have on land are present in its coastal waters.

The biological productivity of the water above Iceland's continental shelf is very high, enabling the country's few hundred thousand inhabitants to prosper in an otherwise harsh environment. Fishing, fish processing and marketing have been the mainstays of the Icelandic economy this century, generating today more than 70 percent of its foreign exchange. No other nation has such a high level of dependence on fishing for generating its foreign exchange. This foreign exchange enables Iceland to import essential raw materials and consumer goods. Changes in the amount of fish landings directly and indirectly affect the health of Iceland's economy. In a speech to the UN General Assembly, Iceland's Foreign Minister, Agustsson, stated that "the fish in the Iceland area continue to form the foundation of our economy – a matter of life or death to our people" (Iceland Ministry for Foreign Affairs, 1976, p. 31).

Cod, *Gadus morrhua*, is the most important demersal species in the waters surrounding Iceland. Other exploited demersal stocks include saithe, redfish, plaice, and haddock. The UK and Iceland were the two major countries landing cod from these waters in the post-World War II period. It has been a popular fish in the UK marketplace and a target of British distant-water trawlers. An idea of the relative proportion of landings is provided by the following average percentages for the years 1954–56 of demersal catches in the waters around Iceland: cod (63.4%), redfish (14.0%), saithe (7.5%), haddock (7.5%). Cod landings from Icelandic fishing grounds between 1950 and 1987 are illustrated in Fig. 12.3.

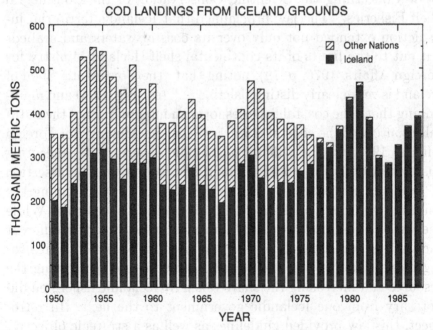

Fig. 12.3 Landings of cod from Icelandic fishing grounds by Iceland and other nations.

Icelandic cod has been considered as relatively independent from other cod stocks in the North Atlantic. Tagging experiments, however, have shown that a significant number of adult cod migrate from southern Greenland to Iceland and admix with Icelandic cod (ICES, 1976).

The abundance and characteristics of Icelandic cod stocks have fluctuated over time. Increases in abundance of cod were related to increases in temperature, noting that such increases were observed (from catch data) during a climate warming in the 1920s and 1930s (Bell & Pruter, 1958). It was also noted that other researchers suggested that the variation in cod landings might be the result of the way that catch information was collected and analyzed, and that cod stocks had in fact been relatively stable over the decade (Bell & Pruter, 1958, p. 645).

When Iceland became independent from Denmark in 1944, it had a three-mile territorial jurisdiction for its coastal waters. That limit had been four miles at the turn of the century but was reduced to three miles, as a result of a 1901 Anglo-Danish agreement that centered on Danish exports to British markets (Jónsson, 1982, pp. 37–9). In 1948 the Icelandic Parliament (*Althing*) passed the "Law Concerning the Scientific Conservation of the Continental Shelf Fisheries." The law proclaimed that Iceland's territorial jurisdiction extended not only over its coastal waters and seabeds but out to the limits of its continental shelf (Iceland Ministry for Foreign Affairs, 1976, p. 19), noting that "the Continental Shelf of Iceland is very clearly distinguishable . . ." (see Fig. 12.4) and maintaining that "the coastal fisheries form an integral part of the natural resources of the coastal state . . ." (Iceland Ministry for Foreign Affairs, 1976, p. 7). Different Icelandic governments, reflecting changes in domestic political coalitions, supported the assertion of Icelandic control over living marine resources surrounding the country, regardless of their conflicting political ideologies. While the various coalition governments might have preferred different tactics to secure control over fish populations on which their fishing industry depended, the strategic objectives of preserving the resource and increasing the share taken by Icelandic fishermen did not vary from one Icelandic government to the next. In retrospect, this law provided guidelines as well as a strategic objective for future Icelandic actions related to unilateral extensions of its jurisdiction over coastal fisheries in its surrounding waters.

Fig. 12.4 Icelandic stamp clearly showing Iceland's continental shelf.

The first Anglo-Icelandic conflict

An important impetus to Iceland's first unilateral extension of its territorial waters began in 1951 with the resolution of a long-standing Anglo-Norwegian dispute (one that originated in 1906) over the methods used to determine territorial jurisdictions in coastal areas. In mid-December 1951 the International Court of Justice (ICJ) supported Norway's methods for determining baselines (the use of rock outcroppings and reefs and not just the mainland) and, therefore, coastal jurisdiction. This decision enabled Norway to control the exploitation of marine resources in its own fjords, thereby excluding UK fishermen from these lucrative fishing grounds. UK fear about changing the traditional methods for establishing baselines in coastal areas stemmed not so much from the loss of Norwegian fishing grounds to its fleet but from the precedent that might be set for other nations to follow suit (see Bilder, 1973, footnote 263). As was the case with Norway, control over indentations in the coastline (e.g., bays) was also a major concern to Iceland. In this regard Iceland had tried to convene a meeting in 1949 to discuss Faxafloi (see Fig. 12.1) in order to establish control over foreign fishing vessels in this and other bays but, as the British refused to participate, no meeting was held (Alexander, 1963, p. 114).

In 1952, Iceland announced that its territorial jurisdiction would be extended to four miles. The UK protested, claiming that such an extension would adversely affect access to traditional fishing grounds by UK trawlers. The UK also feared that such extensions would follow elsewhere. While the UK's national economy was dependent on its fishing sector only to a very marginal extent, the economies of regions within the country, such as Humberside, would be affected in a major way.

Although some authors suggest that UK fishermen have fished cod on the Icelandic fishing grounds since the 1300s, it was not until the end of the 1800s that UK fishermen began to exploit Icelandic waters year-round. Technological changes in the fishing industry (e.g., the development of the steam trawler, improved deep-water trawling techniques, the use of ice to preserve catches) strengthened British involvement in distant-water fishing activities based primarily in Hull, Grimsby, Fleetwood, and Aberdeen (e.g., Barston & Hannesson, 1974). These ports were adversely affected by changes in access to cod in waters around Iceland. Although the fishing industry at the time made up only a small percentage of UK's national work force (0.09% in 1979), it had an inordinate degree of political power for the following major reasons: (1) employment and fishing revenue is concentrated in small, depressed, high-unemployment areas; (2) a small change in fishing revenue in these areas creates a large change in ancillary industries, such as shipbuilding, equipment manufacturing and supply, and fish processing activities; and (3) in UK's single-member, winner-take-all electoral system, fishing seats in Parliament can affect the outcome of an election (*The Economist*, 16 August 1980, p. 41).

In response to Iceland's 1952 extension, UK trawler owners based at Humberside ports led an "unofficial" UK boycott against Icelandic trawlers and fish products. As a result of the ban, Iceland could not sell its fresh and unprocessed fish in the UK. In response, it chose to expand its quick-freezing industry, increasing the value of its fish products, and to locate new markets (Grondal, 1971, p. 61). Thus, Iceland developed trade ties with the Soviet Union and Eastern Europe. By 1955 the USSR had become the largest single importer of Icelandic fish, replacing the UK as Iceland's second largest trading partner (Mitchell, 1976, p. 128). Recall that 1952 was in the middle of the Cold War era between the East and the West. Iceland, a founding member of the North

Atlantic Treaty Organization (NATO), developed a good trade relationship with the USSR and other Eastern European countries such as Poland and East Germany. A NATO ally dependent on trade with Communist countries created some concern among other members of the alliance.

The first Anglo-Icelandic dispute was settled in 1956, when the UK accepted Iceland's four-mile fishing jurisdiction, ending the four-year ban on Icelandic trawlers (Jónsson, 1982, p. 65). This particular unilateral extension did not adversely affect fish catches by UK's distant-water fleet, as British catches and catch per unit effort apparently increased shortly thereafter (see, for example, Iceland, 1958, and Fig. 12.3).

During this period representatives of developing countries were voicing a general concern that customary international law had been developed to support the *status quo*, that is, the dominant position of the major seafaring countries. One approach to redress this imbalance was a call for a UN Conference on the Law of the Sea (UNCLOS). The first UNCLOS conference was convened in Geneva in 1958. Iceland had hoped, if not expected, that lingering questions about fishing jurisdictions would be addressed and resolved in this forum. Inconclusive results, however, meant further delay on addressing the fishing jurisdiction issue, with major powers like the UK and Japan seeking to maintain the *status quo ante* (e.g., Akaha, 1985). The failure to resolve fishing jurisdiction concerns displeased the Icelandic government, prompting it to consider unilateral action once again. Iceland's representatives had said such action would be inevitable if the results of the Geneva conference were inconclusive.

The first Cod War

In 1958, the Icelandic government (apparently after considerable interparty debate) announced a new extension of its fishing jurisdiction, banning all fishing vessels from trawling within 12 miles of its coast. Passive gear, however, such as gill nets, long lines, and hand lines could still be used. Iceland, citing the "special case" doctrine of preferential rights of coastal states, claimed that its action was motivated by fear of the adverse impacts on its fisheries of the rapid growth in size and efficiency of foreign fishing fleets (mostly as a result of technological changes in the distant-water fishing industry). According to Barston and Hannesson,

> *Post-war developments in the field of electronics, hydroa-*
> *coustics, net and hull construction, as well as freezer*
> *techniques have not only made possible sustained, larger*
> *catches for traditional trawlers but facilitated the con-*
> *struction and operation of factory type vessels* (Barston
> & Hannesson, 1974, p. 559).

Icelandic leaders decided to take another step to protect increasingly endangered fish stocks in waters adjacent to its coast. (For a detailed discussion of the internal Icelandic political situation surrounding the 1958 unilateral extension, see Davis, 1963.)

The UK (along with a few other European countries) protested against the second unilateral Icelandic extension. This extension precipitated a conflict which came to be known as the "First Anglo-Icelandic Cod War," under great pressure from UK fishing interests (Gilchrist, 1978). While other European countries such as France, The Netherlands, and West Germany initially protested but later acquiesced, British Royal Navy vessels were used at considerable cost to escort its fishing vessels to the fishing grounds off Iceland. Iceland, however, remained firm with its demands and used its Coast Guard Service (all seven vessels) to harass UK vessels off its coast, seeking to gain UK acceptance of the new jurisdiction by making the economic cost of long-distance fishing prohibitive. Iceland once again increased its trade with the USSR and indirectly threatened the solidarity of the NATO alliance. The UK wanted to avoid the appearance of being Goliath in a David-and-Goliath situation and to avert a schism in the alliance (Mitchell, 1976), and pursued legal arguments in its attempt to roll back the 12-mile jurisdiction, claiming that Iceland's action was contrary to international law. Following a few years of diplomatic maneuverings by the UK and Iceland, with Iceland steadfastly sticking to its original demands, the issue was resolved between themselves in Iceland's favor in March 1961. Three years later, in 1964, Great Britain established its own 12-mile fisheries jurisdiction (Jónsson, 1982, p. 110).

In the mid-1960s, concern about the fate of another Icelandic fishery emerged. The northern Icelandic herring stock collapsed abruptly. Whereas in 1960 more than 700,000 metric tons of her-

ring were caught in eight months, by 1968 landings had dropped
to 65,000 metric tons. Iceland's economic situation drastically
changed, as 40 percent of its export earnings came from this fishery. High unemployment occurred, and Iceland's per capita gross
national product fell from third place behind the US and Sweden
to about thirtieth (Jakobsson, 1988, p. 24). Jakobsson has noted
that "The loss of the herring fishery was without a doubt the
underlying reason behind Iceland's demand for a wide, exclusive
economic zone in order to secure a higher proportion of the groundfish resources on the Icelandic continental shelf" (Jakobsson, 1988,
p. 26).

An assessment of cod landings at that time showed that foreign
trawlers were taking about half the cod catch. Concern was raised
about "foreign vessels" taking Iceland's "natural" resources away
from Icelanders. Interest was heightened not only in protecting
the cod stocks but in providing an increased share of landings
to Icelandic fishermen (many of whom became unemployed as a
result of the herring fishery collapse).

The second Cod War

While the Icelandic government had been relatively stable from
1959 onward, a leftist coalition took power as a result of the 1971
elections. This change in government prompted Iceland to an-
nounce in the fall of 1971 its intention to conserve the cod fishery
and to increase the share of catches by Icelandic fishermen by once
again unilaterally extending its fishing jurisdiction – from 12 to 50
miles (Hart, 1976, p. 19). This particular extension caught the
UK off guard. While many states at that time already claimed 12
miles, and a handful had claimed 200 miles, few countries claimed
any jurisdictions in between. Also, the UK believed that their previous Anglo-Icelandic agreement in 1961 had called for submission
of any dispute to the ICJ for resolution. The new Icelandic government argued at the end of 1971 that the 1961 agreement had
become null and void and "that [its] provisions do not constitute
an obligation for Iceland" (Jónsson, 1982, p. 123).

While the UK blamed the change in policy on Iceland's leftist
government and on the appointment of its new fisheries minister,
a member of Iceland's Communist party, it appears that within
the Icelandic government there was debate over *how* and *when* to

extend the fishing jurisdiction, but not *whether* to do so. The UK
and West Germany protested to the ICJ, but Iceland refused to
be a party to the Court's proceedings (Dickey, 1974).

Following the passing of a 1 September 1972 deadline set by Iceland to enforce the new unilateral extension, UK Navy ships were
sent to escort UK trawlers onto the contested fishing grounds.
Again, the Icelandic response was to use its few Coast Guard Service vessels to harass UK vessels so that fishing in these distant
waters would become unprofitable. This was accomplished, in
part, by causing the UK to protect its trawlers and by Iceland un-
leashing a secret weapon: the trawlwire cutter, the sole purpose
of which was to sharply increase the cost of fishing by severing
the expensive trawls from the trawlers. This weapon was successfully used in 1972 and 1973 against UK and West German trawlers
(Jónsson, 1982, pp. 134–7).

In addition, Iceland used its NATO affiliation and the NATO
airbase at Keflavik as a bargaining chip to bring outside pressure
on the UK to accept Icelandic demands. By threatening to close
the Keflavik airbase, and by denying landing rights to the UK,
Iceland was able to convince the US and other NATO allies that it
was serious about pulling out of NATO (*US News & World Report*,
1973). Iceland's geographic location in the North Atlantic fills a
gap in strategic defenses against Soviet naval activities. Without
Keflavik, a serious flaw in NATO defenses would exist.

Support for the 50-mile fishing jurisdiction became a unifying
theme in Icelandic domestic politics. After several conflicts at sea
and bans on port visits of each other's ships, the second Cod War
ended with the British once again giving in to Iceland's original
demands. (For additional information on this conflict, see Hart,
1976.)

The third Cod War

In the early 1970s the Law of the Sea deliberations focused increasingly on the development of a 200-mile exclusive economic
zone (EEZ). In the midst of discussions about this trend toward
an EEZ, Iceland once again unilaterally extended its fishery jurisdiction to 200 miles. This was done months before the Anglo-
Icelandic agreement stemming from the second Cod War was to

have expired. West Germany and Belgium accepted the extension. The UK opposed it.

Iceland and the UK relied on their old arguments to support their positions: Iceland pointed to the depletion of its demersal stock, about 50 percent of which was being captured by foreign trawlers, and underscored the importance of fishing to its livelihood; the UK cited the ICJ interim ruling prohibiting Iceland from extending to 50 miles. Thus began the third Cod War. To Iceland, "the British demands in effect amounted to expecting the Icelandic people to share their only resource with British trawlers and thus jeopardise their own economic survival" (Iceland Ministry for Foreign Affairs, 1976, p. 9).

Cod in the North Atlantic had been well studied for several decades. During the 25-year period that spanned the Anglo-Icelandic conflicts, both UK and Icelandic fisheries scientists relied on essentially the same catch data and stock assessments to support what politicians turned into diametrically opposed positions on the issue of overfishing (United Kingdom, 1958; Alexander, 1963; Burton, 1973; Iceland Ministry for Foreign Affairs, 1976, pp. 57–68). Despite the high quality of scientific information on cod, uncertainty exists in the way data are collected and in the nature of information about cod stocks (e.g., a reliance of proxy data to determine the standing stock). Thus, different conclusions could be drawn from the same pool of information, depending on how one emphasized the existing scientific uncertainties. In this instance, it became clear that scientific information, no matter how reliable, could be used by those who wished to do so to bolster nationalistic (political) positions on seemingly value-free scientific issues (Martin, 1979, *passim*).

A close look at the scientific arguments of the British and Icelandic scientists shows why the issue could not be resolved for the policymakers by additional scientific research on cod. To policymakers neither side could be proven wrong. As a result, politicians used the uncertainty inherent in scientific information to essentially challenge their opponents' views (whether or not they really disbelieved them), allowing politics, neither scientific objectivity nor the need for conservation, to determine the outcome of what had become a political (not scientific) conflict.

The British Navy was again sent into the contested zone to protect British trawlers from harassment by Icelandic Coast Guard vessels. Again, trawlwire cutters were used by Iceland. The mutual ramming of boats occurred (Fig. 12.5), and shots were fired. Iceland severed its diplomatic relations with the UK and threatened to withdraw from NATO, if hostile British actions did not cease. Iceland argued that the alliance was not protecting it from threats to its national security by another NATO member. What, many asked, was the value to Iceland of NATO membership and of the US-operated air base at Keflavik (Jónsson, 1982, p. 173–8)? The conflict was resolved between the protagonists on 2 June 1976, once again in Iceland's favor.

Fig. 12.5 The 2,200-ton *Lloydsman* at full speed about to ram the 693-ton
         Thor with considerably reduced speed. (Caption and photograph
         from Iceland Ministry of Fisheries, 1975.)

The third Cod War (but the fourth Anglo-Icelandic fisheries dispute) clearly differed from earlier conflicts over fishing jurisdictions in Icelandic waters. For example, British resolve to oppose the 200-mile extension appears to have been considerably lower than its opposition to the extensions that led to the other Cod Wars. In fact, British fishing interests were calling for a 200-mile EEZ for the UK at the same time that they were opposed to Iceland's 200-mile extension (Jónsson, 1982, p. 181). In addition, the cost for protection of UK trawlers by the Royal Navy during the last Cod War, estimated at over £40 million, was more than double the value of the fish being landed from contested Icelandic waters. Also, revenue derived from these fisheries constituted a relatively small percentage of the UK's national revenue, although the impacts on individual UK ports were quite large. This point was highlighted in a UK workers' union report on fishing (Transport and General Workers' Union, 1980, p. 8) which noted that

> *Whilst it is the case that the fishing industry plays only a minor role in the UK's gross domestic product, it nonetheless plays a most important role in a number of peripheral areas of Scotland, England, and Wales, as well as being crucially important to a number of towns and cities, e.g., Peterhead, Aberdeen, Fleetwood, Hull, Grimsby and Lowestoft.*

Ironically, several of the arguments used by Iceland against the UK during the cod wars were used by the UK in its negotiations for the development of a common fisheries policy for the European Economic Community (EEC) in the late 1970s.

Summary

Officially, there were three Anglo-Icelandic Cod Wars (1958, 1971, 1976). In fact, there were four, if one includes the 1952 dispute over extending the fishing jurisdiction from three to four miles. Each of these wars had its own dynamics, its own history, and its own international as well as domestic political and economic settings. In addition, each conflict took on a different degree of importance to the protagonists. In retrospect, the first unilateral extension from three to four miles in 1952 was perhaps the least important to the UK distant-water fishing industry, while the third one, from 12 to 50 miles in 1971, caught them and their government by surprise.

Table 12.1 graphically depicts how these linear fishing limit extensions translated into areal expansion of Iceland's fishing grounds and, therefore, gives a better idea of what such extension actually meant with respect to the area of fishing grounds lost to UK vessels (and gained by Icelandic authorities).

Table 12.1
Areal expansion of Iceland's fishing grounds

Year of fishing jurisdiction extension	1952	1958	1972	1976
Change (nautical miles)	3→4	4→12	12→50	50→200
Areal Coverage (km²)	24,530 to 42,905	42,905 to 69,809	69,809 to 216,000	216,000 to 758,000
Added Area (km²)	18,400	26,900	146,000	542,000
Change (from preceding limit)	75.1%	62.7%	209.5%	251%

Compiled from information presented in Jónsson, 1982, pp. 3–5.

Many publications have appeared about various aspects of these conflicts and the issues that they have raised: the historical development of the Law of the Sea; the role of international scientific, political, and economic organizations in the management of an international conflict between allies; the relationship between national interest and natural resources; the use of uncertain scientific information to resolve resource-related conflicts, and so forth. This chapter clearly cannot do justice to each of these issues. Nevertheless, despite the relative uniqueness of each of these Anglo-Icelandic conflicts, some general themes recur from one conflict to the next.

1. Icelanders (as well as many other observers, including many segments of the UK public) believed that Iceland's national survival was closely tied to the preservation and expoitation of living marine resources that surround their island nation. If anything were to happen to that resource (e.g., an adverse change in availability or abundance), Iceland's existence would be fatally threatened. Thus, the resource had to be (a) protected and (b) made available to Icelanders for their exclusive exploitation. Clearly, the

strength of the Icelandic economy improved or deteriorated with the ups and downs of the fishing sector.

2. The available scientific information about fish populations (e.g., cod), regardless of the quality of the existing scientific data, was interpreted differently (one might argue nationalistically) by policymakers from the two countries in conflict. Thus, the political use of scientific information about the cod fishery made the search for a resolution of the Anglo-Icelandic conflicts more difficult, because the uncertainties inherent in catch statistics, regardless of how good Icelandic catch statistics and stock assessments might have been, permitted many (sometimes opposing) interpretations of the same data. (For an analogous situation with respect to the interpretation of available scientific information pertaining to the environmental impacts of a hypothetical fleet of high-flying SSTs, see Martin, 1979. In addition, the global warming scientific and political debates can also shed light on the political use of scientific uncertainties inherent in scientific research. See also Glantz, 1988.)

3. In each instance, Iceland was able to use the disparity in power and resources between the UK and itself to its advantage. Iceland, with a population of less than a quarter million, with no military establishment except for a few patrol boats, was determined to defend its right to control the exploitation of resources in its coastal waters. The UK used the Royal Navy in a show of force to intimidate Iceland but were unwilling to unleash their full military power against a NATO ally. The disparity in resources between the protagonists aided Iceland, which fostered the image of a David taking on a Goliath. International sympathy generally flowed toward Iceland (see Fig. 12.6).

4. Law of the Sea discussions in the United Nations from the late 1950s onward provided an important backdrop to the Anglo-Icelandic conflicts. Over the centuries Iceland's fishing jurisdiction had changed from 32 miles to 24 miles to four miles (in 1859). As the result of a 50-year agreement signed in 1901 between Denmark (then in control of Iceland) and Great Britain, Danish exports of ham and butter to the UK were to receive favorable tariff arrangements in return for a reduction from four to· three miles of Iceland's fishing jurisdiction. This arrangement became a lingering grievance for Icelanders.

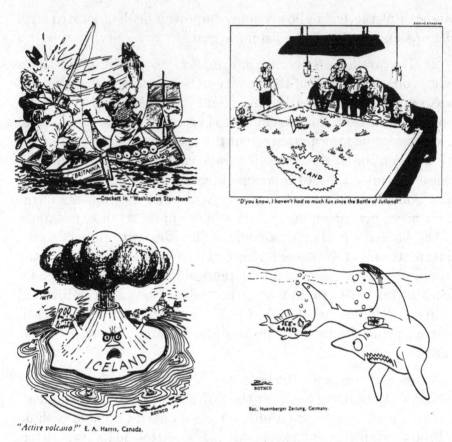

—Crockett in "Washington Star-News"

"D'you know, I haven't had so much fun since the Battle of Jutland!"

"*Active volcano!*" E. A. Harris, Canada.

Bac, Nuernberger Zeitung, Germany.

Fig. 12.6   Upper left: Cartoon by Crockett, *Washington Star News*, reproduced from *US News & World Report*, 11 June 1973, p. 33. Upper right: cartoon from *Evening Standard*, reproduced from *Time*, 29 December 1975, p. 25. Lower left: cartoon by E.A. Harris, Canada, reproduced from *European Community*, March 1976, p. 16. Lower right: cartoon by Bac, *Nuerenberger Zeitung*, reproduced from *European Community*, March 1976, p. 14.

5.   From Icelandic independence (in 1944) onward, successive governments sought to protect the nation's living marine resources. With the passage of a conservation law in 1948, Iceland sought to regain control over its adjacent waters congruent with its isolated continental shelf which it considered a natural extension of its land. Each of the four unilateral extensions was embedded in a mix of already-existing fishing jurisdiction claims made by other coastal nations. For example, in 1948 Peru, Ecuador, and Chile had already claimed 200 miles, at a time when most states claimed only three miles. Ten years later, at the first UN Conference on the Law of the Sea, a vote to extend territorial waters to 12 miles (six

miles for territorial jurisdiction and six miles for fishing jurisdiction) failed to pass by one vote. According to Akaha (1985, p. 38), "... following the unsuccessful second UNCLOS [Geneva,1960], one coastal state after another resorted to unilateral actions to expand their territorial sea and fishery jurisdictional claims."

Thus, Icelandic territorial and fishing jurisdiction claims, with the exception, perhaps, of the 1971 extension (from 12 to 50 miles), should not have been viewed as unprecedented (Table 12.2).

Table 12.2
Territorial sea and fishery zone claims by period

Claim	Pre-WWII	Pre-UNCLOS II (1945–60)	Post-UNCLOS II (1960–70)	Total
12-mile territorial sea	2	10	33*	45
12-mile fishery zone	1	1	16	18
25-mile territorial sea	0	0	1	1
130-mile territorial sea	0	0	1	1
200-mile territorial sea or fishery zone	0	5	5	10

* Including five states claiming fishery jurisdiction beyond and in addition to the 12-mile territorial sea limit.
Source: Kenzo Kawakami, cited in Akaha, 1985, p. 40.

By the early 1970s there was a general movement in the international community toward approval of 200-mile EEZs. Despite its opposition to the 200-mile extension by Iceland, the UK was considering the enactment of its own 200-mile jurisdiction in an attempt to protect its inshore fisheries and to improve its bargaining position in negotiations for a common fisheries policy for the European Community.

6. With successive unilateral extensions by Iceland, British Humberside ports, where distant-water trawlers unloaded and processed their cod catches, opposed Iceland, claiming severe injury to their industry and economy (loss of revenue, loss of jobs in the fishery and in ancillary industries, etc.). They initiated multi-year boycotts against the unloading of Icelandic fish in their ports.

The initial claims of damage were apparently unfounded in the two earlier disputes, as UK landings actually increased following the exclusion of UK vessels. The increase was due in part to a focus on new fishing grounds in the area as well as improvements in fishing technologies and techniques. Such was not the case with the last two Cod Wars (1972 and 1976). There were no new equally productive fishing grounds for UK vessels to exploit. Only overcrowded fishing grounds, less preferred species, and newer grounds were available beyond the travel range of many of the existing vessels (Fig. 12.7) (British Trawler Federation, 1958; British House of Commons (5th Series) 20 July 1971, pp. 1416–17).

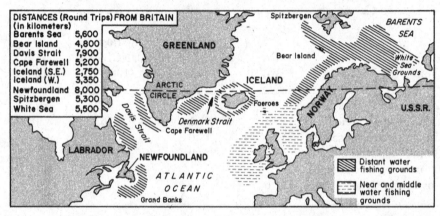

Fig. 12.7 Fishing grounds of British trawlers. (From Dalgleish, 1970, p. 13.)

Concern was voiced about the future of Humberside ports. British trawler owners raised another aspect of the equity issue, when they noted that the population of the port of Hull was totally dependent on the landing of Icelandic cod and that its population was greater than that of the whole of Iceland. In a 1971 House of Commons debate on the unilateral extension of Icelandic fishing limits, a member of Parliament noted that the 12 to 50 mile extension

> *brought an immediate and angry reaction from the whole industry: from the fishermen and their union ... ; from the trawler owners and the merchants, from Fleetwood, 62.2% of whose total catch in 1970 came from Icelandic waters; from Grimsby with 41.1%; and from Hull, where 25% of last year's catch of demersal fish was caught in Icelandic waters* (British House of Commons (5th Series), 20 July 1971, p. 1407).

Another speaker also noted that "the exclusion of our vessels [from fishing grounds in the Icelandic area] would deprive us of between 20–25% of all British landings of such species as cod, haddock and plaice ..." (British House of Commons (5th Series), 20 July 1971, p. 1416). Figure 12.8 depicts the declining number of fishermen at major UK distant-water ports.

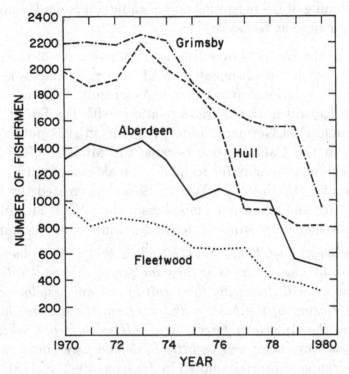

Fig. 12.8 Numbers of regularly employed fishermen in UK distant-water ports, 1970–80. (From Wise, 1984, p. 50.)

7. Clearly, the impacts of Iceland's unilateral extensions of its fishing jurisdiction on the British distant-water fleet were major. However, other factors had adverse impacts on the industry at various times throughout the period of conflict between 1951 and 1976: the sharp increase in oil prices as a result of the Arab oil embargo in the early 1970s, the negotiations concerning UK entry into the Common Market, and the development of a common fishing policy for the European Community (Buchanan & Steel, 1977).

8. It was inevitable that NATO should be drawn into a protracted conflict between two member states (founding members at that), during the Anglo-Icelandic conflicts. Two important issues

were raised with respect to NATO and the Cod Wars: (a) continued Icelandic membership in NATO, and (b) NATO's ability to deal with conflict among its members. With respect to Iceland's membership in NATO, there had always been some degree of opposition within the country to its abandonment of neutrality by signing a 1951 defense agreement with the US, and by permitting the stationing of US personnel on Icelandic soil and a US-operated NATO air base at Keflavik.

Each of the conflicts over fishing jurisdiction enabled Icelandic opponents of its involvement in NATO to raise the issue of the value of this relationship to Iceland. As a result of the first conflict in 1952, Iceland initiated trade relations with the Soviet Union, Poland, and East Germany. Later, elections in 1971 put two communists in the Cabinet (one became the Minister of Fisheries) and heightened pressure for Iceland to break away from NATO. It was, however, the last Cod War in 1976 that created the biggest threat to Iceland's continued membership in NATO. The following statement succinctly sums up Icelandic thinking at the time:

> We ought to tell the NATO leaders that if they do not yield to our demands in this matter of our vital interests, we will, following the recall of our ambassador and the closing of the base, withdraw from the alliance and send the American forces home. If we lose the fishery limits war, there will be little to defend here (moderate newspaper editorial, quoted in Jónsson, 1982, p. 178).

With the resolution of the last of the Anglo-Icelandic Cod Wars in mid-1977, popular Icelandic concern about continued membership in NATO faded away.

NATO's ability to cope with a serious natural resource-related issue between its members was essentially negligible. Whereas NATO did serve as a forum for the protagonists to lobby for international support for their positions, it was unable to directly affect the outcome. Within NATO, members such as Denmark, Norway and Canada sympathized with Iceland's fishing jurisdiction claims (Grondal, 1971, p. 64). In most instances, it was clear that Iceland was willing to quit the alliance while Britain, in part under US pressure, was most concerned to save it (that is, to protect the strategically important NATO North Atlantic surveillance system operating out of Keflavik). While fisheries were seen as vital to the economic as well as cultural survival of Iceland, the UK government apparently was not as concerned about the potential impact of the demise of its distant-water fleet on its national economy.

Lessons

The analogy of the Anglo-Icelandic Cod Wars to the potential impacts of a global warming on marine resources lies in societal responses to changes in the abundance or the availability of resources. Why did Iceland make these unilateral extensions? How did the UK respond to the loss in availability of cod? How did Iceland react to UK responses? What lessons might be drawn from this analogy for possible responses to the regional impacts of a global warming? Such lessons include the following:

- *The economic, political, and social settings at the time of a change in resource availability will be a crucial determinant to the outcome of a resource-related conflict situation. Depending, for example, on the state of the economy or the health of the fishing sector, different (even diametrically opposed) responses could be justified.*

The Icelandic economy during the period of the conflicts with the UK (1952–76) was almost totally centered on its fishing sector. Its foreign exchange was (and still is) in large part derived from this sector; its workforce was tied directly and indirectly to the well-being of this sector; its domestic political situation was very susceptible to changes in this sector. Also, in this period there was a movement within the community of nations to bring some order (and law) to the seas. While Iceland was very much in favor of changing the existing rules of the sea, the government of the UK was steadfastly seeking to maintain the *status quo*. Had Iceland been more successful in diversifying its economy, had the UK been more flexible about changes in the *status quo* with respect to fishing jurisdictions, had the Icelandic herring fishery not collapsed in the late 1960s, had the US–Canadian proposal for a 12-mile limit for national fishing jurisdictions been accepted at the UNCLOS, the entire situation regarding the exploitation of Icelandic cod could have been totally different (especially for those dependent on the fishing sector in Humberside ports).

- *The level of commitment to achieve objectives will be very important. The Icelandic government was deeply committed to take responsibility for managing its marine resources, while the UK appeared less committed to protecting its fishing sector.*

Because of the high level of dependence of the Icelandic economy on its fisheries, any real or perceived threat would be responded to with great dogmatism. Iceland saw the conflict over cod as a life–death struggle. Without the fish, there could be no Iceland. Iceland's commitment to participation in the North Atlantic alliance was seen as subservient to its national economic survival. As a result, it was willing to forego participation in NATO for the sake of defending from outside exploitation, cod populations above its continental shelf. On the other hand, the UK put a much higher value on NATO (and Iceland's participation in it), given that the Cold War was in high gear throughout the period of the Anglo-Icelandic conflicts over fishing jurisdictions. In addition, the UK's national economy was much less dependent on its fishing sector, although regional dependence was demonstrably very high.

- *Traditional international law might not be relied on in times of environmental change. For example, the International Court of Justice was not useful in the resolution of these disputes and, in fact, may have exacerbated conflicts in the short term.*

When a country believes that its existence is threatened, it will react to ensure its survival. Iceland did not believe that its existence should be subject to adjudication by the International Court of Justice. Iceland interpreted the situation and its responses as "right makes might," contending that existing international law was in fact laid down by colonial powers in an earlier (and outmoded) historical epoch. The UK, of course, took an opposing point of view, citing the need to use existing mechanisms to resolve a binational dispute over fishing rights.

- *Countries will resort to "tie-in" tactics (e.g., Lall, 1968), bringing in other issues in order to strengthen the chance for success. For example, Iceland brought NATO into the Cod Wars conflicts; it also brought in the Cold War fears by trading with the USSR, East Germany, and Poland.*

Iceland could not have successfully taken on the UK, if the conflicts had been focused solely on fishing rights. With a population of about 250,000 and a Navy of seven patrol boats, Iceland had to resort to other issues more important to UK interests than Iceland's well-being. Therefore, Iceland used its NATO membership and the existence of the NATO airbase at Keflavik as an important bargaining chip against the UK. At the time of these conflicts over cod the Cold War was at its height. During the first conflict that began in the early 1950s, Iceland, in response to a boycott of its fish products in the UK, began to trade with the Soviet Union and other Eastern European countries. This put the UK (and other NATO allies, especially the US) on notice that Iceland would go to unusual limits to protect what it considered to be its marine resources.

- *When a nation's economy is threatened, long-standing agreements and "habits" may no longer hold.*

The UK cited as support for their actions, that UK vessels had fished Icelandic waters since the 1300s. They argued that these traditional fishing grounds off Iceland's coast were as much theirs to exploit as Iceland's. Again, with its vital resources being depleted (i.e., the fear that foreign vessels would overexploit Icelandic cod), Iceland argued that it had the right to protect its national existence by claiming ownership over fish populations in waters which it defined as territorial. Thus, centuries-old customary practices were ended in the face of a threatened national existence.

- *Regional arrangements should maintain flexibility in the face of changes in environmental conditions and must not become too "bureaucratized" to act. Regional political or military organizations are not good at dealing with environment- or natural resource-related conflicts.*

Despite some attempts to resolve these conflicts, NATO leaders were unable to do so; although Icelandic policymakers were able to use NATO as a forum to generate support for their position while alienating other members from the UK position (Spaak, 1971). In this regard, Iceland played on the appearance (if not reality) of being a David in a David-and-Goliath-like situation.

- *Because perceptions of resource levels, especially in the marine environment, will vary, if not oppose each other, scientific research may not be useful in resolving such disputes (until after the living marine resource has collapsed!).*

Uncertainties inherently abound in all scientific assessments of living marine resources. The responses of regional fisheries management organizations were too slow and inconclusive and, in some instances, national positions were taken on the assessments of the status of cod populations. Politicization of scientific information tended to minimize the role of scientific organizations as potential arbiters in resolving the Cod Wars. They were, however, useful in providing a forum for the protagonists. If the global climate does change, resources will shift at the local and regional levels. Assessing the responses of regional organizations to the Anglo-Icelandic Cod Wars, one can only surmise that they were not up to the task of resolving conflicts among members over living marine resources. Therefore, the operations of regional scientific and resource management organizations must be re-evaluated and when necessary restructured so as to deal more appropriately with regional-scale resource shifts. (On the importance of regional organizations, see Simonis et al., 1989).

- *Resolution of an ecological problem in one location could generate new problems for fish stocks and fishing fleets in other areas.*

As fish populations collapse in one area, trawlers move on to other areas. Usually those areas are already under some degree of exploitation and their fish populations cannot carry the pressures of additional exploitation. In lieu of further burdening already heavily exploited areas, other options include the removal of fishing capacity from the fishery (scrapping inefficient vessels) or the identification of as yet-unexploited fish populations.

- *In light of changes in the marine environment that might accompany a potential climate change, might there be a need in future decades for a UN sea-keeping force?*

It is generally assumed that with a global warming there would be a poleward shift (or expansion) of some living marine resources (e.g., UN WMO, 1988). Given that resources could shift across state borders and national fisheries jurisdictions, there would need to be a mechanism for resolving disputes that might arise from those resource shifts. Similarly, the straddling stocks problem could also lead to conflict. Given such possibilities, the future need for a UN peacekeeping force at sea (this already exists for land-based conflicts) might be warranted.

Concluding comments

Analogies must be used with caution (Jamieson, 1988). With any given analogy there are dissimilarities as well as similarities. With regard to the Anglo-Icelandic Cod Wars, for example, the adverse change in availability of cod to UK trawlers was not the result of an act of nature but of Icelandic political decisions. Also, with a climate change all other factors would not be equal; that is, abundance and availability of resources in other fisheries would also change. The option of shifting trawlers from one fishing ground to another may not be available. The combined effect of all these simultaneous changes would most likely create many additional unforeseen problems. In addition, with a global warming, there may be a change in abundance of cod as well as in availability.

Nevertheless, the Anglo-Icelandic Cod Wars analogy can be instructive, providing a first approximation of how societies deal with change in the marine environment. It can provide useful lessons about how we (our societies) might better prepare for the impacts of the possible occurrence of a global warming. In addition, conflicts over fish populations have not disappeared even with the establishment of the 200-mile EEZ. "Cod Wars" as a type of conflict continue to occur. In the recent past we have had tuna wars and lobster wars. Recently, we have heard about the Franco-Canadian cod wars. For example, media headlines reminiscent of the Anglo-Icelandic conflicts, such as "Gunboat alert in the cod war" (*Maclean's*, 12 April 1982, p. 26) and "Heating up the cod war," (*Maclean's*, 23 February 1987, pp. 12–3) describe Franco-Canadian conflicts over cod quotas in the northwest Atlantic.

Today we also have potential fishing conflicts centered on the exploitation of pollock in the Bering Sea "donut hole," involving the US, USSR, Japan, Korea, Poland, and Taiwan (e.g., US Senate, 1988). This represents the broader problem concerning straddling stocks in other locations (Burke, 1989). Such conflicts over ownership of living marine resources suggest that any change in location of traditionally exploited species could (in theory at least) lead to further extensions of a nation's EEZ.

The method of forecasting by analogy identifies an important role for the use of historical information in understanding the possible societal responses to regional climate impacts on the marine environment. If carefully used, such information can provide insights into how well societies might be prepared to cope with changing environmental conditions. Whether the global climate ultimately becomes warmer, cooler, or stays the same as it is today, societies will still have to learn to cope with climate variability and extreme meteorological events, and hence with resource variability. By assessing how well (or how poorly) we have coped in the recent past, we can provide a first approximation of how well we may be able to respond to such impacts in the near future.

Acknowledgments

I would like to acknowledge the assistance of Maria Krenz, whose research and editorial support was invaluable to the preparation of this chapter. I would also like to thank Dr. Jakob Jakobsson, Director of Iceland's Fisheries Research Institute. His critical and thought-provoking review of an earlier draft of this chapter was extremely useful. The views and interpretations herein are mine.

In collecting information on the Anglo-Icelandic Cod Wars, several people provided their kind assistance. I want to thank Dr. Ray Beverton, who responded to my queries about that period in UK fishing history and to several librarians in the UK who heeded my pleas for printed material on the Cod Wars.

References

Akaha, T. (1985). *Japan in Global Ocean Politics.* Honolulu: University of Hawaii Press.

Alexander, L.M. (1963). *Offshore Geography of Northwestern Europe.* Monograph Series of the Association of American Geographers. London: John Murray.

Barston, R.P. & Hannesson, H.W. (1974). The Anglo-Icelandic fisheries dispute. *International Relations (London),* **4**, 559–84.

Bell, H. & Pruter, A.T. (1958). Climatic temperature changes and commercial yields of some marine fisheries. *Fisheries Research Board of Canada,* **15**, 625–83.

Bilder, R.B. (1973). The Anglo-Icelandic fisheries dispute. *Wisconsin Law Review,* **37**, 37–132.

British Trawlers Federation (1958). Report produced for the delegates to the General Assembly of the United Nations on behalf of the United Kingdom fishermen. November 1958. London: British Trawlers Federation.

Buchanan, N. & Steel, D.I.A. (1977). Meaningful effort limitations: the British case. In *Fisheries of the European Community,* p. 7. Edinburgh: Fishery Economics Research Unit.

Burke, W.T. (1989). Fishing in the Bering Sea Donut: straddling stocks and the new International Law of Fisheries. *Ecological Law Quarterly*, **16**, 285–310.

Burton, R. (1973). Tip of the iceberg. *Oceans*, **6**, 56–63.

Dalgliesh, N.A. (1970). *The British Fishing Industry*. London: Nelson.

Davis, M. (1963). *Iceland Extends its Fisheries Limits*. Oslo: Universitetsforlaget.

Dickey, D.P. (1974). The Cod War: Iceland, England and the International Court of Justice. In *Some Aspects of International Fishery Law*, prepared for National Oceanic and Atmospheric Administration, pp. 66–73. Washington, DC: National Technical Information Service.

Gilchrist, A. (1978). *Cod Wars and How To Lose Them*. Edinburgh: Q Press.

Glantz, M.H. (1988). Politics and the air around us: international policy action on atmospheric pollution by trace gases. In *Societal Responses to Regional Climatic Change: Forecasting by Analogy*, ed. M.H. Glantz, pp. 41-72. Boulder: Westview Press.

Glantz, M.H. & Thompson, J.D. (Eds.) (1981). *Resource Management and Environmental Uncertainty: Lessons from Coastal Upwelling Fisheries*. New York: Wiley Interscience.

Grondal, B. (1971). *Iceland: From Neutrality to NATO Membership*. Oslo: Universitetsforlaget.

Hart, J.A. (1976). *The Anglo-Icelandic Cod War of 1972-1973: A Case Study of a Fishery Dispute*. Berkeley: Institute of International Studies, University of California.

ICES (International Commission for the Exploration of the Seas), 1976. Report of the North-Western Working Group, 8–12 March 1975, Charlottenlund, Denmark. Copenhagen: ICES.

Iceland Ministry for Foreign Affairs (1976). *The Fishery Limits Off Iceland: 200 Nautical Miles*. Reykjavik: Ministry for Foreign Affairs.

Iceland Ministry of Fisheries (1975). *Cod War III between Iceland and Great Britain*. Reykjavik: Ministry of Fisheries.

Jakobsson, J. (1988). The importance of the herring fishery. *Modern Iceland*, **4**, 24–6.

Jamieson, D. (1988). Grappling for a glimpse of the future. In *Societal Responses to Regional Climatic Change: Forecasting by Analogy*, ed. M.H. Glantz, pp. 73–93. Boulder: Westview Press.

Jónsson, B. (1972). Our fight for survival: Iceland's case for extending its fishery limits. *Free Labour World*, March. UK: ICTFU.

Jónsson, H. (1982). *Friends in Conflict: The Anglo-Icelandic Cod Wars and the Law of the Sea*. London: C. Hurst & Company.

Lall, A.S. (1968). *How Communist China Negotiates*. New York: Columbia University Press.

Martin, B. (1979). *The Bias of Science*. O'Connor, Australia: The Society for Social Responsibility in Science (ACT).

Mitchell, B. (1976). Politics, fish, and international resource management: The British–Icelandic Cod War. *Geographical Review*, **66**, 127–38.

Roemer, M. (1970). *Fishing for Growth: Export-Led Development in Peru, 1950-1967*. Cambridge: Harvard University Press.

Simonis, U.E., von Weizsäcker, E.U., Hauchler, I. & Böl, W. (1989). The crisis of global environmental demands for global politics. Report of the Foundation for Development and Peace, F.R. Germany. *Interdependenz*, **3**.

Spaak, P.-H. (1971). *The Continuing Battle: Memoirs of a European, 1936–1966*. London: Weidenfeld and Nicolson.

Transport and General Workers Union (1980). *Fishing: The Way Forward*. Aberdeen: Transport and General Workers Union.

United Kingdom, Government of (1958). The problem of the fisheries around Iceland. Memorandum submitted to the General Assembly of the United Nations, November 1958.

UN WMO (World Meteorological Organization) (1988). *Proceedings of the Conference on The Changing Atmosphere: Implications for Global Security, Toronto, Canada, 27-30 June 1988*. WMO Publication 710, pp. 333–5. Geneva: WMO.

US News & World Report (1973). Where a 'Cod War' is threat to Nato. 11 June, p. 33.

US Senate (1988). Fisheries management and enforecement in the Bering Sea. Hearing before the National Ocean Policy Study of the Committee on Commerce, Science and Transportation, 100th Congress, 2nd Session. Washington, DC: US Government Printing Office.

Wise, M. (1984). *The Common Fisheries Policy of the European Community*. London: Methuen.

13

Adjustments of Polish fisheries to changes in the environment

ZDZISLAW RUSSEK

Sea Fisheries Institute
Gdynia, Poland

Introduction

This assessment of some dramatic changes in the Polish fishery system has been inspired by the general assumption that a warming of the global climate may turn out to be unavoidable over the next several decades. The consequences of increasing CO_2 emissions as a result of industrial activities are highly complex. The rise of average atmospheric temperatures by about 2–3°C could have numerous implications for marine ecosystems. Primary production and natural fish habitats will be disturbed to some, as yet unpredictable, extent. Consequently, the distribution of some living marine resources could shift, adversely affecting some fishing communities while benefiting others.

It is not yet known whether, or how, different sectors of societies will be able to respond to the environmental impacts of a global warming. Hence, the question arises whether past experiences of some fishing nations could be used to anticipate impacts on society that could take place in the future. The approach taken in this assessment is based on the concept of a case scenario. The character, qualities and range of applicability of the case-scenario approach have been comprehensively discussed for this project by, for example, Glantz (1988) and Jamieson (1988). Thus, this chapter presents basic facts regarding what actually took place in the Polish fishery. It also considers whether these facts bear any significant resemblance to those which might occur in other national fisheries as a result of global warming in the future. The kind of "analogical reasoning" that could be viewed as either strained or misplaced must be avoided and the extent to which the use of a specific analogy is limited by the level of technology and/or the political system in force in a given country, must be identified.

The disparity in research

Awareness of the necessity to think about, as well as to research, possible societal responses to a global warming is at present quite high among both scientists and policymakers. To date, however, this progress focuses mainly on terrestrial impacts. Fishery scientists and administrators have only been marginally involved – with a few exceptions – in elaborating measures which could prevent or mitigate or adapt to the impacts of a global warming on marine ecosystems. They have been quite occupied addressing an increasing number of pressing issues affecting their national fisheries. Many national stocks are overfished and few zones (EEZs) are exploited wisely in terms of optimum effort. In addition, there is a lack of expertise and appropriate enforcement of fishery regulations in many developing countries. According to the UN FAO, substantial improvement in national self-reliance and skill through training, along with the transfer of appropriate technology, is a key need facing many national fisheries today (UN FAO, 1984). The global warming issue does not yet contain a "dread factor." Since global warming is expected to take place some decades in the future, the real importance of the issue is barely perceived by fishery administrators.

Three decades of spectacular development

The take-off period

The Polish fishery industry started with modest catches in the southern Baltic. The International Council for the Exploration of the Sea (ICES), established in 1902 in Copenhagen, Denmark, reported the total landings in 1903 of all countries surrounding the Baltic to be 50,000 metric tons. By 1908, landings had increased to 113,000 mt. It would be a very speculative (and arbitrary) task to separate out the pre-World War I catches of Polish fishermen, as Poland was not independent at that time. Polish landings were included either in German or in Russian data reported to ICES. The first officially reported Polish marine catches in 1921 amounted to 1,307 mt. The average annual Polish catch in the Baltic Sea, between 1921 and 1930, accounted for 2,490 mt, compared to about 120,000 mt taken by other countries. Initially,

the activities of Polish fishermen were entirely restricted to inshore areas. In the 1930s, however, the introduction of motorized
cutters expanded their operations with otter trawls to offshore
grounds and, as a result, catches of herring, sprat and cod gradually increased (Thurow, 1978). Simultaneously (from 1931), some
Polish enterprises began to operate in the North Sea. As a consequence, during the four years prior to World War II, the first signs
of progress in the Polish fisheries were noted. From a global or
regional viewpoint, this progress was rather inconspicuous. The
average annual Polish catch in the years 1935–38 accounted for
18,000 mt compared to 177,000 mt taken by seven countries in the
whole Baltic Sea. Polish national catches did not keep up with
the constantly growing local demand for fish products, mainly
for salted herring and canned sprat or sardines. Because of the
scarcity of distant-water fishing vessels and the relatively poor fishing grounds in the Baltic Sea, demand was met mostly by imports
ranging from 61,000 to 104,000 mt (product weight) per year, at
a cost of US$6 to 13 million (1920 dollars). So, fish consumption
in Poland – measured in product weight – exceeded by six to ten
times the levels of national catches.

The ambitious plans of some Polish companies aimed at expanding the national fishery were destroyed during World War
II (1939–45), together with the entire fleet of 180 offshore fishing
boats that was in operation in 1939, in addition to the loss of more
than 500 small motorized boats.

When the war ended, Poland inherited a southern Baltic sea-
coast which resembled a scorched desert. Fishing vessels were
sunk, processing plants ruined and harbors devastated. The reconstruction of the fishing sector and the shipbuilding industry
took two years. In 1947, Polish catches in the Baltic Sea had
doubled, compared to the pre-war level. From 1948, the shipbuilding industry was steadily growing, which was very important
for the development of the rudimentary Polish fishery. Before
Poland mastered the construction of ocean-going vessels, the then-
fledgling shipyards in Gdansk and Gdynia were operating at their
fullest capacity to support, first and foremost, the national fishery.
Within the three-year period from 1947 to 1949, there were 383
small side-trawlers with 80–150-HP engines completed in Polish
shipyards to be operated on the Baltic Sea. To exploit the North
Sea resources, 24 bigger, medium-sized trawlers with 900-HP en-

gines were constructed locally and 40 second-hand trawlers were
acquired from the UK. In this way, during the late 1940s, Poland
suddenly came into possession of a relatively excessive fishing fleet
in comparison to the number of skilled crews of Polish nationality.
The inadequate capacity for training sea-going personnel became
a bottleneck. For this reason, more than 300 Dutch skippers and
engineers were hired for a three-year period (from 1948 to 1950) to
catch fish and simultaneously to train unskilled Polish fishermen.
This on-the-job training turned out to be extremely successful and
much needed before the first three schools for fishermen were established in 1950–53. The total catch jumped from 39,000 mt in
1947 to 66,000 mt in 1950 to 107,000 in 1955. From then on, the
Polish fishery advanced rapidly during the next 30 years; an increase in tonnage, catches and market supply was evident almost
every year. It appears that the years 1947–50 could be considered
a take-off period for this sector of the economy, to use a classic
economic term. As a matter of fact, this take-off dates back to
the late 1930s, but it was brutally destroyed by World War II. Its
development was thus delayed by ten years.

Growing catches and market supply

Following the first boost into the late 1940s, the uninterrupted
flow of newly built fishing vessels and growth of landings continued into the 1950s and 1960s. It is worth noting that in 1960, only
one-half of the total catch was extracted from the southern Baltic,
while the other half was taken from the North Sea. Afterward,
however, the share of catch taken outside the Baltic Sea increased
at a considerable pace. The resources of the Baltic Sea were considered by planners to be poor and fishing grounds of the North
Sea were overcrowded with different foreign fishing vessels operating there. So, the Polish fishing fleet began to search for new
fishing grounds on the Atlantic Ocean and to evaluate the feasi-
bility of exploiting them. For some time, the managers of Polish
enterprises were quite interested in the results of the impressive
British venture (in the late 1950s) involving the operations of the
first fishing and factory vessels *Fairtry I* and *Fairtry II* (owned
by Salvesen Co.). The ability to catch and process fish on remote
fishing grounds was attractive. The same type of venture as that
of Salvesen Co., although on a larger scale, had been launched by

the *Tralflot*, a Soviet enterprise in Murmansk. In the late 1950s the Polish shipyards put forward a tender to construct the first 15 sophisticated fishing and factory vessels for Polish fishing companies. The shipyards were, at that time, primarily export-oriented, constantly designing new vessels for both shipping and fishing. They were very anxious to use the Polish fishery as an experimental ground for testing their ability to construct good vessels which would make them more competitive against other European companies (mainly UK, West Germany, and Spain). The economic justifications of the projects that were carried out pointed to the likelihood of a beneficial expansion of the Polish fishery to the Newfoundland and Labrador (Canada) fishing grounds.

Eventually, all three parties (government planners, shipyards, and fishing enterprises) came to accord, and the first fishing factory trawler, *Dalmor I*, completed its maiden voyage at the end of 1960. This began the noticeable presence of Polish vessels in the Northwest Atlantic. From 1960 to 1988, as many as 300 different sizes and types of fishing and factory vessels were constructed in long or short series. They became involved in the exploitation of vast – at that time – underutilized resources in the Atlantic and Pacific Oceans, achieving high catch rates and a good quality of processed and frozen fish. Their operating distances ranged as far as 8,000 nautical miles from the ports of registry. It is no wonder that from 1960 until 1980 Polish catches grew almost five times from 168,000 to 781,000 mt. Figures 13.1 and 13.2 illustrate the dependence of the catches on the tonnage of fishing vessels.

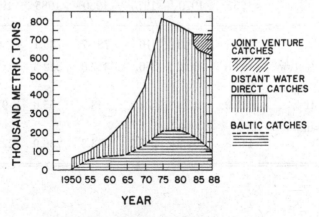

Fig. 13.1 Total Polish catches.

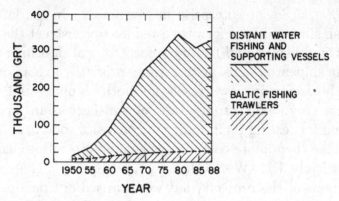

Fig. 13.2   Polish fishing fleet in the years 1950–88 (in thousand gross registered tonnes).

The increase of demand in the local market was actually lower than the increase in catches counted in live weight. The reason for this disparity lies in the worsening economic position of the Polish fishery and, in particular, in the dramatic increase of that part of the costs which have been expended in foreign currency. Until 1963, imports of fish products exceeded exports. From 1965 onwards, the surplus of export over import grew continuously. Its magnitude, calculated in five-year intervals, together with the catch in live weight, is shown in Table 13.1.

Table 13.1
Fishery production and local market supply 1965–88
(in thousands of metric tons)

	1965	1970	1975	1980	1985	1987	1988
Catch	230.1	451.3	816.7	781.7	735.1	750.4	732.4
Final production	152.8	230.2	256.8	317.0	312.5	316.2	310.6
Surplus of export over import	6.0	42.3	33.1	43.7	55.6	93.5	110.1
Domestic market supply	146.8	187.9	223.7	273.3	256.9	222.7	200.5

The Polish fishery had been subsidized by the government. The subsidies generally accounted for six to 19 percent of the value of domestic market supply. Beginning in 1987, the companies were left entirely to their own devices and had to deal with steep increases in unit cost resulting from three factors: (1) sharp increases in fuel prices in the 1970s, (2) increasing distances to the fishing grounds, and (3) higher royalties assessed by foreign exclusive economic zones. Thus, the Polish distant-water enterprises were compelled to export each consecutive year more fish straight from the fishing grounds (using fish carriers) in order to earn enough foreign exchange to cover their operating costs.

As shown in Table 13.1, in 1988 about 35 percent of the total final net production (surplus of export over import) had to be exported to keep the Polish fishery machinery working, whereas in 1980, the figure was 14 percent. In the late 1980s, catch composition and the profile of the final production had changed considerably. In brief, the Polish distant-water trawlers searched for species of fish which brought the best export price instead of those for which there was local demand (squid is exported and herring is imported). The terms of trade for the Polish distant-water fishery in the 1980s were much worse than in the 1970s, to say nothing of the 1960s. The development strategy of the Polish fishery was reviewed by the 15th session of FAO's Cooperative Oceanic Fisheries Investigations (COFI) in the following manner:

> *The opportunities for greater catches are generally offshore, where catch rates may sometimes be higher. Some of the largest unexploited stocks are to be found in remote areas of the ocean far from centers of human population. It often pays to exploit more distant fishing grounds if the catch rates are sufficiently high, but comparison of near- and distant-water options by straightforward cost-and-earnings calculations are valid only if the skill, supporting services and infrastructure necessary for the operation of off-shore fleets are already in place or if the time and cost of establishing them are accepted as justifiable in terms of general development* (UN FAO, 1983).

For Poland, the cost of distant-water fishing was entirely justified in the national economy of the 1960s and 1970s. It still appeared favorable as a whole in the 1980s, although some operations in some areas brought losses. The 1990s will provide an answer to the crucial question about whether and to what degree it pays to roam the remote seas under conditions restricted by the establishment of the EEZs.

Factors of development

On the basis of a number of analyses, we can identify four explicit factors which have been contributing to the development of the Polish fishery within the last decades, and they are as follows: (1) demand for fish, (2) well-trained and cheap human labor, (3) high input of technology, and (4) open access to all living resources. Each is briefly discussed in the following paragraphs.

(1) Demand. A strong potential demand has its roots not only in increasing personal income and in a very high income elasticity of demand but also in the sociocultural features of the nation. Ethnic background and religious beliefs, generating consumer preferences, converged with the government propaganda launched in the 1950s through to the 1970s, advocating high consumption of fish for purely dietary reasons. This propaganda was combined for some time with subsidies for the fishery, particularly when meat was scarce on the local market. Fish was rated as a substitute for meat, the supply of which often fell short of the government's plans. Furthermore, most of the species caught by Polish distant-water trawlers were also attractive for export opportunities. All these circumstances generated a wider market outlet (local and foreign) for Polish fishing enterprises than they were in a position to fulfill, due to limited access to fishing areas.

(2) Trained manpower. An extensive education system was set up in the 1950s and 1960s by the government to train a large number of sea-going and other personnel. First of all, it embraced occupations required to operate fishing vessels, such as master fishermen, mates, engineers, radio operators, etc. Likewise, thousands of food technologists, plant foremen, and freezer and cannery operators were trained in secondary schools. Separately, managers, government administrators, and development personnel were educated in four-year courses in the College of Fisheries, or other schools. As a result of this training, about 18,000 fishermen and 23,000 inland workers held relevant certificates of proficiency or competency. Over 1,400 employees were ranked as diploma-level personnel. Most of them were attached to state-owned enterprises,

earning only a small remuneration compared to that paid in the US or West German fishing industries.

(3) Technology and capital. Ambitious development plans obviously called for heavy investments, mainly in the form of large and sophisticated vessels. They were financed chiefly from the government budget and contained technical innovations from all over the world. It is worth mentioning briefly the following cornerstones of development: (a) the use of synthetic fibers in the manufacture of nets, (b) the introduction of processing and freezing at sea, (c) the use of electronic aids facilitating the location of the concentrations of fish shoals, (d) the mechanical hauling and stern trawling permitting use of large nets, and (e) an impressive increase in the size, versatility and operational range of fishing vessels (UN FAO, 1987). In the 1980s, however, investment declined seriously. The construction of new vessels was lagging behind the scrapping of old ones. The total number of distant-water catchers in operation dropped from 115 in 1980 to 86 in 1988. The number of freezer carriers supporting the fishing operations increased because of the growing distances to fishing grounds. Eventually, the total net value of fixed assets in fisheries declined as shown in Table 13.2. This process of the decline of investment propensity which started in the 1980s may continue in the 1990s.

Table 13.2
Net value of fixed assets involved in the Polish fisheries 1970–88
(in millions of US dollars)

Component	1970	1975	1980	1985	1988
Fishing and auxiliary vessels	180	271	405	360	325
Inland processing plants and cold stores	43	95	137	120	115
Total net value of capital out	223	366	542	480	440

(4) Open access. For more than three decades, starting from the late 1940s, the Polish fisheries enjoyed practically free access

to all abundant living marine resources in the world oceans which, according to the economic calculations of the planners, were worth exploiting. The only restrictions as to the rate of stock exploitation were those imposed by the recommendations of international fishery regulatory bodies. This comfortable period of open entry ended definitely in some areas in the late 1970s and in others in the early 1980s. The establishment of EEZs were acknowledged internationally in December 1982, when most governments signed the UN Convention of the Law of the Sea (UNCLOS). The freedom of the seas was gone and the new legal regime constituted a serious challenge to the very foundations of the then-existing model of the Polish fishery.

The turbulent fourth decade

The first big shift in fishing operations

In the preceding section a profile of the Polish fishery was presented, casting some light on the scale and factors of its development. In this section, it is necessary to clarify the character of changes and their settings. The deployment of fishing effort, measured in standard fishing days, is presented in Fig. 13.3 for the years 1970–88. Furthermore, the distribution of catches over nine fishing areas is shown in Fig. 13.4. Looking at both Figs. 13.3 and 13.4, we can easily see that the fishing operations have either distinctively declined or were stopped in three important areas:

- The Northeast Atlantic (the North Sea and Norwegian Sea)
- The Northwest Atlantic (the Newfoundland shelf)
- Eastern Central Atlantic (the Mauritanian coast)

In the fourth area, that is, the Southeast Atlantic (along the Namibian coast), Polish operations gradually waned from 1978 until 1987, and eventually ended in 1989. The reason for the withdrawal of Polish vessels was the establishment of EEZs in those regions and its effect seemed to be devastating for the enterprises concerned. It meant relinquishing large quantities of fish which these enterprises had been extracting with no problem. The magnitude of losses incurred by the Polish fishery can be assessed from Table 13.3. Column 2 of this table shows the last years when the vessels were allowed to operate freely. Column 3 contains the quantities of fish caught during these last years. Column 4 presents the

Table 13.3
Quantities of fish caught in some regions by Polish
distant-water vessels prior to and following the
establishment of the exclusive economic zones
(in thousands of metric tons, live weight)

Region	The last year prior to the establishment of the EEZs	Catch in the last year prior to the est. of EEZs	Catch extracted in 1983	Opportunities lost annually (average)
1	2	3	4	5
Atlantic Northeast	1976	124.5	0.9	123.6
Atlantic Northwest	1975	187.5	13.3	174.2
Atlantic Eastern	1977	203.4	–	203.4
Atlantic Southeast	1978	164.1	114.1	30.0
TOTAL	–	679.5	128.3	551.2

quantities extracted in these regions following the establishment of EEZs, and Column 5 shows the opportunities lost in terms of catches.

During this first wave of compulsory withdrawal from the regions specified in Table 13.3, Poland lost a potential catch of 551,000 mt or approximately 70 percent of the total national catch at that time. Obviously, the vessels could not have been put aside to stand idle in the harbors. Luckily, the managers of Polish fishing enterprises received official notice two to three years ahead of the date when the EEZs went into effect. During this transition period, they were able to find new fishing grounds either within the EEZs of other countries or on the open sea. The exploration and scouting for new fishing grounds involved considerable but unavoidable additional costs.

In some areas, equity joint ventures were founded, the biggest one being with Peru, which operated during a 10-year period. The shifts which were carried out before the end of the 1970s are shown

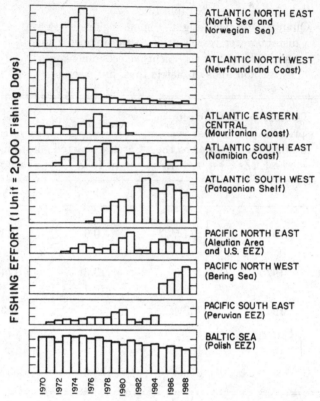

Fig. 13.3 Deployment of fishing effort by fishing areas.

in the set of five maps (Figs. 13.5 to 13.9). In particular, when one compares Fig. 13.5 with Fig. 13.6, one can see to what degree the deployment of fishing effort was reshaped between 1970 and 1980. The fishing grounds on the east and west shelves of the North Atlantic were almost entirely abandoned by Polish vessels. New ones, mainly in the southern part of the Atlantic and East Pacific came to the fore. The distribution of catches in 1970 in contrast with that in 1980 is shown in Figs. 13.7 and 13.8. Figure 13.9 presents the overall picture of the redeployment of fishing effort and displacement of catches. It shows the magnitude of the first big shift of the Polish fishing fleet over the world's oceans. This first big shift eventually turned out to be a success. None of the distant-water vessels had been idled or scrapped before their normal life span. Until 1981, the catches did not drop seriously. Nonetheless, everyone concerned with the Polish fishery became aware of the fact that open access to the living resources of the world's oceans had ended and that more problems were still ahead.

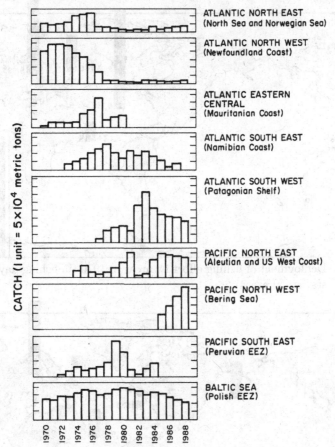

Fig. 13.4 Distribution of catches over fishing areas, 1970–88.

The second big shift in fishing operations

After the first shock in the late 1970s, the Polish fishery reclaimed its ability to operate in distant waters in the 1980s. However, the institutional and economic terms of catches started, again, to deteriorate. First, the terms of cooperation with the Peruvian enterprises were becoming more costly. When the royalties became unacceptably high, Poland declined to renew the annual contract. The joint venture was declared nonoperational in 1984 and was dissolved in 1985.

Second, the stocks close to the Namibian coast (Atlantic Southeast area), where no EEZ existed, were heavily exploited by numerous vessels of different countries. The size of fish and its price were gradually declining. As signs of overfishing became evident, scientists recommended cutbacks in catches. Poland thereby sus-

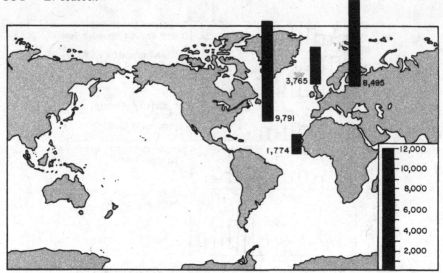

Fig. 13.5 Deployment of fishing effort by areas in 1970 (in fishing days in three statistical areas).

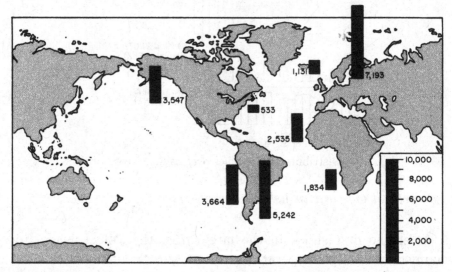

Fig. 13.6 Deployment of fishing effort by areas in 1980 (in fishing days in seven statistical areas).

pended its fishing operations in the area for a period of five years or until the time when an EEZ would be declared by an independent Namibia. The overall scope of withdrawals by the Polish fishery during the so-called second wave is shown in Table 13.4.

As it appears from Table 13.4, the loss of potential catches in the Pacific Southeast and the Atlantic Southeast was estimated at 284,000 mt. It was compensated for by the redeployment of fishing effort to the North Pacific area (see Figs. 13.10 and 13.11).

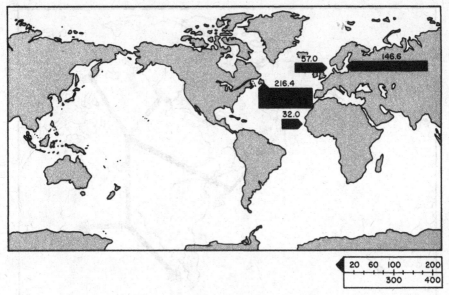

Fig. 13.7 Distribution of Polish catches over main fishing areas in 1970 (in thousand metric tons of live weight).

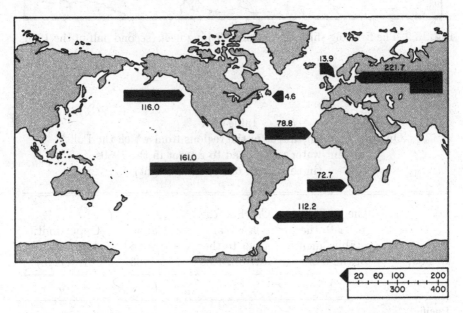

Fig. 13.8 Distribution of Polish catches over main fishing areas in 1980 (in thousand metric tons of live weight).

Figure 13.12 illustrates the general direction of the so-called second big shift of the Polish fishing fleets. The second shift, as did the first, called for a number of additional expenses on the part of Poland in the form of: (1) direct input for fish scouting, and (2)

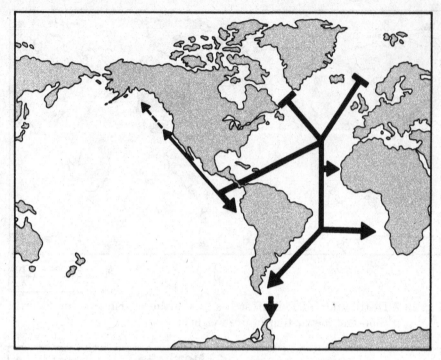

Fig. 13.9 The first big shift of the Polish fishing fleet (second half of the 1970s).

Table 13.4
Quantities of fish caught in the regions from which the Polish
distant-water vessels had to retreat in the 1980s
(in thousands of tons live weight)

Region	The last year prior to the establishment of the EEZs	Catch in the last year prior to the est. of EEZs	Catch extracted in 1983	Opportunities lost annually (average)
1	2	3	4	5
Pacific Southeast	1980	161.0	–	161.0
Atlantic Southeast	1981	123.0	–	284.0
TOTAL	–	284.0	–	284.0

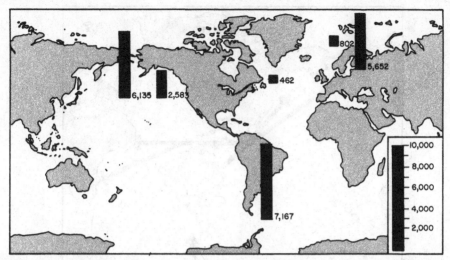

Fig. 13.10 Deployment of fishing effort by areas: 1988 (in fishing days upon three statistical areas).

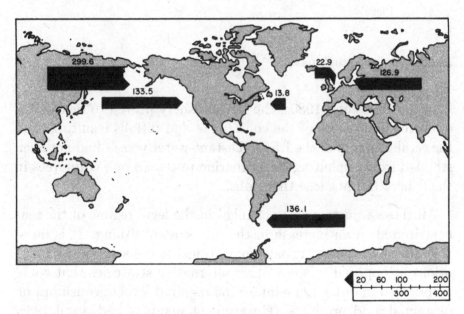

Fig. 13.11 Distribution of Polish catches over main fishing areas, 1988 (in thousand metric tons of live weight).

negotiations coupled with payments for the right to fish in foreign EEZs.

It is difficult to envision the many additional shifts that must be carried out in the future and other measures that must be taken to keep the existing fleet operating.

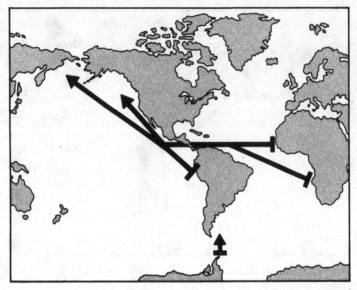

Fig. 13.12 The second big shift of the Polish fishing fleet (second half of the 1980s).

Economic implications

In the 1970s and 1980s, the Polish fishery proved to be capable of overcoming many of the constraints and pitfalls resulting from the establishment of the EEZs. Distant-water vessels had not been stopped from exploiting living marine resources and employees in the fishery did not lose their jobs.

All these adjustments to changes in the legal regime of the seas contributed to an increase in the unit cost of fishing. It is questionable now whether it pays to continue with distant-water operations as opposed to some other alternative strategies that could be pursued in order to maintain the required level of consumption of animal food products. Government planners had, until 1988, a ready answer at hand: in their opinion, the distant-water fishery should be supported as long as the unit cost of net animal protein derived from fish remained lower than that derived from meat. This yardstick was "sacred" in centrally planned economies for making decisions related to the allocation of national resources between agriculture and fishery. Thus, from the national viewpoint, the relevant comparative evaluation in this respect (in local currency, of course) was carried out every year. An attempt to

Table 13.5
Costs of animal protein production in Poland
(in US dollars per 1 kg of net protein contained in
three different groups of products in the period 1966–87).

Year	Meat	Fish	Dairy products	Fish as percentage of meat
1966	6.85	4.20	3.32	61
1970	6.97	4.25	3.16	61
1971	6.59	5.08	3.46	77
1972	6.50	5.88	3.75	90
1973	6.28	5.94	3.98	94
1974	6.41	5.82	4.14	90
1975	7.19	6.22	4.79	86
1976	8.43	8.10	5.10	96
1977	8.60	8.48	5.42	98
1978	9.20	8.73	5.17	94
1979	9.82	8.42	5.20	85
1980	11.24	9.50	6.17	84
1981	13.10	9.70	8.62	74
1982	15.78	12.09	8.86	76
1983	17.76	13.94	9.01	78
1984	18.10	16.13	10.04	89
1985	19.06	16.03	9.08	83
1986	19.48	16.52	10.28	84
1987	20.09	17.49	11.10	87
Index in 1987*	293	416	334	

*Assuming 1966 as 100.

convert it into US dollars is shown in Table 13.5 and illustrated by Fig. 13.13.

One must note, however, that the use of unit cost of protein as a measurement for decision making has been questioned for many reasons. First, pure protein is never the object of trade, local or international. Its value varies from country to country depending on the composition of fish species traded and on the income elasticity of demand plus some social factors. In practice, individual investment decisions which ultimately constitute macroeconomic resource allocations are based on financial criteria and not on protein measurement.

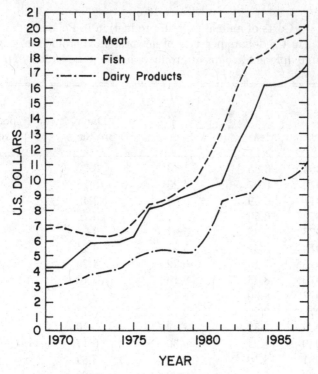

Fig. 13.13 Protein production costs in US dollars per 1 kg net protein in the years 1969–87.

Second, all detailed figures of prices and costs of fishery and agricultural production are available in Poland in the local currency which is not convertible, and the official rate of exchange is inadequate for carrying out any sensible assessment for international comparative use.

The national economy as a whole is a muddle (even for the top government planners), and the figures produced in Table 13.5 may contain a large margin of error (up to 50%). So, one may conclude that the response of the Polish fishery administration to the establishment of EEZs was successful in terms of technological capacities, although disputable in terms of economics.

Deceptive fish protein concentrates

The issue of functional fish protein concentrates (FFPCs) could have been left unmentioned, had it not caused so much damage to the Polish fishery between 1974 and 1977. At a time when Polish managers were having trouble adjusting to the constraints stemming from the creation of EEZs, a certain dubious inventor appeared on the scene, offering a miraculous solution by converting to human consumption enormous quantities of by-catch and trash fish usually discarded by Polish vessels when catching target fish, or fish which was being reduced to fish meal. Since he was unable to get fishery managers to accept his technology of producing FFPC, he turned to the two top politicians in the country. To the despair of scientists, these politicians believed him and as a result ordered the hasty construction of a special laboratory and, simultaneously, of two big processing plants for FFPC, with a capacity of 30,000 mt each. This was done before the proposed technology was worked out in detail and tested in a pilot plant.

The FFPCs were to be added at a ratio of 20 percent to wieners and luncheon meat in an attempt to alleviate severe meat supply shortages which Poland was experiencing at the time. Unfortunately, all attempts to work out a technology for manufacturing FFPCs from full (un-deboned) trash fish failed. No single trash fish was utilized as promised. On the other hand, the gel and other shapes of FFPCs obtained from the deboned, filleted, and minced flesh of edible fish turned out to be more expensive than the traditional fish products. In addition, the production of FFPC competed in terms of raw material requirements with the existing processing plants and canneries. Thus, the scarcity of raw material was aggravated. Furthermore, the wieners stuffed with FFPC were not accepted by consumers, and the ratio of rejection was enormous. And so, after four years of futile struggle, the big processing plant was disassembled and transformed into a cannery (the second proposed FFPC processing plant had not as yet been constructed).

The harm done to the Polish fishery by the interest in the FFPC concept stems mainly from the fact that other more promising alternative opportunities to gain more raw material were neglected. Some options to establish joint ventures with access to foreign EEZs were lost. Such mistaken decisions – as in the case of FFPC – always have a considerable chance to crop up in centrally planned economies where the politicians have more control over final decisions than do the entrepreneurs who take into account market forces. In this specific case, the politicians wanted to find a detour to escape from the agriculture and fishery squeeze. In fact they achieved the opposite, making a bad situation even worse.

*Search for resources on the lower trophic level
(the Antarctic krill)*

On the areas of the Atlantic and Pacific Oceans, Polish fishing vessels had traditionally been extracting mainly secondary and tertiary carnivores. Because of imminent constraints in exploiting the EEZs, it was decided in the mid-1970s to investigate the possibilities of shifting operations down by two trophic levels and to start catching the Antarctic krill (*Euphasia superba*). The task was difficult due to the lack of knowledge about the distribution of remote stocks and puzzling technological properties of the would-be food products. The first three years of research were financed by the government through the Marine Fisheries Institute in Gdynia and, later, starting in 1980 by some enterprises. The following are the results of their investigations:

E. superba occurs in the belt of Antarctic waters between the minimum and maximum annual range of ice cover, south of the Antarctic Convergence, as shown in Fig. 13.14. Its distribution is circumpolar and biomass varies greatly within this area, fluctuating in different years. From the 1960s until the 1980s, there were at least seven controversial assessments of the standing biomass ranging from 44.5 to 5,000 million mt with a potential yield from 25 to 2,000 million mt annually (Budzinski et al., 1985). Like other pelagic animals living in shoals, krill migrate vertically to the upper water layer from 5 to 50 m, although some shoals have been caught at depths of up to 300 m. Fishing techniques for catching krill were mastered by skippers within a year, since they are generally similar to those used in pelagic trawling. Relevant hydroacoustic equipment is necessary to spot the shoals and distinguish them from other organisms and to plan the daily catch rate. Ice around the Antarctic determines the duration of the fishing season, which lasts on average from mid-November to mid-April. The catch rate obtained after two years of experimental fishing proved to be highly satisfactory, on average around 60–80 mt per day per boat during the five to six months of the fishing season.

The real obstacle for undertaking industrial production has been caused by an inconsistent daily fluctuation of catches combined with the very peculiar biochemical properties of krill meat. The main obstacle in processing krill into edible products is the active system of protetic enzymes and the large amount of water-soluble

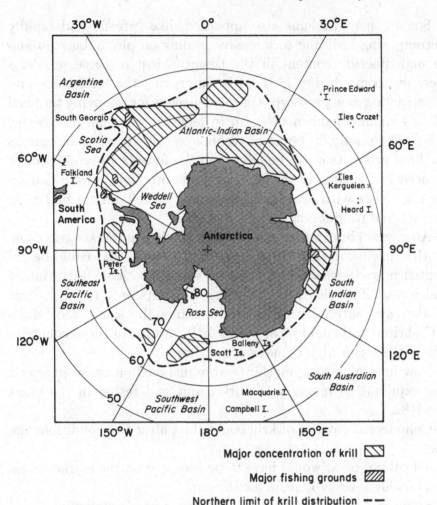

Fig. 13.14 Areas of concentration of the Antarctic krill and the major fishing grounds exploited during the three Polish krill expeditions in the years 1976–79.

protein in krill muscles. Hence, after catching, the storage time on deck in air temperatures of 0–5°C cannot exceed three to four hours. After that time, the decomposition of tissue is advanced and the raw material can be directed only to reduction to fodder meal, provided that the overall storage time takes no more than eight to 10 hours. After that time, only empty shells remain. To keep within these deadlines a vessel must be properly designed, so that its capacities to catch, store, process (peel), and freeze are in harmony with one another. The time and frequency of trawling must be wisely regulated, corresponding to the processing and freezing capacities during the daily operation schedules.

Some other problems also appeared, like "green" and rapidly decomposing krill due to intensive feeding on phytoplankton and to high fluoride content in the tissue. Most of these problems were overcome during 1981–84 by improving the technology and constructing new prototypes of roller-peelers for removing the shell of this small organism (4–5 cm in length), weighing on average 1.5 g. Eventually, in terms of technology, a wide range of options for final products could be offered in the form of frozen, dried, or minced krill. Some hopes for large-scale production were tied up for some time with coagulated krill paste. They did not find a big market and production did not develop.

After years of experiments, Polish scientists and managers came to the conclusion that the appropriate product that could be accepted nowadays by the consumers would be "a shell-free intact tail-meat." The yield, after peeling, amounts only to 16 percent of the raw material caught. It resembles the small size "shell-off" shrimp and could most probably be traded in those countries where there is a high demand for shrimp.

New lines of peeling machines were installed on two vessels and the exploitation model was worked out and tested in the years 1984–87.

Commercial catches of krill could start after two conditions are met:

- krill tail-meat would have to be accepted on the market as an analogue or substitute for shrimp,
- the ex-vessel price paid for this frozen product should exceed US$3,500 per metric ton (at 1987 prices) to cover the unit costs of products.

If any other less efficient model of exploitation, different from that devised by Polish enterprises is adopted, or if the crew's remuneration is as high as it is on US vessels, the minimum selling price would probably need to amount to US$5,000 per metric ton.

The model of krill exploitation discussed here addresses the high-quality, low-volume production for human consumption, causing no disturbance in the ecosystem of the Antarctic. It is believed that within the next 10 years, the crucial issue of consumers' acceptance of the analogue of the small shrimp will finally be resolved either positively or negatively.

Societal response to climatic change

The rationale of the response

In an attempt to make use of some past occurrences, when organizing a wide societal response to climatic change, we should also accumulate maximum information about the latter. What is known up to now is this (put very briefly): If the emission of greenhouse gases remains constant at the 1985 level, carbon dioxide concentration would reach about 440 parts per million by volume (ppmv) by the year 2100, as compared with the 350 ppmv of today and 290 ppmv a century ago (US EPA, 1989). The doubling of CO_2 would contribute to an increase of the global average temperature by some 2 to 3.8°C.

Because of a number of uncertainties as to the intensity of future emissions and several natural feedbacks occurring in the biosphere, the expected average warming would lie within the range of 1.2 to 5.5°C (US EPA, 1989). Furthermore, "The temperature increase would not be distributed uniformly among regions: if there is a 2°C warming near the equator there might be a 6–10°C warming (annual average) near the poles" (Glantz & Ausubel, 1988, p. 120). The latter occurrence alone would have serious consequences. Oceanic circulation is mainly wind-driven – if the meridional gradient of air temperature drops, the oceanic circulation might become less intensive. The horizontal motion of water masses could be disturbed, deviating from its usual direction or changing the geographical range. This would also lead to less vertical mixing of water. Some coastal upwellings may become weaker or may shift, re-emerging in other places. In consequence, the distribution and intensity of the inflow of nutrients to the surface could change. Nutrients constitute the critical component of oceanic productivity and are particularly scarce at lower latitudes. The boundaries of some ecosystems could expand or shrink and this may lead to the genetic mixing of some populations. Warmer water could pass beyond the thermal tolerance of some finfish or shellfish. By changing the temperature or other properties of the ecosystem, some species may vanish from a niche and others may enter that ecological niche (Glantz & Feingold, 1990).

In general, one may assume that global warming may shift fisheries productivity northwards in the Northern Hemisphere and southwards in the Southern one. It is most likely that governments will react in two (conflicting) ways in response to the change in distribution of living resources in the world's oceans. In particular, some fishery administrators would be very keen to learn, first, whether the fishermen in their countries will gain or lose something as a result of the shift of fish stocks before they initiate any action in response to the global warming. Others will hold views that a premature division between those who gain and those who lose will do nothing but inhibit the internationally consolidated effort in combating the increasing CO_2 emission. For example, as Glantz states: "This perception about potential benefits as well as adversities associated with the yet-to-be-identified regional impact of a global warming will clearly serve to constrain international cooperation on the CO_2 issue" (Glantz, 1988, p. 47). The last viewpoint appears to be correct. The issue, as a whole, is awesome, and there are no reasons whatsoever to hesitate or postpone an internationally regulated response to climate change. The strengthening of international cooperation, by its very nature, does not hinder the intensification of national and international research, the results of which should be accessible to all cooperating parties. A tiny example of a national attempt, in this respect, made *ad hoc*, is presented below.

Changes in primary production

Comprehensive changes which occur in the entire flow of organic matter and energy through oceanic food webs, as a result of rising air temperature, cannot be estimated. Some processes are poorly understood and, therefore, are not reliably quantifiable. On the basis of existing data, one could crudely assess the changes in one parameter, that is, primary production, assuming that the air temperature would increase by 2.5°C on average and by 7°C in the polar regions. This kind of rough assessment was carried out especially for this chapter by Dr. B. Wozniak, oceanologist. Data collected from 1,000 monitoring stations were processed and analyzed. They include, among others, the following: underwater irradiance, temperature in the euphotic zone, and content of organic and inorganic nitrogen. Phosphorus was ignored so as not to complicate the matter, as well as due to the lack of data. Two forecasting models were applied in the course of this assessment. The first one relates to changes in primary production caused by temperature only. The second one, apart from the increase in temperature, allows for a simultaneous reduction of the content of total nitrogen. The outcomes of the computations have been produced in the form of tables and diagrams. They distinguish the standard climatic zones used by oceanologists. In each zone three types of areas have been singled out: eutrophic, mezotrophic, and oligotrophic. The smallest change ranging from zero to minus 10 percent in primary production would be expected in the equatorial zone (from 10°N to 10°S), if the single factor model (temperature) turns out to be credible. If both models which allow for changes in the temperature and total nitrogen content are used, a reduction of primary production ranging from 5 to 21 percent in the eutrophic areas of the equatorial zone could be expected. Consequently, in oligotrophic areas of the equatorial zone, a reduction from 0 to 23 percent could likely occur. In the sub-Antarctic zone, on latitudes from 40° to 50°S, which is mostly oligotrophic, the increment of primary production could amount from 6 to 12 percent or 6 to 18 percent, respectively, depending on the model. The biggest growth of primary production could be envisaged between latitudes from 50° to 70°N and from 60° to 70°S.

The models are far from accurate, and some reservation regarding the very concept might also be expressed. Nevertheless, for the purposes of an exercise, we wish to submit the results of computations related to the two selected areas, out of a total number of 10 with which the Polish distant-water fishery is concerned.

The first area constitutes a part of the Bering Sea shown on the map of Fig. 13.15. The second area stretches around the Falkland Islands (Malvinas). Figure 13.16 points to the British Zone, to the median line dividing the Falklands from Argentinian waters, the zone claimed by Argentina in order to incorporate the Falklands into its territory, and in addition presents the UK–Falkland interim conservation zone, which is explicitly defined and well monitored. The outcomes of the exercises, completed with the use of both models for the two areas, are shown in Table 13.6.

The assessments and the models do have some faults and deficiencies. The distribution of monitoring stations has not been adequate to secure the required accuracy. The quality of data is

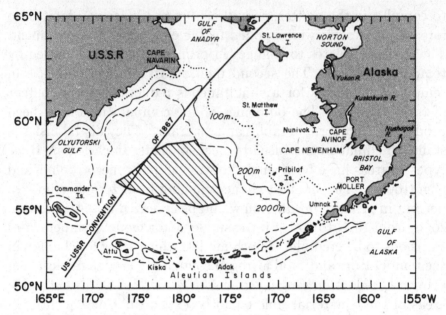

Fig. 13.15 Fishing grounds exploited by Polish distant-water vessels, Bering Sea.

not perfect in regions of very small or very high concentrations of nitrogen (below 0.3 and above 3 mg/liter). The models could not include other factors of a local character that influence primary production – factors which are often inextricable. Furthermore, one may also object that the assessment of primary production is not the same as the estimates of biomass of the secondary or tertiary carnivores that the fishery administrator would like to have on his desk for an easy decision in defining the total allowable catch. It really is not an assessment of biomass of a target fish we want to catch, but only a small step forward in constructing a general picture of what may occur in the future in the marine environment. It should be mentioned that similar computations carried out for some other regions in low latitudes are rather disappointing; the overall picture is far from optimistic. Based on this first, provisional, crude assessment of the primary production of the world's oceans, one would be inclined to assert that nobody will gain from global warming.

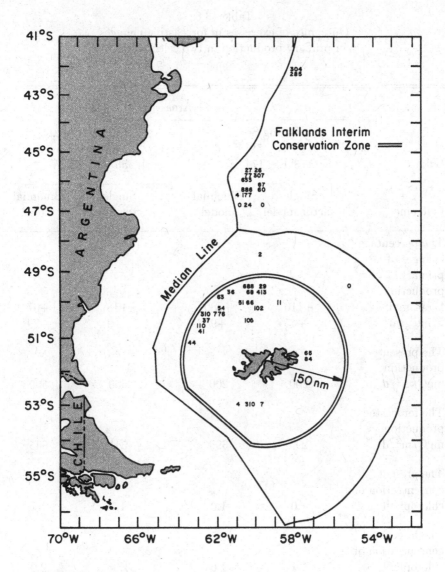

Fig. 13.16 Fishing grounds exploited by Polish distant-water vessels, around the Falkland Islands. (Numbers denote metric tons of catches extracted in May 1987.)

Table 13.6
Outcomes of exercises in forecasting changes
in primary production in two selected areas

	Area			
	Bering Sea 55°N – 59°N 173°E – 174°W		Falklands area 45°S – 54°S 55°W – 63°W	
Outcome	Single-factor model	Binomial model	Single-factor model	Binomial model
The percentage changes of primary production				
E optimal*	+110	+280	+18	+67
E integral**	+45	+95	+9	+29
The present production mgC/m²/d	200	200	500	500
The forecasted production mgC/m²/d	290	390	541	645
The present concentration of chlorophyll	1.0	1.0	1.7	1.7
The forecasted concentration of chlorophyll	1.1	1.6	1.55	2.0

*on the optimal depth
**in water column

Findings and comments

Three case studies have been presented in this paper with the aim of possibly using them in future scenarios of societal response to global warming.

1. As a consequence of establishing EEZs, the managers of the Polish distant-water fishery decided to find new accessible fishing grounds. Two big and dramatic shifts of fishing effort turned out for some time to be a success, in terms of nominal catches. This experience could serve as an analogy for future dealings with external climate-related stress/distress in fisheries, provided two conditions are met:

- *First, decision-makers should have available the appropriate level of technology required for catching and processing different species in remote areas. Second, political arrangements should permit entry into the appropriate EEZs.*

- *The transferring of investment capital to other parts of the world, after the collapse of the Californian sardine fishery (see Ueber & MacCall, this volume), bears a resemblance to the Polish shifts of effort. A good number of specific investment projects could be developed under the heading "shift" or "transfer," when the need arises, drawing on some past events in fisheries.*

2. A different approach to the one mentioned above was the experimental catching and processing of herbivores, specifically the Antarctic krill. The experiment, still in progress, if fully successful, would mean a sort of technological breakthrough in using the fauna on the lowest trophic level for direct human consumption. The experiments carried out during more than 10 years became year-by-year more selective and concentrated, focusing eventually on low-volume, high-quality production. The issue rests not so much on the technology of quick peeling and freezing but on the consumers' acceptance of krill-tail-meat as an analogue for small shrimp. This crucial issue will most likely be clarified during the next decade in the course of market surveys.

- *The lesson arising from the krill study may turn out to be instructive: in the face of global warming, people should not reject outright endeavors aimed at utilizing the lower trophic levels of marine resources, which are enormous as compared to the upper levels that we utilize and consume at present.*

3. The story of FFPCs in Poland had a very disappointing ending. This project implemented in the 1970s collapsed and the idea was completely abandoned. The impediments in utilizing trash fish for direct human consumption instead of for fish meal for animal feed supplement are really vast. The decision-makers responding to global warming must be able to distinguish between fake inventors and genuine explorers, and misplaced analogies and valid ones.

Commenting on the above findings, it is important to emphasize that none of these three cases (and many others by different authors) could be transformed through analogical reasoning to workable measures in responding to global warming if we ignore: (a) the characteristics of marine fisheries, and (b) the fact that not only the specific elements of interest are affected but the entire setting is subject to change.

Changes in the setting (i.e., the marine environment) will gradually be recognized. As to fishery characteristics, one cannot forget that the resources of the world's oceans are limited and catch per unit of effort is declining. Common property resources within the EEZs and poor management in many areas too often lead to either overcapitalization or overfishing. The level of technology is extremely differentiated in many countries. The level of investment uncertainty, already high in the past, is increasing nowadays, as world catches approach their biological limits.

So far, more attention has been paid to uncontrolled development than to appropriate fisheries management and the evolution of regulatory measures toward more efficient ones. Quite often, management practices concentrate on individual stocks, assessing only the ratios of recruitment and mortality with little regard to fish habitat. The fishery policies of specific countries are usually regulated by three main factors: (1) type and dynamic of the fish stocks, (2) the importance of the fishery production in the national economy, and (3) the prevailing socioeconomic environment in a given country. Thus, we cannot expect homogenous views and actions from different regions around the globe in coping with global warming. Some governments will act according to a "nothing to lose" belief. Others will wait indifferently because of the lack of a "dread factor." Still others might simply panic.

In alleviating regional differences and making societal responses more effective, two kinds of activities would be most welcome.

- *International cooperation should be strengthened through relevant institutional arrangements. The existing 29 international fishery bodies should be geared to act in a unified manner. Their statutory capacities and responsibilities should be extended. Perhaps the bodies concerned could establish a coordinating committee.*

- *Knowledge about the consequences of global warming must be widely disseminated among fisheries administrators and managers. And finally, it would be beneficial if a world-wide pressure group of fishermen's communities were established to combat the increasing emission of greenhouse gases.*

References

Budzinski, E., Bykowski, P. & Dutkiewicz, D. (1985). Possibilities of processing and marketing of products made from the Antarctic krill. FAO Fisheries Technical Paper No. 268. Rome: UN FAO.

Glantz, M.H. (1988). Politics and the air around us: International policy action on atmospheric pollution by trace gases. In *Societal Responses to Regional Climatic Change: Forecasting by Analogy*, ed. M.H. Glantz, pp. 41–72. Boulder: Westview Press.

Glantz, M.H. & Feingold, L.E. (1990). *Climate Variability, Climate Change and Fisheries*. ESIG/NMFS/EPA Study. Boulder: National Center for Atmospheric Research.

Jamieson, D. (1988). Grappling for a glimpse of the future. In *Societal Responses to Regional Climatic Change: Forecasting by Analogy*, ed. M.H. Glantz, pp. 73–93. Boulder: Westview Press.

Thurow, F.R. (1978). *The Fish Resources of the Baltic Sea*. FAO Fisheries Circular, 708. Rome: UN FAO.

UN FAO (1983). Objectives, policies and strategies for fisheries development. 15th Cooperative Oceanic Fisheries Investigations, COFI/83/3. Rome: UN FAO.

UN FAO (1984). *Report of the FAO World Conference on Fisheries Management and Development*. Rome: UN FAO.

UN FAO (1987). World Fisheries Situation and Outlook. 17th Cooperative Oceanic Fisheries Investigations, COFI/87/2. Rome: UN FAO.

UN FAO (1989). Trends and prospects for future capture: fisheries and aquaculture in the next 25 years and the role of the FAO. COFI/89/2. Rome: UN FAO.

US EPA (1989). Policy options for stabilizing global climate. Draft Report to US Congress. Executive Summary. Washington, DC: US EPA.

14

Climate-dependent fluctuations in the Far Eastern sardine population and their impacts on fisheries and society

TSUYOSHI KAWASAKI

Faculty of Agriculture
Tohoku University
Sendai 981, Japan

Historical overview of the Far Eastern sardine fishery

Trends in catch

The Japanese sardine fishery has had a long history since the beginning of the Tokugawa era (1600–1867). There have been six peaks in sardine catches since the seventeenth century; 1633–60, 1673–1725, 1817–43, 1858–82, 1930–40 (Kikuchi, 1958) and the 1980s. In the Tokugawa era, sardines were caught primarily by coastal beach seines and eight-angle lift nets along the coasts. Changes in the availability and abundance of sardine stocks caused the development as well as collapse of various coastal fishing villages.

Figure 14.1 shows trends in sardine catches in the Northwest Pacific between 1894 and 1988. The sardine catches began to increase in the 1910s and peaked in the 1930s. This major increase in sardine catches resulted from an increase in stock size, caused by an enormous spatial expansion of the range of sardine. When the sardine stock was abundant, its range was broad and it was distributed throughout the Sea of Japan and as far east as 173°W. When it was at a low level, the stock was confined to a small, coastal area along southern Japan. In the years of the most abundant sardine stocks, a large quantity of sardine was caught along the eastern coast of the Korean Peninsula and the Coast Range of the USSR, as well as along the Sea of Japan side of Japan. As seen in Fig. 14.1, sardine fishing along the Korean coast began in 1925

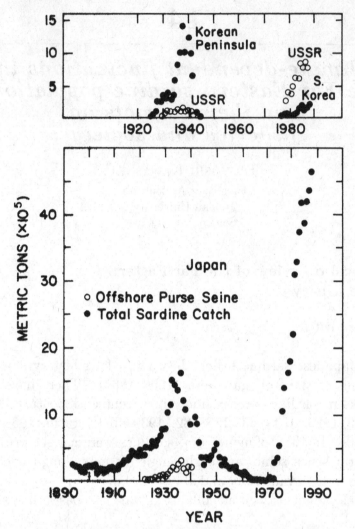

Fig. 14.1 Trends in catch of the Far Eastern sardine, 1894–1988. Statistics of
DPR Korea are not available.

reaching a peak catch of about 1.2 million metric tons (comparable to that along the Japanese coast). This peak occurred mainly
along the northern Korean coast in 1937. All fishing ceased there
in 1943, when sardine schools were no longer found in the area. On
the other hand, sardine fishing by the USSR continued between
1930 and 1946.

In the 1940s the sardine catch declined sharply and reached a
low of only 9,000 mt in 1965. Around 1970, the sardine catch
began to increase once again. At present it is at its highest level.
Substantial catches have been obtained along the coasts of South

Korea since 1976 and the USSR since 1978. North Korean catch records have not yet been made public.

Pre-war changes of fisheries

The sardine fishery during the Taisho era (1912–26) has been inseparably linked to the development of purse seines. The annual catches of sardine by offshore purse seiners larger than five metric tons in Japan was under 10,000 mt between 1915 and 1921, but increased to over 20,000 mt between 1922 and 1925. After the beginning of the Showa era (1927–88), catches increased rapidly from 50,000 mt in 1927 to 300,000 mt in 1936. These increases were made possible by targeting the sardine for the purse seine (Fig. 14.2).

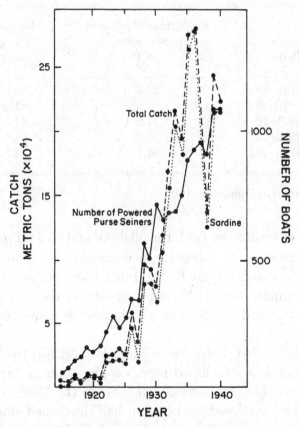

Fig. 14.2 Offshore seine fisheries around Japan, 1915–40 (Ounabara, 1980).

Between 1915 and 1940, most of the sardine catch was in the coastal fisheries; catches by offshore purse seines made up less than two percent of the total sardine catch until 1921 and only four percent each year between 1922 and 1927. As powered purse seiners were developed and brought into the fishery, the percentage exceeded 10 percent after 1928, but it was not until 1940 that it exceeded 20 percent. Average tonnage of the offshore purse seiners was 15 mt and the total number of boats, each of which was manned by a 15-person crew, somewhat exceeded 1,000 around 1940. About one-fourth of the sardine catch in those days would be ascribed to this fleet. As the sardine stock expanded, the average tonnage as well as the proportion of power-driven purse seiners in the fleet also increased (Table 14.1).

Table 14.1
Change in proportion of powered purse seiners to the
total number of offshore purse seiners in Japan

Period	Proportion of powered purse seiners (%)	Average tonnage (mt)
1915–19	20.4	60.8
1920–24	62.6	83.9
1925–29	77.6	92.3
1930–34	96.3	98.1
1935–39	97.7	98.7

Source: Ounabara, 1980

Catches of sardine by both the offshore and coastal purse seiners in 1941 were about 750,000 mt, accounting for 77 percent of the total sardine catch (about 970,000 mt). The purse seine fisheries in pre-war times, after 1926, were the core of the sardine fisheries and in this regard the former was almost synonymous with the latter.

In pre-war Korea (then under Japanese control) the purse seine fishery for sardine developed more rapidly than in Japan proper. The sardine fishery off the eastern coast of the Korean Peninsula, especially the northeastern portion, had developed since the sudden appearance of sardine there in the fall of 1923. Subsequently, catches peaked in 1939 and 1940 and declined rapidly thereafter.

Powered purse seiners, operating along the eastern Korean coast in 1927–41, increased in number from 61 to 598, or about one order of magnitude (Table 14.2). Needless to say, the tonnage and horsepower of fishing vessels also increased during this period.

Table 14.2
Number of powered purse seiners on the eastern coast of
the Korean Peninsula, 1927–41

Year	Number
1927	61
1928	112
1929	123
1930	166
1931	199
1932	139
1933	174
1934	227
1935	269
1936	435
1937	589
1938	616
1939	632
1940	621
1941	598

Source: Ounabara, 1980

In the heyday of the fishery, around 1940, fleets were composed of 50-ton vessels, equipped with an engine of 130 HP and wireless, and manned by a crew of 50 men; sometimes airplanes joined them to locate schools of fish. In 1936, 70 percent of the total sardine catch along the eastern Korean coast was obtained by the powered purse seiners. It has been said that the number of fishing boats steaming from Japan proper amounted to 110 in the fishery's heyday. This huge fishing power, however, declined with the fall of sardine stock.

Development of the fisheries in post-war times

Japan entered the period of high level economic growth in the 1960s and the purse seine fishery was developed as part of that economic growth. The fishery improved due to the invention and dissemination of labor-saving devices, by the structural reform policy for the elimination or reduction of labor, and as a result of the increase in availability of chub mackerel and the recovery of the sardine stocks.

The development of the purse seine fishery is characterized by increases in the number of, and the catch by, one-boat purse seiners. Target species of purse seiners operating around Japan are tunas, such as bluefin and yellowfin, and smaller pelagic species such as jack mackerel, chub mackerel, and sardine. The number of large- and medium-sized one-boat purse seiners increased until 1977 (Fig. 14.3), while at the same time there was an increase in

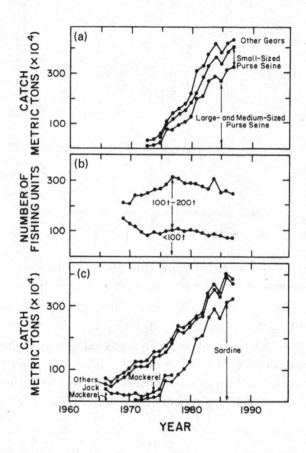

Fig. 14.3 (a) Sardine catch by fishing categories, 1973–87, (b) number of large- and medium-sized one-boat purse seiners by tonnage classes, 1969–87, and (c) catch by large- and medium-sized purse seiners by species, 1966–87.

larger boats (between 100 and 200 mt) as well as a decrease in smaller ones (under 100 mt). Since then, there has been a decrease in the number of both larger and smaller boats. This shows that the fishing capacity of each purse seiner became more powerful as a result of progress in the development of modernized fish school-locating techniques and electronic fishing devices.

Catches by the large- and medium-sized one-boat purse seiners have increased rapidly from 300,000 mt in 1967 to four million mt in 1987, a change of more than one order of magnitude. During the late 1960s, while the stock size and, hence, catch of jack mackerel were still large, there was an increase in the abundance and catch of chub mackerel. Moreover, the sardine stock began a dramatic increase in the 1970s. The catch has skyrocketed since then, reaching a level of more than three million metric tons in 1987 (Fig. 14.3). Thus, most of the recent increases in fish catches by large- and medium-sized one-boat purse seiners has been ascribed to the Far Eastern sardine. Trends in sardine catches by respective fishing categories show that most of the sardines have been caught by large- and medium-sized purse seines (Fig. 14.3).

Impacts on society of changes of the sardine fishery

A case study: Kushiro

As described earlier, a great increase in the sardinestock resulted from a major expansion of their spatial range. Impacts on society of fluctuations in sardine abundance are found most markedly on the fringes of the sardine stock's distribution. Kushiro, a city located on the eastern Pacific Ocean side of Hokkaido, a base of fishery operations, where total landings had been the highest in Japan in this period, is used as a case study (Fig. 14.4).

Figure 14.5 shows trends in landings at Kushiro since 1964 in terms of major landed species; Alaska pollock, chub mackerel, and sardine. Alaska pollock has been caught by trawlers in the Bering Sea as well as around Hokkaido. The size of the pollock catch is dependent to a large extent on the international regulations of the sea. Since about 1960, when a method to produce frozen minced fish meat (*reito surimi*) was invented, the landings of Alaska pollock at Kushiro had increased year by year until it reached about 600,000 mt in 1974, when regulations for limiting pollock catches by foreign vessels in the Bering Sea were strengthened by the US. On 1 March 1977, a regulation went into effect governing fisheries resources in the 200-mile exclusive economic zone of the US and

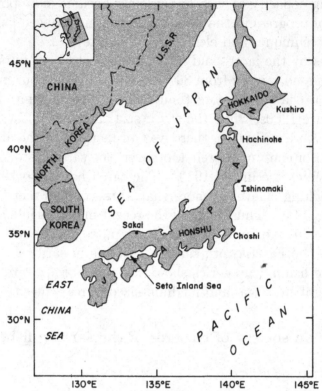

Fig. 14.4 Locations of landing ports where a large quantity of sardine is landed.

of the USSR based on their domestic laws. Landings of Alaska pollock at Kushiro dropped to about 200,000 mt in 1978 and have remained more or less at this level since then.

Chub mackerel landings caught by purse seiners at Kushiro began to increase in the early 1960s and peaked between 1970 and 1974. In 1976 a large quantity of sardine suddenly appeared in the waters off southeastern Hokkaido, replacing chub mackerel. As shown in Fig. 14.5, the sardine landings at the Kushiro Fish Market increased rapidly thereafter and by 1987 made up 65 percent of the total landings of all species. The proportion of large-quantity fish such as Alaska pollock, chub mackerel, and sardine to total landings, has risen from 42 percent by weight and 28 percent by value in 1964 to 91 percent and 56 percent, respectively, in 1987 (Fig. 14.6). Thus, the constitution of the Kushiro Fish Market has become more dependent on species with large quantity landings. The total value of landings, however, has declined recently in spite of the increase in total catch, as a result of the increase in

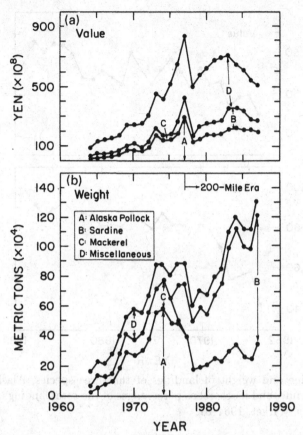

Fig. 14.5 Year-to-year change in landings at the Kushiro Fish Market in terms
of value (a) and weight (b), 1964–87. This figure shows that fishing
for mackerel finished in 1977, but the sardine fishery began in 1976.

low-value landings. Changes in utilization categories and prices of
landings at Kushiro (Fig. 14.7) also occurred for various species.

In 1964–66 the uses for Alaska pollock were categorized as
"fresh" and "miscellaneous." In 1967 their use was expanded to
include "minced fish meat" and in 1968 to include "fish oil and
fish meal." Use for "minced fish meat" has been expanding annually, while the use for "fish oil and fish meal" ended in 1978, as
the catch of the low-value sardine increased. The use of pollock as
"fresh" fish also declined and, since 1977, it has been used almost
exclusively for "minced fish meat" (*surimi*).

Between 1965 and 1968 only part of the chub mackerel landings
were used for "fish oil and fish meal" but after that time most
of their landings were used for "oil and meal," until the mackerel
population was replaced in the waters southeast of Hokkaido by

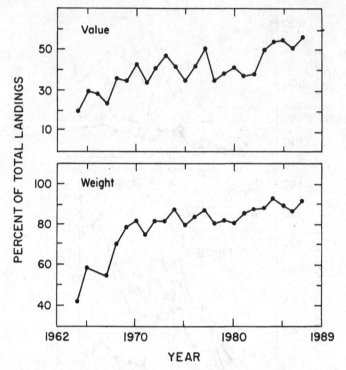

Fig. 14.6 Value and weight of landings of the three species, Alaska pollock, sardine, and mackerel as percentages of the total landings at Kushiro Fish Market, 1964–87.

a newcomer – the sardine. Most of the sardine landings which have been caught in place of mackerel have been reduced to "oil and meal." Thus, uses of large-volume but lower-value fish have become very unvaried.

Trends in mean landing prices at the Kushiro Fish Market are also interesting. Between 1964 and 1975, when fishing for mackerel off Hokkaido ended, landing prices of mackerel and Alaska pollock were almost the same – low and fluctuating between 20 and 35 yen/kg. Prices were also the same for Alaska pollock and sardine in 1976, when fishing for sardine began. The price of Alaska pollock, however, jumped sharply in 1977 to twice that of the previous year (73 yen/kg) because of the fear that its supply might decline as a result of the imposition of the 200-mile fishing jurisdiction. The price of sardine, however, remained unchanged (36 yen/kg). Since 1977, the price of sardine declined to a low of 9–10 yen/kg as its landings drastically increased. The tendency of the price of Alaska

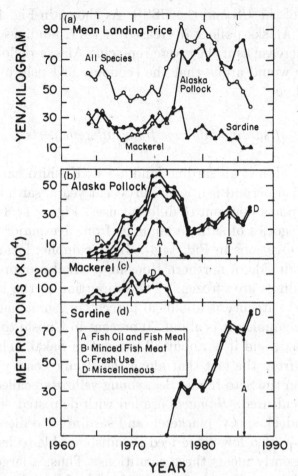

Fig. 14.7 Mean landing prices at the Kushiro Fish Market for all three species, Alaska pollock, mackerel, and sardine (a), and breakdowns of landing to different utilization categories for each species (b–d), 1964–87.

pollock has also been downward as its supply gradually increased; 56 yen in 1987, down from 79 yen/kg in 1980 (Fig. 14.7).

Thus, Kushiro has become a city which depends on large amounts of low-value sardine and Alaska pollock for its economic well-being. There are many fish processing and reduction plants in Kushiro, and many people work in these plants. Not only are facilities directly related to fisheries dependent on large-quantity fish landings but so also are many ancillary industries, such as transportation.

People worried that the economy of Kushiro would decline as a result of the possible (even likely) decrease in landings of Alaska pollock caused by the regulation of Japanese fishing in the 200-

mile zones of the US and the USSR. As shown in Fig. 14.7, when landings of Alaska pollock decreased in 1977, landings of sardine began to increase as if they were replacing Alaska pollock. In this regard, one would argue that the economy of Kushiro was saved by the sardine.

Diverse use of sardine landings from different ports

Although almost all sardine landings at Kushiro have been reduced to "fish oil and fish meal" in recent years, sardine landings from other ports were put to different uses. Figure 14.8 shows utilization categories of sardine landings from five major ports (see geographic locations in Fig. 14.4). In Hachinohe, located on the Pacific Ocean side of northern Honshu, a small proportion of the sardine landings are "frozen." The proportion of the landings in the "frozen" category is about 40 percent in Ishinomaki, located south of Hachinohe; it is about 70 percent in Choshi to the south of Ishinomaki; and it is about 30 percent in Sakai. These differences stem from the fact that along the Pacific coast of southern Japan and in the Seto Inland Sea, young yellowtail called *hamachi* are widely cultured. *Hamachi* are fed with defrosted "fresh feed" such as sandeel, saury, mackerel, and sardine. Needless to say, a stable supply of a low-value feed is indispensable to fish culture; sardine presently meets these conditions. Thus, a larger proportion of sardine landings is frozen as feed for *hamachi* at southern ports than is the case at the northern ports.

Dependence of hamachi *culture on sardine production*

Hamachi culture in Japan began in 1962. Its production had risen steeply until it reached about 160,000 mt in 1979, after which it remained almost unchanged (Fig. 14.9). This figure also shows trends in the amount of fresh feed for *hamachi* culture and trends in the production of frozen sardine. In 1973 frozen sardine made up 20 percent of fresh feed; this proportion rose to 77 percent by 1987, indicating that *hamachi* culture has become primarily dependent on the sardine.

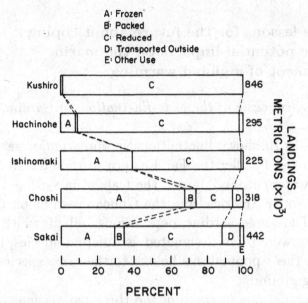

Fig. 14.8 Percentages of utilization categories for sardine landings at major fish
markets in 1987.

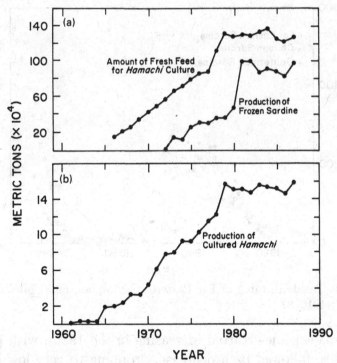

Fig. 14.9 (a) Amount of fresh feed for *hamachi* culture and production of frozen
sardine, 1966–87, and (b) production of cultured *hamachi*, 1961–87.

Possible lessons for the future about coping with the potential impacts on the marine environment of a global warming

The biological basis of the wide fluctuations in sardine

Figure 14.10 shows fluctuations in three major sardine populations in the Pacific; the Far Eastern sardine distributed in Far Eastern waters around Japan, the California sardine off the west coast of North America, and the Chilean sardine off the Chilean coast.   The three sardine populations, all of which belong to *Sardinops*, were often designated as different species, but the notion that these populations belong to the same species, *S. sagax*, is gaining ground.

The fluctuations in catch of the three populations are quite in phase with one another (Fig. 14.10). Catches of Far Eastern and

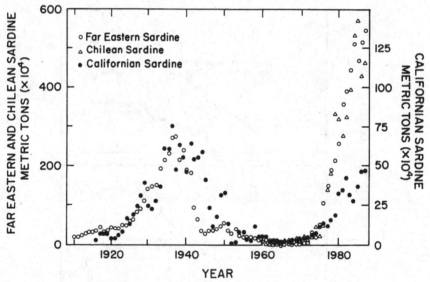

Fig. 14.10 Trends in catch of Far Eastern, Californian, and Chilean sardines, 1910–87.

California sardines started increasing in the 1920s, with peaks in the 1930s, followed by a decrease, dropping to very low levels in the 1960s.   The three populations began to increase in the 1970s

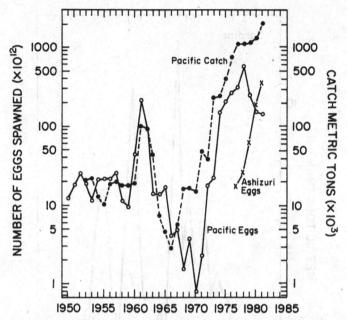

Fig. 14.11 Annual change in the estimated total number of eggs spawned and
the nominal catch of the Pacific sub-population of Japanese sardine
during 1951–81. The change in the former in the Ashizuri sub-
population is also shown for 1977–81 (Watanabe, 1983).

and have been at very high levels in the 1980s. Yet, catch differs
from abundance. In order to establish a relationship between catch
and abundance, Fig. 14.11 (Far Eastern sardine) and Fig. 14.12
(California sardine) are shown. Figure 14.11 shows estimates of
the total amount of eggs produced from Pacific and Ashizuri sub-
populations in the period 1949–82. The total amount of eggs can
be regarded as a measure of the spawning biomass. As one can
see, the trends in the amount of eggs accord well with the trends
in catch.

The estimates of the spawning biomass of the California sardine between 1790 and 1970 are derived from the scale-deposition
rate in two anoxic basins (Fig. 14.12). The trend in spawning
biomass after 1910 is basically similar to that in catch as shown
in Fig. 14.10. Figure 14.12 also suggests the important point that
even before 1910, when there was no sardine fishery, the biomass
of sardine showed a wide fluctuation. This implies that the population change in sardine resulted from natural causes and not from
fishing pressure.

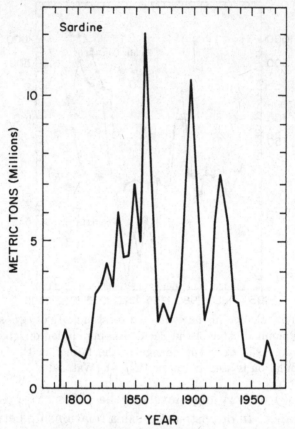

Fig. 14.12 Revised estimate of the spawning biomass of the California sardine
between 1790 and 1970 (180 years at five-year intervals) derived from
the scale-deposition rate in two anoxic basins (Soutar & Isaacs, 1974).
The Pacific sardine biomass remains the same as in Smith, 1978
(Smith & Moser, 1988).

The pattern of large-scale fluctuations
in sardine populations

Large increases in the sardine catch are the result of an increase
in biomass which is caused by a great extension of the spatial range
of sardine. In the 1980s when the sardine stock was abundant, its
geographic range was broad throughout the Sea of Japan and as far
east as 173°W (Fig. 14.13). Around 1965, when the stock was at
a low level, it was confined to a small coastal area along southern
Japan (Fig. 14.14).

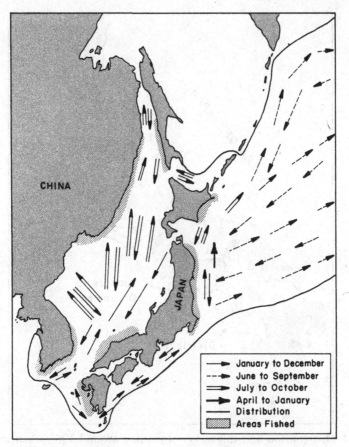

Fig. 14.13 Distribution and migration of sardine in the 1980s when the population level was very high. Areas fished are dotted.

In the period of low levels, the sardine was truly a coastal species, remaining in coastal areas and managing to maintain the species' survival. In those days it was even called the "illusional fish." In recent years with the population at its highest level, however, the range of the sardine extends throughout the Sea of Japan and eastward, well into the Pacific Ocean. According to experimental surveys, the distribution of sardine in a broad area of 35°–50°N and 155°E–173°W between June and August in 1977 to 1984 was confirmed by drift net, purse seine, and angling (Tokio Wada, 1988, personal communication). The sardine has made a dramatic transformation from a real coastal fish to a widely migrating pelagic fish. It was as if it had become a different species.

In 1988, locust outbreaks occurred in broad areas from North Africa to the Middle East, and swarms flew to many distant loca-

Fig. 14.14 Distribution and migration of sardine around 1965 when the population level was very low.

tions, devastating vegetation. In terms of biology, this event was
the result of a "phase variation." This morphologically, physiologically, and ecologically variational phenomenon is observed in
insects such as locusts and armyworms. Such insects are inactive, not making long-distance migrations, and their populations
are usually very small when their environment has deteriorated
(solitary phase). If, however, their environmental setting becomes
ameliorated, not only do morphological characteristics (including
body color) vary, but they also become very active, resulting in
their swarming over hundreds of kilometers (gregarious phase).
Entomologists once believed that the two types of locusts belonged
to separate species because they were quite different in appearance.
The outbreak of sardine populations might be considered as a fish
population version of a phase variation.

Another characteristic strategy of the sardine is its feeding niche. As Hyatt (1979) stated, most of the marine teleosts are carnivorous; there are few herbivores, in particular planktivores, among them. The sardine is one of the very rare species that can take phytoplankton with its elaborate filtering apparatus – gill-rakers. It can be said that when the population level of sardine is low, therefore, phytoplankton in broad open oceans are not directly utilized by fish. When the environment takes a more favorable turn, however, the sardine is transformed into a widely-migrating pelagic fish, dramatically extending its range by grazing over the open oceans. This leads to sharp increases in sardine populations.

A marked characteristic observed in the phase variation of the sardine is a great change in its growth rate as well as the expansion and contraction of its spatial range. As seen in Fig. 14.15, strong

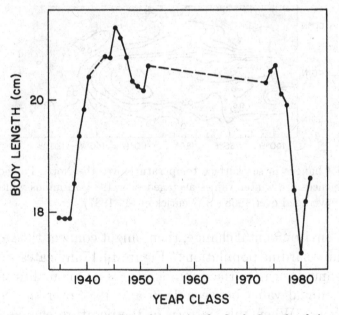

Fig. 14.15 Year-class-to-year-class change in the body length of three-year-old Far Eastern sardine (dashed line signifies lack of data).

year classes of the Far Eastern sardine, such as 1935–37, exhibit slow growth rates and three-year-old fish were very small, with a mean length of less than 18 cm. As the stock size declined thereafter, fish became larger year by year and three-year-old fish of the poor 1941–76 year classes were very large, with a mean length of more than 20 cm. The sardine population once again increased and the three-year-old fish became very small.

Causes of large fluctuations in sardine populations

It is difficult to explain fluctuations in each sardine population by changes in the local or regional environment in which it lives. Winter values of sea surface temperatures (SSTs) averaged over the years 1980–85, minus winter values averaged over 1968–73 in the North Pacific (Fig. 14.16) show that, although SSTs decreased in the middle and western part of the mid-latitudes, they rose in the eastern areas. However, the two sardine populations in the Far Eastern areas and off the California coast increased rapidly between the two periods cited above. It is difficult to believe that different populations of a single species of sardine would react in opposite ways to the change in SSTs.

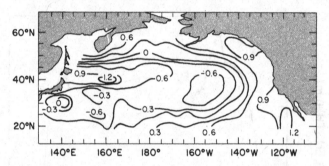

Fig. 14.16 Changes in sea surface temperature over the North Pacific. Differences are winter values averaged over 1980–85 minus winter values averaged over 1968–73 (Venrick et al., 1987).

What environmental change, then, might connect the variations in the three sardine populations? Figure 14.17 indicates variations in the annual mean surface temperatures as anomalies from the average annual value between 1876 and 1987 over the Northern Hemisphere. When this pattern of temperature change is compared with variations in catch, hence in abundance, in Fig. 14.10, one can see that the temperature changes are quite in phase with the population change. To examine the relationship somewhat more quantitatively, Fig. 14.18 shows a clear relationship for the Far Eastern sardine – when the temperature anomalies were positive, catch (abundance) was high. If variation in temperature is taken as a measure of climatic change, this could suggest that population change is affected by global climatic change.

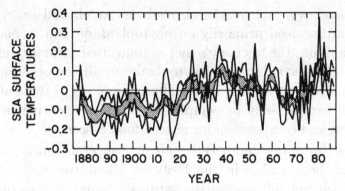

Fig. 14.17 Trends in anomaly of average surface temperatures in the Northern Hemisphere, 1876–1987. Two thin (thick) lines represent upper and lower bounds on annual (five-year running) average surface temperature anomaly (Meteorological Agency of Japan, 1989).

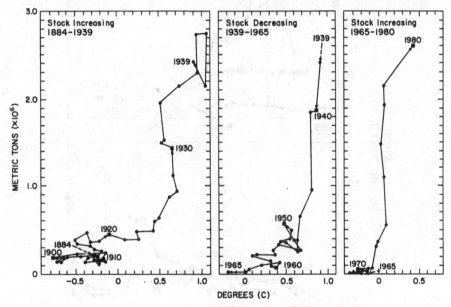

Fig. 14.18 Relation of Far Eastern sardine catch to anomaly of mean surface air temperatures. Circles indicate mean values for respective year and squares indicate values 10 years apart (Kawasaki & Omori, 1988).

Dependence of variations in other
pelagic stocks on sardine variations

There is a pelagic fish community composed of several large biomass plankton-feeding fish in each temperate area of the world's oceans. In Far Eastern waters the pelagic community comprises sardine, anchovy, jack mackerel, chub mackerel, and saury. Although sardine feed primarily on phytoplankton and secondarily on zooplankton, the anchovy's diet is comprised primarily of zooplankton and secondarily of phytoplankton; all the other species feed exclusively on zooplankton. Thus, the foods of these species overlap either partially or completely, possibly resulting in close interaction and competition for shared food resources.

Catches from the temperate pelagic fish community around Japan had been stable in the 1960s at about two million metric tons. During this period the sardine population had been at a very low level (Fig. 14.19). In the 1970s, however, the sardine

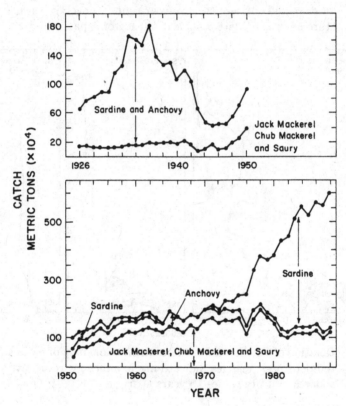

Fig. 14.19 Trends in catch of sardine, anchovy, jack mackerel, mackerel, and saury around Japan, 1926–88.

population began to rise steeply and, as a result, the catch from the community increased drastically, approaching six million metric tons in 1982. Through this process, a decrease in catch of fish other than sardine (from 2 million mt to 1.3–1.5 million mt) occurred, possibly indicating that it was caused by population pressure imposed by the sardine on other species. This suggests that the sardine may be the key species in the temperate pelagic fish community around Japan. It is evident from Fig. 14.19 that such variations of the pelagic fish community also took place from the 1930s to the 1940s, implying that the temperate pelagic fish community is strongly influenced by the sardine qualitatively as well as quantitatively.

As seen in Fig. 14.20, the total catch of all marine animals from the Pacific had increased from about 24 million mt in 1965 to over 46 million mt in 1985. At the same time, the sum of the catches of the three sardine populations increased from 80,000 mt, only 0.3 percent of the total catch from the Pacific in 1965, to over 10 million mt, one fourth the Pacific total in 1985, indicating how great an effect the variation in sardine populations can have on fisheries production in the Pacific.

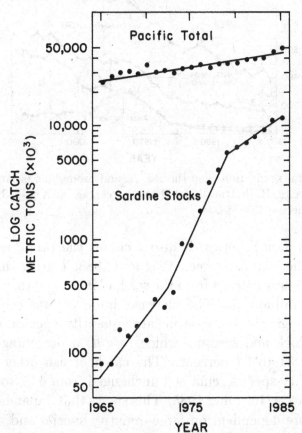

Fig. 14.20 Trends in catch of all marine animals and that of sardine in the Pacific Ocean, 1965–85.

Possible lessons for the future

- *How might we forecast the future of sardine populations?*

As shown earlier, the increase in the recent catch of sardine has had an enormous impact on Japanese fisheries. Figure 14.21 indicates the production of large-quantity species such as Alaska pollock, sardine, and mackerel, and the total production from the

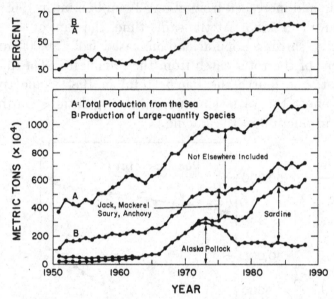

Fig. 14.21 Total production from the sea, A, and production of large-quantity species, B, (bottom); and B as a percentage of A (top) by Japanese fisheries, 1951–88.

sea by Japanese fisheries; it also includes the proportion of the former to the latter, between 1951 and 1988. During this 38-year period, Japanese catches from the sea had increased threefold from 3.8 to 11.3 million mt. The increase, however, was achieved primarily through the increase in large-quantity species, especially Alaska pollock and sardine, which rose as a percentage of total catch from 31 to 64 percent. The catch of fish other than the large-quantity species remained unchanged from 4.07 to 4.05 million mt between 1961 and 1985. This shows that Japanese fisheries became more dependent on large-quantity species and, therefore, vulnerable to societal and environmental changes. If, at present, the stock of Far Eastern sardine starts declining quickly, the total catch of Japan's fisheries could possibly be reduced to as little as

7 million mt in about 10 years, down from 11.3 million mt in 1988. This would be similar to the years from 1936 to 1945, when the total catch decreased from 4.2 to 1.8 million mt.

To protect the fisheries and society from dislocations caused by a sudden decline in sardine landings and in order to try to mitigate its impacts, it is very important to know the future of the sardine stock as far ahead as possible. To this end, it is essential to raise the level of fisheries science and to continue to collect key data about fisheries and the global environment.

- *How can we absorb the shock of sudden drops in sardine landings?*

If the landings of sardine drop considerably in Japan, needless to say, Japanese society as well as its fisheries will be seriously adversely affected. In major sardine landing ports, such as Kushiro, Hachinohe, Ishinomaki, Choshi, and Sakai, many plants and facilities dependent on sardine landings would face economic hardships. Fish culture, especially of *hamachi*, would lose its supply of low-cost feed; clearly local, regional and national economic problems would ensue. Measures to cope with problems such as these must be sought in the event that the sardine population collapses once again.

Reorganization of the purse seine fisheries
after World War II

As shown in Fig. 14.10, an abrupt decline in sardine catch began in the 1940s, after its peak at 1936 and 1937, but the degree of decrease in catch varied from region to region. The rate of decrease was higher in more peripheral regions of the range of sardine, which was centered on the area west of Kyushu where the sardine seemed to spawn and develop (Table 14.3). The big drop in sardine migration in the waters around northern Japan and the increase in the East China Sea caused a great change in the regional distribution of purse seine fisheries, the major target of which was the sardine.

An absolute shortage of food in Japan immediately after the end of World War II brought about a temporary prosperity of fisheries. As a result, the purse seine fisheries, which had been heavily damaged in wartime, recovered quickly. As seen in Table 14.4, as early

Table 14.3
Regional trend in sardine catch

Statistical sub-area	Average annual catch (× 1,000 mt)		
	between 1934–36 (A)	between 1948–50 (B)	B/A (%)
Around Hokkaido	368	14	3.8
Pacific side			
North	396	17	4.2
Intermediate	295	63	21.4
South	79	46	59.0
Japan Sea side			
North	335	28	8.5
West	39	20	51.3
East China Sea	155	227	146.5
Seto Inland Sea	47	54	115.2
TOTAL	1714	469	31.5

Pacific North: between Aomori and Ibaraki
  Intermediate: between Chiba and Mie
  South: between Wakayama and Miyazaki
Japan Sea North: between Aomori and Ishikawa
  West: between Fukui and Yamaguchi
  Source: Ounabara, 1980

Table 14.4
Change in number of fishing units of purse seiners
by tonnage categories

Year	smaller than 10 mt	10–20 mt	larger than 20 mt	Total
1939	629	537	125	1291
1947	805	470	112	1387
1950	1587	702	606	2895

Source: Institute for Fisheries Research, 1953a

as 1947 the number of purse seiners exceeded what it had been in 1939, reaching about twice that number by 1950.

The number of purse seiners larger than 20 mt in 1950 had increased to four times as many as that in 1939, indicating that the average size of boats got larger (Table 14.4). Although the total number of purse seiners throughout Japan almost doubled between 1939 and 1949, when we look at the regional breakdowns, their numbers in statistical sub-areas (around Hokkaido, Japan Sea North and West) decreased, those in Pacific North remained nearly unchanged, but those in Pacific South and East China Sea increased markedly, strongly affected by the change in geographical distribution of the sardine (Table 14.5).

Table 14.5

Change in number of purse seiners by tonnage categories
(regional breakdown)

Statistical sub-area	Year	Smaller than 10 mt	10–20 mt	Larger than 20 mt	Total
Around Hokkaido	1939	6	42	1	49
	1949	0	2	–	2
Pacific					
North	1939	16	160	95	271
	1949	67	93	137	297
Intermediate	1939	154	125	28	307
	1949	115	166	71	352
South	1939	151	16	–	167
	1949	525	70	7	602
Japan Sea					
North	1939	3	–	–	3
	1949	–	1	1	2
West	1939	16	5	–	21
	1949	–	3	1	4
East China Sea	1939	178	176	1	355
	1949	237	73	266	626
Seto Inland Sea	1939	110	8	–	118
	1949	200	1	73	274
TOTAL	1939	634	532	125	1291
	1949	1194	484	451	2159

Source: Institute for Fisheries Research, 1953b

Another important issue relates to the reorganization of purse seine fisheries after World War II. Huge enterprises that had prevailed under national protection in the overseas and colonial fishing grounds in pre-war times, and had seldom competed with domestic intermediate and small fishery enterprises, came into direct conflict with these enterprises in coastal fishing grounds after World War II (as a result of restriction of their operating areas). In those days, each purse seiner was permitted to operate only within a small local area in accordance with fisheries regulations. Top-ranking ones among the intermediate and small enterprises as well as the huge enterprises demanded revisions of the fisheries regulations then in effect, so as to allow a purse seiner to fish unrestricted throughout the coastal areas around Japan. This demand was in conflict with the desires of the traditional, native intermediate or small enterprises, which tried to continue their fishing activities by rejecting newcomers, since their economic power was too small to compete with the huge enterprises of markets.

In 1951 the Japanese government's Fisheries Agency issued a new regulation for the purse seine fishery, which was to:

- change the system of fishing areas from regional to national, allowing any purse seiner to fish anywhere around Japan;
- permit the building of additional large-sized boats;
- permit purse seiners to operate at night in the East China Sea;
- establish closed areas along the coasts to protect native fisheries; and
- permit the building of new boats if smaller enterprises are managed jointly.

As shown in Fig. 14.22, more purse seiners were built in the statistical subarea Pacific North after World War II, with a peak in 1947. As the sardine stock declined there (Table 14.3), however, new shipbuilding declined rapidly. Measures to establish a financing system to assist the purse seine fishery in this area were discussed in the Japanese Diet at the end of 1948. The Fisheries Bill System for the purse seine fishery was established in January 1949. This led to the nationwide Fisheries Bill System six months later. The subareas of Japan Sea West and the East China Sea had their peaks of shipbuilding later in 1950, indicating that investment had continued until then (Fig. 14.22).

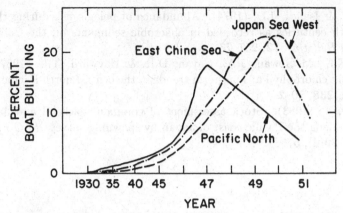

Fig. 14.22 Regional differences in trends in fishing boat building. (Institute for Fisheries Research, 1953a.)

Acknowledgments

I wish to thank Professor K. Ounabara, Tokyo University of Fisheries, for his kind cooperation.

References

Hyatt, K.D. (1979). *Feeding Strategy, Fish Physiology III*, pp. 71–119. San Diego: Academic Press.

Institute for Fisheries Research (1953a). *Structural Change in the Japanese Fisheries after World War II*. Mimeo. Tokyo: Institute for Fisheries Research.

Institute for Fisheries Research (1953b). *Purse Seine Fishery and Its Running*. Tokyo: Institute for Fisheries Research.

Kawasaki, T., & Omori, M. (1988). Fluctuations in the three major sardine stocks in the Pacific and the global trend in temperature. *International Symposium on Long-Term Changes in Marine Fish Population* in Vigo, Spain in 1986, pp. 37–52.

Kikuchi, T. (1958). A relation between the alternation between good and poor catches of sardine and the establishment of Shinden and Naya villages. Memorial Works dedicated to Professor K. Uchida, pp. 84–92.

Meteorological Agency of Japan (1989). *Recent Anomalous Weather and Climatic Change: Its Actual State and an Outlook.* Tokyo: Meteorological Agency of Japan.

Ounabara, K. (1980). A short history of purse seine fisheries. In *History During the Decade*, ed. All-Nation Purse Seine Fisheries Association, pp. 1–211. Tokyo: All-Nation Purse Seine Fisheries Association.

Smith, P.E. & Moser, H.G. (1988). CalCOFI time series: an overview of fishes. *California Cooperative Oceanic Fisheries Investigations Reports*, **29**, 66–78.

354   *T. Kawasaki*

Soutar, A. & Isaacs, J.D. (1974). Abundance of pelagic fish during the 19th and 20th centuries as recorded in anaerobic sediments off the Californias. *Fishery Bulletin*, **72**, 257–73.

Venrick, E.L., McGowan, J.A., Cayan, D.R. & Hayward, T.L. (1987). Climate and chlorophyll a: long-term trends in the central north Pacific Ocean. *Science*, **238**, 70–2.

Watanabe, T. (1983). Stock assessment of common mackerel and Japanese sardine along the Pacific coast of Japan by spawning survey. *FAO Fisheries Report*, **29**(2), 57–81.

15

The Peru–Chile eastern Pacific fisheries and climatic oscillation

CÉSAR N. CAVIEDES and TIMOTHY J. FIK

Department of Geography
University of Florida
Gainesville, FL 32611, USA

There is little doubt that the fisheries of the Humboldt (Peru) Current (Fig. 15.1) represent the largest in the world, not only if one considers the primary production (Paulik, 1971) and catch potential of its waters (Sharp, 1987), but also in view of record-setting catches. During the 1960s and early 1970s, these catches comprised nearly 20 percent of the world's landings (Fig. 15.2). Yet, the Humboldt Current that supported such high volumes of fish and nourished an enormous marine ecosystem has been constantly beset by oceanic–climatic oscillations – known as El Niño–Southern Oscillation (ENSO) – that noticeably depressed biological productivity and yield levels.

There have also been ecological perturbations in the historical or geological past, as documented in the pioneering work of Schweigger (1959) which offers revealing details of fish, bird, and mammal mortality resulting from pre-1972 El Niño events. DeVries (1987) also provides information about geological findings that hint at past catastrophic perturbations along the western coast of South America. Recent ENSO events have resulted in remarkable variations of fish production in the coastal waters of Peru and Chile, measured in terms of fish landings and biomass estimations. It is certain that the depressed Peruvian catches since 1972 reflect the simultaneous occurrence of overfishing and El Niño (Instituto del Mar del Perú, 1981).

Considering the intensity of ENSO events in recent times, particularly in 1972–73 and 1982–83, a dominating thought has begun to develop among ecologists and earth scientists that ascribes to changing global environmental conditions the marked changes in primary productivity and fish population levels that have occurred in eastern boundary current fisheries (see Sharp, 1987; Steele,

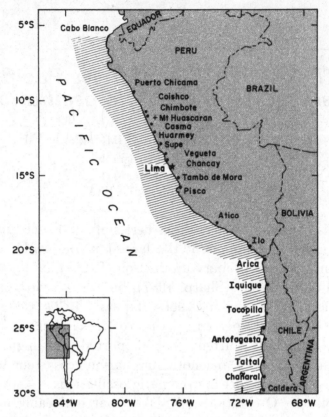

Fig. 15.1 Major fishing ports and biological production areas along the coasts of Peru and Chile.

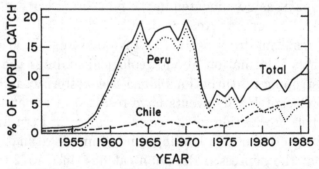

Fig. 15.2 Southeast Pacific fisheries output, 1951–86.

1989). This hypothesis, although supported by some initial climatic/oceanic indicators, requires additional testing in order to verify the assumption of a secular change of oceanic–climatic conditions in the Humboldt Current system.

A first step, when proposing that the dwindling fisheries of Peru and Chile are related to oceanic–climatic changes, is to determine whether a steady change of conditions has occurred. It could be that only a long-term fluctuation, by no means of permanent character, has upset the system during the last decade and that reduced fishery capacity is a reversible situation. The collapse of the anchovy fisheries in the waters of Peru and Chile and the surrounding circumstances, other than just the oceanic–climatic variables, must also be carefully assessed. In this sense, the study of time series of catches illustrating the dramatic growth of the eastern tropical Pacific pelagic fisheries during the 1950s and 1960s might uncover disruptive processes which may not have been entirely induced by nature. An answer to a third question to be addressed requires the availability of a long time series to check if we might not perhaps be witnessing the short segment of a biological cycle with a long periodicity in some pelagic species, the total effect of which cannot be uncovered using short time series (i.e., from the mid-1950s to the present). Research suggests the existence of natural cycles greater than 50 years in the abundance of Pacific clupeids (Kawasaki, 1983; Sharp et al., 1984; see also Kawasaki, this volume). Therefore, the possibility that the contraction of some species may possibly originate from these cycles and not from irreversible environmental changes cannot be ruled out.

Using statistical methods, this chapter examines the trends of Peruvian and Chilean fisheries since the introduction of industrial fishing activities. The procedure allows for the recognition of models that can be fitted to that growth and reveals the existence of natural events ("interventions") that may have affected the observed trends. A similar treatment is applied to available oceanic and meteorological data to detect periodicities, changes, or oscillations that might account for the dwindling fish stocks.

Development of Peruvian and Chilean fisheries

Until 1950, most of the fish extracted from the waters of the Humboldt Current along the western coast of South America was destined for consumption. Artisanal fishermen provided the local markets with mostly neritic fish, since transportation from the fishing ports to centers of consumption was very limited. Still, some canning plants were scattered along the Peruvian and northern Chilean littoral, specializing mostly in tuna, bonito, and sardines for domestic consumption. In 1950 the volume of fish captured by Peru and Chile was 86,723 and 73,500 mt, respectively. Hake accounted for 38 percent of the Chilean captures, while herring made up 13 percent of the Peruvian landings. When the first fish-meal processing plants opened on the coast of central Peru, in 1953, landings rose to 111,700 mt, with anchovies accounting for 40 percent of all captures. That same year, Peruvian captures surpassed those of Chile by 4,900 mt. By 1962 Peru had become the number one fishing nation in the world (by volume, not value, of landings).

The impressive growth of Peruvian industrial fisheries was made possible by the major concentration on anchovy (*Engraulis ringens*), which constituted 90 to 98 percent of all Peruvian commercial landings between 1957 and 1976 (Fig. 15.3). The concentration of fishing effort on this species can be best understood if one considers that from 1958 to 1959, as speculative growth boosted the installation of fish-meal plants along the coast of Peru, anchovy landings more than doubled. In addition, from 1959 to 1971, the yearly increases varied between 15 and 25 percent. Only in 1965 and 1969, when relatively minor El Niño episodes occurred, did the catches drop below those of the previous year.

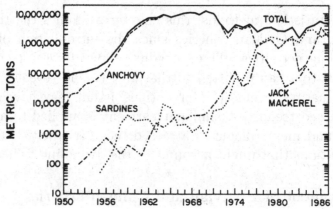

Fig. 15.3 Development of Peruvian fisheries, 1950–87.

The uncontrolled fashion in which anchovy were extracted from Peruvian waters and the contractions of the standing stocks observed in 1965 and 1969 worried international fishing authorities and conservationists. At a meeting held in Lima in 1970, a panel of experts (Instituto del Mar del Perú, 1981) critically reviewed the situation created by the frantic multiplication of fishing plants and the high volume of anchovy extraction. The panel issued a warning that, if the harvest remained above the maximum sustainable yield (MSY) of 9.5 million mt a year – including nearly 2 million mt estimated to be consumed annually by seabirds – the anchovy fishery would risk a collapse similar to those of the Sakhalin and the California sardine fisheries. These warnings were heeded neither by the fishing industry, which was fast amassing big fortunes, nor by a government that enjoyed the tax revenues and foreign exchange generated from the rising exports. Moreover, the warnings about resource depletion were also ignored for fear of reducing employment opportunities in coastal communities that had been flourishing since 1955. The extraction of anchovy continued to increase reaching 12.4 and 10.5 million mt in 1970 and 1971, respectively. Under these societal circumstances, a major El Niño developed in 1972.

Although relatively mild when compared to the global effects of the 1982–83 El Niño, the impact of the 1972–73 El Niño was devastating and long-lasting (e.g., Hammergren, 1981). In 1973, total landings dropped to 2.29 million mt. It was not until 1984 that the volume of fish landings rose above 4 million mt. The regeneration of the anchovy stock in Peruvian fishing waters has proven a painfully slow process, despite hopes and expectations to the contrary.

In Chile, the story of industrial fisheries growth has been shorter and different. The expansion from artisanal to industrial fisheries was restricted to the northern Chilean littoral, where a few fishmeal plants were established in Arica (Fig. 15.4), Iquique, Tocopilla, and Antofagasta (Salinas, 1973), in contrast to over 16 Peruvian localities with industrial processing plants, during the halcyon days of the Peruvian fishery. The increase in Chilean landings was never spectacular and reflected the contractions associated with El Niño occurrences as well as the severe fishing restrictions such as closed seasons (*vedas*) imposed by the administrations of Presidents Frei and Allende. Restrictions were imposed whenever fishing officials noticed declines in the reproduction of certain species. The 1972–73 El Niño did not spare Chile; the total level of landings, which in the 1960s hovered around 1.2 million mt, stayed below one million from 1972 through 1975 (Fig. 15.5).

In the early years of Chile's industrial expansion, anchovy was the main fish input with proportions oscillating between 80 and

Fig. 15.4 Fish-meal plants of Arica, Chile.

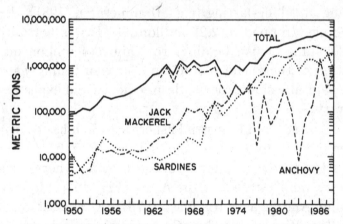

Fig. 15.5 Development of Chilean fisheries, 1950–88.

95 percent of the total catch. When the 1972–73 El Niño struck, the Chilean fisheries industry adapted to the conditions and re-structured itself around an alternative fish supply that Peru had not exploited – jack mackerel (*Trachurus murphyi*). Of secondary value to consumers, this fish had been overlooked by industrialists. It proved, however, to be a valuable resource for northern Chilean fisheries. In 1976–77, when anchovy fishing was discontinued to let the dwindling stock recover from the effects of overfishing, the jack mackerel resource was available to replace anchovies (Caviedes, 1981). This development in Chilean fisheries was accompanied by

a secondary emphasis on sardines (*Sardinops sagax*), the capture of which had been steadily increasing since 1977. By 1988, sardine had become the most exploited species, accounting for 48 percent of Chile's total landings.

Ecosystem variations induced by El Niño events

Shortly after the beginning of the development of industrial fisheries in the coastal regions bordering the Humboldt Current, it was discovered that oceanic variations along the coasts of Peru and northern Chile had a noticeable impact on living marine resources. El Niño phenomena in 1972–73 and 1982–83 revealed the severity of their impacts on fish availability and accessibility as well as on the vulnerability of certain species and the resilience of others. Furthermore, the relatively minor ENSO events of 1977 and 1986 had the marine scientists focus on the long-term effects of oceanic–climatic variabilities on fish population dynamics. Before the question of climate-induced change is addressed, it is necessary to identify the impacts of previous variations on the ecosystem of the Humboldt Current, as recorded in reliable statistics.

Prior to 1972, the areas of major anchovy concentrations along the coast of Peru coincided with upwelling centers located at latitudes 8°, 10°, 11° and 12°S. Since 1973, the centers of concentration have shifted to the south of latitude 14°S and into Chilean waters south of latitude 18°S (Santander & Flores, 1983). At the same time, there has been a noticeable increase in the sardine populations between 5°S and 12°S, an increase which Santander and Flores attributed to a warming of coastal waters and to the migration of high-seas sardines into this warmer than usual coastal environment. In terms of capture distribution, sardines have steadily grown into the mainstay of Peruvian fisheries, except for a temporary contraction between 1983 and 1985 (Fig. 15.6). Furthermore, along with the decrease of anchovies and bonito (*Sarda chilensis*) after 1973, Peruvian industrial fisheries began to rely increasingly on jack mackerel (*Trachurus murphyi*) and hake (*Merluccius gayi*), but at levels far below the traditional high landings of anchovies and sardines (Barber & Chavez, 1983), or the catches from central and southern Peruvian ports.

The spatial and volume variations of captures in northern Chilean waters differ from their Peruvian counterparts, particu-

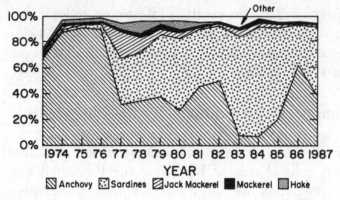

Fig. 15.6 Species contribution to Peruvian catch, 1973–87.

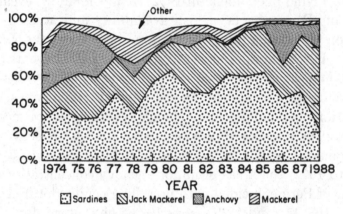

Fig. 15.7 Species contribution to Chilean catch, 1973–87.

larly after the 1972–73 El Niño (Fig. 15.7). The dominance of anchovies in pre-1972 landings was replaced by a surge in jack mackerel, a fish that thrives in habitats with water temperatures between 10° and 18°C, and in sardines that live under the same ocean conditions as jack mackerel, their major predator (Méndez & Neshiba, 1976).

In the early 1980s, the biomass of northern Chilean waters (specifically, the fishing districts of Arica and Antofagasta) was estimated at above 13 million mt, with jack mackerel accounting for 6 to 8 million mt and sardines for about 5.5 million mt (Castillo et al., 1986). This situation was drastically altered by the 1982–83 El Niño which reduced the stock of both jack mackerel and sardines to 3.5 million mt each. By 1986, the captures of jack mackerel had dropped to 100,000 mt, and those of sardines to 600,000 mt. Coincidentally, anchovy catches rose to 1.3 million mt, and surveys of anchovy eggs and larvae revealed a surprising rise in the anchovy stock, the capture of which had been prohibited in 1983 to allow for its recovery (Martínez, 1987).

According to Chilean oceanographers and fisheries specialists, since the warming episode of 1982–83 off the coast of the northernmost fishing district of Arica and Antofagasta, a persistent intrusion of subtropical waters with high salinity and low oxygen content has created strong saline and thermal fronts extending 30–60 km from the coast. Upwelling foci have been restricted to coastal pockets (Martínez, 1987), and the general conditions have been favorable for sardines and jack mackerel, as shown by the volume of their landings. South of this district conditions were normal as reflected in the stability of the fishery's output. This situation continued until the moderate 1987 El Niño warmed the tropical and subtropical Pacific, altering once again the oceanic conditions of the Humboldt Current (Pizarro, 1987). This El Niño, however, did not have serious consequences for the development of the fisheries. Anchovies spawned normally between latitudes 11° and 18°S and adult exemplars were caught between latitudes 10° and 14°S. Sardines were also captured at levels considered normal for the winter season, and spawning occurred mostly at latitudes 13° and 15°S (Zuzunaga et al., 1988).

These oceanic and biological events ceased in the late fall of 1987 (ERFEN, 1987), when surface water along the coast of western South America began to be affected by a severe cooling episode (i.e., an anti-El Niño) that lasted through the spring of 1988, and continued into the southern summer of 1989 (ERFEN, 1988, 1989). This cooling episode terminated the warming trend of the Southeast Pacific that had been observed since the early 1980s.

There is some debate about whether the anti-El Niño of 1988–89 was only a temporary interruption of the assumed warming trend or if such a trend was just one of the many oscillations that characterize ocean–atmosphere systems in the Southern Hemisphere (Ramage, 1987). Disregarding for a moment the discussion about the transitory or long-term nature of such change, it seems evident that the 1972–73 El Niño introduced an element of disruption into the thermal regime of the Humboldt Current and that a return to "normal" (i.e., to pre-1972 conditions) has not yet occurred because of the warming episodes of 1976–77, 1982–83, and 1986–87. The 1972–73 disruption triggered an alteration in the latitudi-

nal distribution of traditional stocks and also in the availability
of different species. In the years following the 1972–73 El Niño,
coastal upwelling species, such as anchovies, have been replaced
by offshore oceanic species, such as sardines and jack mackerel.
Confined to isolated pockets of upwelling along the coast from
the Gulf of Guayaquil to Valdivia (Chile), anchovies have weathered unfavorable warming episodes and appear to be making a
comeback (Sharp & McLain, 1991). A successful resurgence of
the anchovy may possibly be enhanced by the generalized cooling
conditions off the western coast of South America that began in
1988 (Caviedes, 1989; ERFEN, 1989). Since the arguments cited
in favor of a permanent change in the oceanic–climatic conditions
(i.e., a warming of the Southeast Pacific) cannot be substantiated
after the occurrence of the 1988–89 cold episode, it is necessary
to look for alternative explanations for the changing patterns of
Peruvian/Chilean fisheries.

Modeling the growth and contraction of Peruvian–Chilean fisheries

Since climate variations alone do not suffice to explain the contraction and resurgence of these fisheries, it is necessary to examine
as well: (1) the burden placed on living marine resources by overfishing, (2) the impact of ecological crises such as El Niño on fish
landings, (3) the observed variations in marine environments, and
(4) the influence of the demand for fish in international markets.

Time series analysis allows the calibration of a model that val-
idates the statistical significance of differentiating between a pre-
and a post-collapse period (i.e., before and after the 1972–73 El
Niño). According to this distinction, the fisheries collapse is considered to have been the outcome of both overfishing and the exac-
erbating effects of the 1972–73 El Niño. The established model can
then be used as a forecasting tool that allows comparison of real
trends with those simulated by the model's estimated parameters.
Subsequently, the economic forces operative in this marine system
(i.e., the influence of the world economy implicit in the demand
for fish products), as well as the physical and temporal constraints
on capacity as a function of past fishing trends (a supply-side argument), can be viewed in the context of the established model. It
must be stressed that this econometric model is applied to a time

series which consists of only one measure of fish availability in an ecosystem, namely, captures. Bax (1989) emphasizes that modeling a marine ecosystem entails more than a primary relationship between catch and population size; it also encompasses biomass volume, primary production, recruitment, internal and external predation, and not the least important, environmental disruptions such as El Niño. This analysis departs from the premise that human intervention (overfishing) brought about the collapse of the fisheries of Peru and Chile during the 1970s and that this intervention was aggravated by major oceanic–climatic variabilities. Other natural elements which influence the stability of fish stocks, such as biomass volume and predation, are not directly considered in this assessment.

The total catches (in metric tons) of each country were converted to a percentage of global landings to illustrate the severity of the demise of Peru's fishing industry in the early 1970s and the consistent expansion of its Chilean counterpart since the mid-1970s. In Fig. 15.2, Peruvian and Chilean catches from 1951 through 1986 and their sum (total for the Southeast Pacific) are expressed as percentages of the world total. The conversion of the total tonnage into percentages of the world catch is done for the purpose of comparison.

The time series data present the usual estimation problems associated with modeling nonstationary processes (Pindyck & Rubinfeld, 1981; Newton, 1988). Preliminary tests on the first-differenced series of total fish captures as a percentage of world catch (TFD) revealed signs of both autoregressive (AR) and moving average (MA) operators. Since the original data showed the interruption caused by the 1972–73 El Niño (which is also conspicuous in the first-differenced data), a combined regression-time-series format was chosen using 1972 as the starting point.

Given the combined influence of AR and MA operators, a multivariate autoregressive moving-average framework (MARMA) or transfer function model was chosen to test the sensitivity of total fish captures to both physical elements (sea surface temperatures and El Niño) and the temporal nature of the fish catches expressed by $TFD(t)$. Further, a (dichotomous) dummy variable ($D(t)$) to be used as a control variable was introduced in the model to distinguish between periods before ($D(t) = 0$) and after ($D(t) = 1$)

the occurrence of the 1972–73 El Niño (see the Appendix of this chapter for calculations).

The results

The MARMA ($p = 1$, $d = 1$, $q = 8$) model with the two explanatory variables – average sea surface temperature ASST(t) and the interrupted time-series dummy $D(t)$ – provided the best fit of all combinations estimated, with a coefficient of determination of 0.9488. With over 90 percent of the explained variation in first-differenced total fish catch accounted for, the model yielded nine statistically significant estimated parameters (at the $\alpha = 0.05$ level). An analysis of the residuals for the model shows that they have been reduced to white-noise – there is no statistical evidence of first- or higher-order autocorrelation in the disturbance vector – as measured by the Durbin–Watson and Box–Pierce Q statistics.

Since the inclusion of the El Niño dummy variable, denoting the absence or presence of the phenomenon, provided no additional explanatory power to the model, it was subsequently dropped from the analysis. Nevertheless, that dummy variable which controlled for parameter estimation in pre- and post-collapse periods (1972) was one of the most influencial regressors, even after first-differentiation.

Independent from the severity of the six El Niño events comprised in the period examined, the most intense of them (1982–83) had a relatively weaker effect on regional fish landings when compared with the 1972–73 El Niño, or even with the moderate El Niño of 1965. In other words, the time series vector of El Niño dummies proved to add no statistically significant explanatory power to the model, yet the interrupted time-series dummy (representing pre- and post-El Niño 1972) was shown to be highly significant. This finding suggests that the prolonged period of overfishing conducted before the 1972 El Niño, between the years 1957 and 1971, appears as a reasonable explanation for the subsequent decline of fish stocks in the southeastern Pacific.

As a corollary, two interpretations can be advanced for the non-significance of the individual yearly El Niño dummies, and the concurrent importance of bifurcating the time series into pre-1972–73 and post-1972–73 El Niño periods: (1) regional fish catches had reached a threshold in which replacement conditions were precarious, and that, (2) simultaneously, the effects of El Niño exacerbated the negative effects on the fisheries, thus preventing the rebound of fish stocks to pre-disruption levels.

It can, therefore, be posited that the dampening effect of oceanic warming and reduced upwelling coupled with a sustained burden placed on the stock by overfishing, unleashed the dramatic decline of the anchovy fisheries in the southeastern Pacific. Indeed, the 1972–73 El Niño played a secondary role in the exacerbation of a crisis that was already underway, yet given its relatively mild characteristics (in comparison to the 1982–83 El Niño), it cannot be blamed – as Peruvian fishing authorities contended – as the sole reason for the anchovy fishery collapse. The 1982–83 El Niño caused a 25 percent reduction in fish catch, whereas the 1972–73 El Niño produced a 75 percent decline in fish landings (as a percentage of the total world catch). If a tolerance threshold was exceeded – as, indeed, was the case if one recalls the 9.5 million mt upper limit (MSY) recommended by the FAO/IMARPE experts in 1969 – the depletion caused by overfishing was clearly exacerbated by the 1972–73 El Niño. As demonstrated by the insignificance of its explanatory role in the model, El Niño alone, and for that matter any other warming event in the Southeast Pacific, cannot solely account for the serious decline of the fishery, since this pattern should have repeated itself, even at reduced levels of annual captures, during the El Niño events that followed the 1972–73 El Niño. Actually, after the subtle decline in catches during 1982–83, there has been a certain upward trend which is evident in Fig. 15.2.

Since El Niño information is both subsumed in sea temperature variations (found to be relevant in the model) and contained in the structural aspects of the AR and MA aspects of the model, its dampening effects have already been included. For this reason, the El Niño dummy offered nothing more than redundant information. The structural form of the model demonstrates that fishery yield is tied to a fairly long MA process and is also affected by a first-order AR scheme.

Conclusions

The fisheries of Peru and Chile offer an interesting example of major eastern boundary current fisheries that reflect in their trends the mixture of internal biological controls, the effects of climatic and oceanic variabilities, and – perhaps more importantly – the national resource policies dictated by the economic imperatives imposed by international trade and world demand. Having expanded during the 1950s and 1960s in an uncontrolled manner with virtually no consideration for the maintenance of the exploited stocks, the fisheries began to show productivity fluctuations triggered by recruitment failures and ecological disruptions associated with El Niño events.

During the exponential growth phase of the fisheries, none of the El Niño events was severe enough to prompt fishing regulations aimed at preventing a possible collapse of anchoveta-based fisheries. Driven by the desire to increase profits and exports, the Peruvian government opposed the implementation of drastic restrictive management measures. When the 1972–73 El Niño occurred, overfishing, and not ecosystem disturbances, was largely responsible for the collapse of the anchovy stocks.

The same trend occurred in Chile, but the ensuing consequences were different. The export pressures felt by the governing military rulers after 1973 led to a maximization of the benefits obtained from the exploitation of living marine resources and a shift of the fishing effort from anchovy to sardines and jack mackerel, which had received a biological stimulation from the ocean warming that followed the 1972–73 El Niño, and continued after the 1982–83 El Niño. Also, the change of fishing location from offshore to coastal waters made catches more profitable. The concentration on these species and the increasing sophistication introduced into the fishing sector in order to generate a convenient flow of hard currency enabled Chile to become the leading fishing nation in the Southeast Pacific after 1980. By shifting the emphasis from anchovy to sardines, from sardines to jack mackerel, and from jack mackerel back to anchovy – when they recuperated in 1985 – Chilean fishery industrialists have been able to maintain relatively large captures even during the ocean-warming episodes that followed the 1982–83 El Niño (Serra, 1987).

In Peru the shifts were mostly restricted to sardines because jack mackerel were less abundant than in Chilean waters; hence, the volume of catches has remained below that of Chile. Recent years have witnessed a resurgence in the captures of anchovy both in Peru and Chile, even during the short-lived warming episode of 1986–87. This fact makes it doubtful that the failure of the anchovy to recover from the 1972–73 El Niño-related collapse was due to continued excessive fishing pressures.

The proposition of tying an established oceanic–climate change in the Southeast Pacific and along the western coast of South America in recent decades to global warming trends caused by heightened levels of carbon dioxide (Takahashi, 1989) as yet offers no conclusive evidence. From a global atmospheric–oceanic perspective, there are alternative ways of explaining this deviation from a possible norm. First to be considered is the large spatial dimension of the Peru–Chile ecosystem, which extends from the equatorial belt to high middle latitudes, encompassing a broad range of climatic variety across and along the Humboldt Current. Such variety is rarely found in other eastern boundary current systems. To alter the climatic–oceanic patterns of a system of such spatial dimensions, the climatic variations would have to be of greater magnitude than those recorded by our instruments.

Second, the spatial dimension of the system also implies the existence of several upwelling foci and numerous primary productivity centers which, in times of ecological crises, can act as temporary refuges for endangered species. Only when overfishing impinges on these sanctuaries (as has happened in the past), is the reproduction capacity of sensitive species, such as anchovies, severely curtailed.

Third, the elongated geographic configuration of the Peru Current allows not only the eastward penetration of equatorial, tropical, and subtropical Pacific water masses (enhancing the climatic variety mentioned above), but it also prevents a uniform reaction of the system to generalized climatic anomalies. In the context of this reality, migration of mobile species is also greater, allowing north–south movements, across maritime political boundaries in times of oceanic–climatic crises such as El Niño events (see the cases illustrated in Pauly & Zuzunaga, 1987).

Thus, the shifting fortunes of fish populations in the waters of Peru and Chile appear to be somewhat independent of global climatic events and more dependent on the pressures that these countries place on the reproductive capacity of fish stocks already under great stress from ecological disruptions of El Niño.

The implications for resource management and conservation policies are obvious. Overfishing makes fish stocks with delicate recovery rates highly vulnerable to El Niño perturbances. Warnings must be voiced against future repetitions of stock exploitation during contractions in this regional fishery whenever ocean–atmosphere conditions resemble those of 1972–73. The emphasis, therefore, must be on restricted use and proper regional management of living marine resources to neutralize, or compensate for, the negative effects of El Niño.

While overfishing and stock conservation at the regional level demand concerted actions by Peru and Chile, they each have developed in the course of their individual economic and political histories their own way of coping with these coastal ocean crises. Indeed, because the economic advantages of one of these countries are not necessarily good for the other, there has been a lack of coordination. The coexistence of two economic and political models – one with clear authoritarian and free enterprise overtones, the other with obvious statist and populist traits – of a common living marine resource has compounded the problem of dealing jointly with ecological crises.

Since society has to operate within the constraints inherent in the ecosystem, limitations must be placed on those parameters which society has the ability to control – total catch. Future inquiries into this issue should discern how the constraints placed on the system by environmental disruptions and human intervention influence – separately or combined – fish availability and stock replacement. Only then can the consequences of society's concentrated fish exploitation strategies be put into a useful perspective. In order to reduce the possibility of fishery devastation, strict biological monitoring needs to be implemented and properly enforced fishing controls must be instituted. In this way it will be possible to keep stock replacement levels stable in the face of the physical stress caused by natural phenomena such as El Niño events, while still meeting the growing world demand for fish meal and other fish products. An international body, consisting not only of coastal countries but also of distant-water fishing nations that harvest fish stocks in the Peru Current, should monitor and regulate the capture volume necessary to safeguard the viability of the stocks. This body should keep a detailed record of the oceanic, climatic, and biological variations that accompany generalized oscillations in the Pacific Ocean. Their findings and propositions should be binding for all parties involved, so that fishing limitations can be implemented quickly at the first sign of ENSO-like events. This goal is not out of reach, as both Peru and Chile have adequate scientific research and record-keeping facilities that permit the simultaneous monitoring of abnormal oceanic–meteorological events. What has been missing to date is the willingness to act in a coordinated fashion to protect a fishing patrimony which is not the sole property of any one country, but a resource that belongs to the world.

Appendix

The MARMA model can be expressed as an autoregressive (AR) and moving average (MA) process with k explanatory variables $(X_{jt}, j = 1, \ldots, k;$ and $t = 1 \ldots \Gamma$ time periods) and is written

$$y_t(d) = \sum_{j=1}^{k} \beta_j X_{jt} + \phi^{-1}(B)\Theta(B)\tau_t \ ,$$

with autoregressive coefficients

$$\phi(B) = 1 - \phi_1 B - \phi_2 B^2 - \ldots - \phi_p B^p \ ,$$

moving average coefficients

$$\Theta(B) = 1 - \Theta_1 B - \Theta_2 b^2 - \ldots - \Theta_q B^q \ ,$$

and a backward shift operator (B). The term $y_t(d)$ denotes the first-differenced $(d = 1)$ dependent variable $(TFD(t))$ and τ_t is the ARMA error term.

In its more general form the model can be expressed as

$$y_t(d) = \sum_{j=1}^{k} v_j^{-1}(B)w_j(B)X_{jt} + \phi^{-1}(B)\Theta(B)\tau_t \ ,$$

where transfer function operations are performed to identify the lag polynomials $v(B)$ and $w(B)$ of degree z_j and m_j (Box & Jenkins, 1970; Makridakis & Wheelwright, 1978).

The AR and MA structures were chosen through a lengthy trial and error process. It was found that the most efficient model demonstrated an autoregressive structure of $p = 1$ and a moving average operator of $q = 8$, for $d = 1$ (the first-differenced data). The following explanatory variables were then used to account for the remaining variation in yt (in various combinations thereof) using the MARMA ($p = 1$, $q = 8$, $d = 1$) model:

X_{1t} = the dummy variable $D(t)$;

X_{2t} = average sea surface temperature $\text{ASST}(t)$; and

X_{3t} = an additional dummy variable $DD(t)$ denoting the El Niño years of 1965, 1969, 1972, 1977, 1982, and 1986, where $DD(t) = 1$; and $DD(t) = 0$ for the remaining years.

These additional explanatory variables are supposed to capture the impact of changing environmental conditions on the availability of fish stock. Nonlinear Gaussian estimation was performed using the econometrics package SORITEC (version 6.3, 1989), and the statistically relevant findings and estimated parameters are listed in Table 15.1. An illustration of the actual and simulated

Table 15.1
Output for MARMA model of Peruvian and Chilean
fishing industries for the period 1951–86

Coefficient description		Estimated coefficient	Standard error	*t*-statistic
X_1 (Dummy)	$\beta_1 =$	−0.0782698	0.016827	−4.6512*
X_2 (ASST)	$\beta_2 =$	−0.0096424	0.004186	−2.3032*
AR-term(-1)	$\phi_1 =$	0.707452	0.174205	4.0610*
MA-term(-1)	$\Theta_1 =$	0.957460	0.197138	4.8568*
MA-term(-2)	$\Theta_2 =$	−0.557181	0.259636	−2.1460*
MA-term(-3)	$\Theta_3 =$	0.354013	0.288278	1.22803
MA-term(-4)	$\Theta_4 =$	−0.428059	0.199701	−2.14350*
MA-term(-5)	$\Theta_5 =$	0.891684	0.187595	4.75324*
MA-term(-6)	$\Theta_6 =$	−0.604302	0.259006	−2.33315*
MA-term(-7)	$\Theta_7 =$	−0.022545	0.284620	−0.07921
MA-term(-8)	$\Theta_8 =$	0.418544	0.202160	2.07036*

*Significant at the 95% confidence level ($|t| > 1.714$)

Summary Statistics:
　　　Sum of Squared Residuals = 0.006054
　　　Variance of Residuals = 0.0002632
　　　Durbin-Watson Statistic = 1.9234
　　　Box-Pierce Q Statistics (30 lags) = 8.73
　　　R-Squared = 0.9488

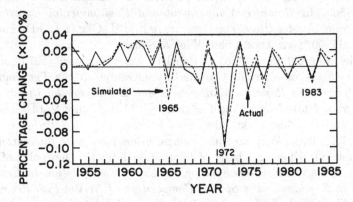

Fig. 15.8 First-differenced time series, actual and simulated fish catch.

first-differenced time series appear in Fig. 15.8, demonstrating the close fit produced by this model.

References

Barber, R.T. & Chavez, F.P. (1983). Biological consequences of El Niño. *Science*, **222**, 1203–10.

Bax, N. (1989). What can models tell us of interactions between the environment and fisheries? Paper presented at *The Environmental Influences on Marine Biological Resources International Seminar*, Murmansk, USSR, 1–8 June 1989 (mimeographed).

Box, G.E.P. & Jenkins, G.M. (1970). *Time Series Analysis, Forecasting and Control.* San Francisco: Holden-Day Publishers.

Castillo, J., Lillo, S. & Pineda, P. (1986). Distribución y abundancia de sardina española, jurel y anchoveta en otoño de 1986. Programa Investigaciones Pesqueras de Recursos Pelágicos – Zona Norte. Santiago, Chile: Convenio Pesquero Coloso-Pesquera Guanaye-IFOP.

Caviedes, C.N. (1981). The impact of El Niño on the development of Chilean fisheries. In *Resource Management and Environmental Uncertainty: Lessons from Coastal Upwelling Fisheries*, ed. M.H. Glantz & J.D. Thompson, pp. 351–68. New York: John Wiley & Sons.

Caviedes, C.N. (1989). The anomalies of South America during the winter of 1988. *Tropical Ocean–Atmosphere Newsletter*, **52**, 1–5.

DeVries, T.J. (1987). A review of geological evidence for ancient El Niño activity in Peru. *Journal of Geophysical Research*, **92**, C13, 14471–9.

ERFEN, (1987). *Boletín de Analisis Climático. Pacífico Oriental Sudamericano*, No. 12.

ERFEN, (1988). *Boletín de Analisis Climático. Pacífico Oriental Sudamericano*, Nos. 13–16.

ERFEN, (1989). *Boletín de Analisis Climático. Pacífico Oriental Sudameri-cano*, No. 17.

Hammergren, L.A. (1981). Peruvian political and administrative responses to El Niño. In *Resource Management and Environmental Uncertainty: Lessons from Coastal Upwelling Fisheries*, ed. M.H. Glantz & J.D. Thompson, pp. 317–50. New York: John Wiley & Sons.

Instituto del Mar del Perú (1981). Panel of experts' report (1970) on the economic effects of alternative regulatory measures in the Peruvian anchoveta fishery. In *Resource Management and Environmental Uncertainty: Lessons from Coastal Upwelling Fisheries*, ed. M.H. Glantz & J.D. Thompson, pp. 369–400. New York: John Wiley & Sons.

Kawasaki, T. (1983). Why do some pelagic fishes have wide fluctuations in their numbers? Biological basis of fluctuation from the viewpoint of evolutionary ecology. In *Proceedings of the Expert Consultation to Examine Changes in Abundance and Species Composition of Neritic Fish Resources*, ed. G.D. Sharp & J. Csirke, pp. 1065–80. Rome: UN FAO.

Makridakis, S. & Wheelwright, S.C. (1978). *Forecasting Methods and Applications*, chapter 11. New York: John Wiley & Sons.

Martínez, C. (1987) Los recursos pelágicos y la actividad pesquera del norte de Chile en relación con las características oceanográficas. *Boletín ERFEN*, **23**, 3–10.

Méndez, R. & Neshiba, S. (1976). Relación entre las variaciones mensuales de temperaturas superficiales y distribución de los recursos pesqueros de sierra, jurel y sardina, en el Pacífico Sur-Este. *Revista de la Comisión Permanente del Pacífico Sur*, **5**, 139–46.

Newton, H.J. (1988). *Timeslab: A Time Series Laboratory*. Pacific Grove: Wadsworth & Brooks/Cole.

Pauly, I. & Zuzunaga, T. (Eds.) (1987). *The Peruvian Anchoveta and its Upwelling Ecosystem: Three Decades of Change*. Manila: International Center for Living Aquatic Resources Management.

Paulik, G.J. (1971). Anchovies, birds, and fishermen in the Peru Current. In *Environment: Resources, Pollution and Society*, ed. W.W. Murdoch, pp. 156–89. Stamford: Sinuaer Associates. Reprinted in *Resource Management and Environmental Uncertainty: Lessons from Coastal Upwelling Fisheries*, ed. M.H. Glantz & J.D. Thompson, pp. 35–79. New York: John Wiley & Sons.

Pindyck, R. & Rubinfeld. D. (1981). *Econometric Models and Economic Forecasts*, 2nd ed., pp. 539–605. New York: McGraw-Hill.

Pizarro, L. (1987). Anomalías oceanográficas a principios de 1987 frente al Perú. *Boletín ERFEN*, **23**, 11–4.

Ramage, C.S. (1987). Secular change in reported surface wind speeds over the ocean. *Journal of Climate and Applied Meteorology*, **26**, 525–8.

Salinas, R. (1973). The fish-meal industry of Iquique. In *Coastal Deserts: Their Natural and Human Environments*, ed. D.H.K. Amiran & A.W. Wilson, pp. 137–46. Tucson: The University of Arizona Press.

Santander, H. & Flores, R. (1983). Los desoves y distribución larval de cuatro especies pelágicas y sus relaciones con las variaciones del ambiente marino en frente al Perú. In *Proceedings of the Expert Consultation to Examine Changes in Abundance and Species Composition of Neritic Fish Resources*, ed. G.D. Sharp & J. Csirke, pp. 835–67. Rome: UN FAO.

Schweigger, E. (1959). *Die Westküste Südamerikas im Bereich des Peru Stroms.* Heidelberg–Berlin: Keysersche Verlag.

Serra, R. (1987). Impact of the 1982–83 ENSO on the southeastern Pacific fisheries, with an emphasis on Chilean fisheries. In *Climate Crisis: The Societal Impacts Associated with the 1982-83 Worldwide Climate Anomalies*, ed. M.H. Glantz, R. Katz, & M. Krenz, pp. 24–9. Nairobi: UNEP.

Sharp, G.D. (1987) Climate and fisheries: Cause and effect or managing the long and short of it all. In "The Benguela and Comparable Ecosystems," ed. A.I. Payne, J.A. Gulland & K.H. Brink, issue of *South African Journal of Marine Science*, **5**, 811–38.

Sharp, G.D. & McLain, D. (1991). Fisheries, ENSO and upper ocean temperature records: an Eastern Pacific example. *Biological Oceanography.* (In press.)

Sharp, G.D., Csirke, J. & Garcia, S. (1984). Modelling fisheries: What was the question? In *Proceedings of the Expert Consultation to Examine Changes in Abundance and Species Composition of Neritic Fish Resources*, ed. G.D. Sharp & J. Csirke, pp. 1177–224. Rome: UN FAO.

Steele, J.H. (1989). The message from the oceans. *Oceanus*, **32**, 4–9.

Takahashi, T. (1989). The carbon dioxide puzzle. *Oceanus*, **32**, 22–9.

Zuzunaga, J., Mariduña, L., Carrasco, S., Arriaga, L., Cornejo, M., Cabezas, E., Cruz, M., & Hurtado, M. (1988). Condiciones oceánicas, climáticas y biológicas en el Pacífico Sudeste durante el segundo semestre de 1987. 3, Condiciones biológicas. *ERFEN Boletín*, **21**, 22–8.

16

Climate change, the Indian Ocean tuna fishery, and empiricism

GARY D. SHARP

Visiting Scientist
NOAA Center for Ocean Analysis and Prediction
Monterey, CA 93943, USA

Introduction

The unique characteristics and opportunities within the western Indian Ocean are finally being exploited, after many decades of conservative commentary and misplaced concern. Although there were several local and regional fisheries for tropical tunas in the western Indian Ocean that have been operating for periods often exceeding written history, none of these had developed to any remarkable degree because of the types of gear being used, and the long distances from landing ports to processing facilities and markets.

The background history behind the development of the western Indian Ocean tuna fishery, particularly within and about the Seychelles Plateau, is unique and worthy of documentation, even though that development is still ongoing. For various reasons this particular fishery thrives, while development efforts in other ocean areas over recent decades have been marginal; some have even regressed. It began in parallel with several similar, but less successful, efforts.

During the last two decades, there have been several national efforts to develop nearshore fisheries for tunas. Projects have mostly been effective only at locating resources, but have not been very effective at transferring technology needed to extend fishing grounds into open ocean areas. Projects have come and gone from Indonesia, Sri Lanka, Thailand, India, Somalia, Zanzibar, Madagascar and, recently, the Maldives and the Seychelles (Fig. 16.1). The latter two have been most successful, but for very different reasons.

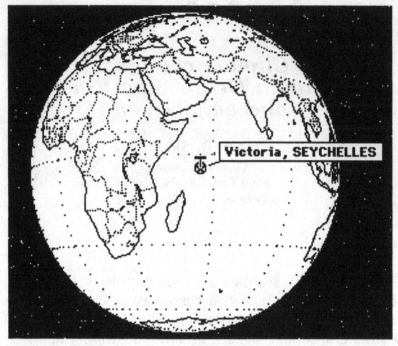

Fig. 16.1 The Seychelles Plateau is one of the world's largest tropical open
ocean shoals, comprising 48,334 km^2 with many small islands. The
capital of the Seychelles is Victoria, Mahé, where the infrastructure
developments related to this fishery have been centered.

The Maldivian and the Seychellois cultures are distinct and
unique. Yet, they have somehow succeeded in accommodating to
the developments of the recent decade's opportunities within the
tough, competitive framework of global tuna markets and infrastructure requirements. This cultural accommodation is in sharp
contrast to many of the other ventures.

Since the early 1970s, the long-line fleets of Japan and other
Asian countries have been in constant flux, as the values of products have changed, and as problems of labor costs and recruitment of fishermen to these fleets have evolved along with national
economies. While the value of both fresh and flash-frozen sashimi-
grade products has increased dramatically, the costs of fishing and
product delivery (e.g., air freight) have also undergone significant
changes.

The political backdrop for
Indian Ocean tuna activities

To better understand the pressures behind the conservative approach to the development of surface fisheries in the Indian Ocean, it is useful to describe a few of the expert reports and attempts to project potential "sustained yields" that were derived from fisheries catch data in the pre-development period. For example, Shomura et al. (1967, p. 2) concluded that:

> the data on the pelagic fisheries suggest that the quantity of pelagic resources now being harvested by the long-line technique may be at the upper limit of rational harvesting ... and that there is an immediate need for basic data that will permit accurate estimates of the maximum sustainable yield for each of the commercially important species.

These comprise five species: yellowfin tuna (*Thunnus albacares*), bigeye tuna (*Thunnus obesus*), albacore (*Thunnus alalunga*), the southern bluefin tuna (*Thunnus maccoyi*), and the skipjack tuna (*Katsuwonus pelamis*).

Catch statistics for the region's long-line fleets were summarized for the year 1964 (surface fisheries at that time were mostly artisanal). Long-line fishing effort was estimated to be 66 million hooks; see Table 16.1 for catches.

Table 16.1
Catch statistics for Indian Ocean tuna

Species	metric tons
Yellowfin tuna	28,000
Bigeye tuna	14,000
Albacore tuna	18,000
Southern bluefin tuna	25,000
TOTAL	85,000

Skipjack tuna catches, not a primary target of long-line fisheries, were estimated to be around 5,000 mt for the entire Indian Ocean.

It is noteworthy that at present the primarily artisanal Maldivian fishery alone annually lands more than 80 percent of these early estimates (i.e., more than 70,000 mt of tropical tunas were landed in the Maldives in 1988). Early attempts at resource assessment were misleading at best, since one of the major reasons for doing them was to justify existing fisheries catch levels, rather than to promote further fisheries development within the region.

Early catch-based stock assessment estimates for sustained yields are summarized in Table 16.2, in contrast to the 1973 and 1983 UN Food and Agriculture Organization (UN FAO) catch statistics (the onset of the major surface fishing effort in the Indian Ocean was 1983), and the FAO Indo-Pacific Tuna Development and Management Programme (IPTDMP) catch estimates for 1986, the most recent complete statistical records compilation for the region (IPTDMP, 1988).

Table 16.2
Early catch-based stock assessment estimates

Source:	Shomura et al. (1967)	Suda (1973)	FAO yearbooks (1973)	(1983)	IPTDMP (1986)
		($\times$ 1000 mt)			
Yellowfin	30–35	30–35	41.3	59.8	114.2
Bigeye	20–25	20–25	16.7	44.4	43.0
Albacore	22–28	30–32	22.8	17.2	25.3
Southern bluefin	38–40	38–40	26.8	36.7	21.9
Skipjack	?	?	Maldives alone – 16.0–45.4		
TOTAL	110–120	118–132	107.6	158.1	286.8

In fact, any in-depth look at the sources of these data will show that, until the implementation of the systematic program of the FAO/IPTDMP in the early 1980s, there were simply no data from which to begin robust assessments of the abundance of the tuna resources of the Indian Ocean, except by analogy.

The conventional resource assessment approach (via the extrapolation of long-line and pole-and-line tuna catch data to ocean potentials) does not adequately represent the existing knowledge in regard to tuna behavior, their distributions with regard to habitat properties, or the wealth of understanding about the relations between tuna aggregation and behavior-related vulnerabilities to different fishing methods under varying climatic and oceanographic conditions (e.g., Sharp, 1976, 1978, 1979, 1982a, 1982b; Sharp et al., 1984).

The recent decade's development efforts

The progress of earlier efforts to develop various surface fisheries within the Indian Ocean has been well documented by Silas and Pillai (1982), Stequert and Marsac (1986), and various reports of FAO and other fisheries development projects (Cort, 1983; UN FAO/UNDP, 1970, 1973, 1981, 1982, 1985; Lee, 1982; Marcille & Uktolseja, 1984; Messieh, 1983; Sharp, 1982b; Sivasubramanian, 1965, 1970, 1972, 1981). Each is informative and contributes to the understanding of Indian Ocean tunas. Recent studies provide improved insights (e.g., Marsac & Hallier, 1991).

Perhaps the underlying causes for the relatively rapid development of the western Indian Ocean surface fisheries were as follows: declining production because of increased competition (overcapitalization) in the eastern tropical Atlantic Ocean fishery, the fall of prices paid throughout the late 1970s and early 1980s for small tunas, and the need to find more productive fishing areas for the larger French and Spanish superseiners. The immediate stimulus, however, came in the form of timely publications describing the likelihood of the development of seasonal surface fisheries within the Indian Ocean based on ocean and climate characteristics.

The first manuscript on this topic was prepared during the summer of 1978. It was an assessment of the areas of potentially successful exploitation of tunas in the Indian Ocean with emphasis on surface methods, and was published by the Indian Ocean Fisheries Development Program (Sharp, 1979). It first summarized the way each of the various types of tuna fishing methods operates, and then described the behavioral responses of a generic tuna to its physical habitat limits, focusing on temperature and oxygen, since these comprise "necessary" conditions within which individual fish must be able to sustain their physiological functions. It describes the basis for prospection, using graphics produced at the Inter-American Tropical Tuna Commission. The basic data was derived from the global oceanographic data files held at the University of California in San Diego, at Scripps Institution of Oceanography.

The Pacific Ocean analogues had just been published (Sharp & Dizon, 1978), along with the basic logic from which the relationship between dynamic oceanographic properties and vulnerability of tunas to fishing gears of various types were deduced. The principles are concise and predictive. The underlying understanding required is about the limiting features for sustained activity within an organism's habitat. The biophysical principles are not complex and, by any standards, should form the basis for studies of living organisms that are free to select their own locations within a continuously varying, heterogeneous environment.

Within a few months of this 1979 publication, the author was visited in Rome by Jacques Marcille and Liné Gery, who had been tasked by the French tuna industry (COFREPECHE), to investigate the possibility of developing a tuna fishery in the Indian Ocean. Marcille had spent many years working in Madagascar- and Indonesia-based tuna fisheries. Gery had lived for several years on La Réunion, in the western Indian Ocean. Both were confident that, if the data summarized in Sharp (1979) were representative, then an effort to develop a regional surface tuna fishery would be appropriate.

Upon their return to France, they began to work on the various problems related to financing prospections, infrastructure development, and making international contacts. Marcille and De Reviers (1981) provided the French industry with an in-depth review of the early developmental efforts, but they focused on providing seasonal descriptions of the oceanography, seasonal climatology, and specific analyses of probable areas of greatest potential for tuna prospection efforts.

Given that Marcille had already been working in the western Indian Ocean, in the Madagascar tuna prospection efforts in previous years, he had clear and direct knowledge of the existence of abundant untapped surface schools that were primarily comprised of tunas smaller than those of interest or vulnerability to long-line fishing techniques.

Gery visited the governments of several Indian Ocean nations to discuss the needed infrastructure requirements. She found little to be optimistic about, because of the morass of complex national legislation that had emerged in response to centuries of imperialism and/or colonialism, which had, in turn, become the anvil against which all international investment negotiations were to be beaten into form. In many cases these obstacles appeared to be insurmountable, at least in the near term.

Due to existing cultural and economic ties, the Seychellois were the most likely recipients of overtures related to the early bases for prospection ventures in the western Indian Ocean, particularly since the expansive Seychelles Plateau provides a natural aggregating function for tropical tunas and their supporting food web.

Meanwhile, the Spanish had already begun negotiations with the Government of the Seychelles for a program of prospection for tropical tunas employing pole-and-line vessels and they were actively working on and about the Seychelles Plateau when the first French purse seine efforts began in December 1980.

The Spanish pole-and-line fishing efforts lasted only a few seasons due to the difficulties of this two-stage fishery related to finding, fishing, and holding baitfish, while also trying to sustain continuous at-sea fishing activities for tunas, then landing them and selling them at appropriate prices. Although this fishery did not thrive then, given the poor state of development of the needed infrastructure, it would possibly be a successful venture under present circumstances. While this fishing method might generate less gross landings than purse seining (due to its lower initial and day-to-day costs), the lower initial outlay and lower day-to-day costs provide ample offset to the initial investment costs and operating costs of modern high seas purse seine vessels above 650 mt or so of capacity.

Identifying potential tuna fishing grounds

There are many published studies for tunas, and all related scombroid fishes, that indicate that there are distinct physical boundaries that, although permeable to varying degrees, appear to inhibit major abundances or aggregations of schools, individuals, or some size groups within a given species (e.g., Sharp & Dizon, 1978).

Acoustic tagging studies (e.g., Carey & Olson, 1982) of the behavior of individual scombroids while free swimming in the open ocean show clearly that they exhibit differential swimming depth and feeding behavior patterns during day and night that provide the backdrop against which all fisheries operate. It cannot be assumed that ocean gradients have their own unique effects on the behavior of these species without regard to the absolute temperatures comprising those gradients, the size and thermal histories of the fish, and their motivations for entering, for whatever periods of time, any variously inhospitable environments. The basic premise is that, wherever there is sufficient oxygen and non-limiting temperatures from the emergent 15° isotherms bounding the poleward regions of the subtropical oceans to the equator, it is nearly impossible to find an area where, fishing long-line or trolling a feather lure, one will not catch at least one species of tuna at some time over the seasonal cycle. Moving toward the equator, the number of likely encounters increases, as do the numbers of species to take the hooks, given that the baited hooks are fished within well-oxygenated waters. Even beyond the thermocline where temperatures at hook levels decrease to well below 10°C in the tropics, larger and diverse species are likely to be caught (e.g., bigeye, albacore and bluefin species as well as billfishes).

I will also identify here the less predictable "sufficient" conditions. These include responses to dynamic habitat characteristics such as salinity, food aggregations, and reproductive opportunities that are scattered over broader spatial and temporal bounds. The difference between the sufficient conditions and the necessary ones is only a matter of the length of time that individuals can tolerate continuous exposure. For example, below certain food encounter rates, an organism cannot continue even at optimal temperature; likewise, even short exposures to very low temperature or oxygen levels can be devastating. However, critical values of oxygen and temperature for sustained exposures change with species, size, and developmental stage.

Tunas and their relatives have evolved characteristics that extend their capabilities to pass beyond these critical physical boundary conditions, but they cannot survive these exposures for long, indefinite periods. That distinction makes their behavior even more predictable than the less-adapted species; that is, because they are so energy-consuming, they are dependent on ocean processes and features which promote the aggregation of the prey resources which they must find within finite time periods, or die. These are the fronts, thermoclines and productive shoal regions of the ocean.

Clearly, even the most primitive models of behavioral responses to such conditions as food aggregations and thermal preference limitations will show that the place to search for such species is at the interfaces of the ocean, where similar processes, on smaller scales, promote aggregations of entire food webs over time.

The proximate causes of these physical features are quite relevant, as these are mostly related to surface winds, current impingements, convergences and divergences related to physical, remotely forced processes. Storms, strong wind events, and current impingements can also be looked upon as being analogous to plowing fields, turning over fertile, productive strata, providing for rapid food web development, and subsequent predator aggregations.

The Indian Ocean is a patchwork of seasonally varying surface wind dynamics, and resulting physical features and processes (Hastenrath & Lamb, 1979). Between the monsoonal seasonal and interannual dynamics, and the longer-term epochal variations of the region, there are many significant sources of ocean current dynamics. This is the key to the success of the Indian Ocean surface tuna fisheries, including the subsurface long-lining for large tunas, billfishes, and other predators which also dominate the region.

Climate-driven ocean variabilities are important on daily, seasonal, up to epochal time scales, depending on whether one is trying to plan for the next decade's investments in infrastructure, next season's orders for canning materials, or tomorrow's fishing day. The patterns of change are only now becoming apparent, and the year-to-year ocean variability in some regions can be as large as the decadal differences in others. Yet seasonality dominates all scales, resetting to some degree the anchor points from which to do short-term, annual comparisons, and relevant economic planning. There are certain dependable properties of each of these systems which provide the basic underpinnings of successful tuna location, if not abundance estimations.

Optimization of surface tuna fishery catch rates depends upon the "local" characteristics of the broader habitat limits related to prey densities and aggregation phenomena, as functions of ocean production features. In response, fishermen (and sea birds) have, generally, learned to operate in the vicinity of frontal discontinuities of any sort, as a necessary adaptation to the difficulties in locating food aggregations in the open ocean. With the onset of satellite-based observations, the location of "features" has become far more efficient. The use of bathythermograph equipment, such as the XBT, along with near real-time satellite imagery, can provide the bases for efficient subsurface feature location and, hence, optimal fishing success.

Characteristics which appear to improve catches for surface tuna fisheries include shoal habitat limits and strong vertical thermal gradients, as well as proximity to the surface emergences of these features in the form of fronts and transition zones. Understanding is often confounded in studies employing only thermocline slope and sea surface temperature (SST) as major proxy variables. The single most abused diagnostic variable is SST, which is nearly irrelevant to tunas since they do not really spend much time near the ocean surface.

Scalar wind records (Fig. 16.2) provide useful integrations of strong forcing functions which affect the upper ocean and, hence, tuna vulnerabilities. Greater surface winds indicate stronger upper ocean mixing, and deeper thermoclines; lesser winds provide less mixing force, and promote shallow tuna habitat, hence greater accessibility of the schools to surface fishing techniques.

COADS GLOBAL SCALAR (WINTER) WIND DEPARTURES

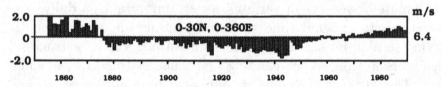

Fig. 16.2 The comprehensive ocean and atmosphere data set (COADS) scalar wind data show the long-term changes in surface wind speed (m/s) and their deviations since 1854 for the Northern Hemisphere from the equator to 30°N. Note the large changes that center around 1875, 1900, 1915, 1925, 1940–45, and 1955–75.

Many still argue the basis for these statements from academic perspectives. Speaking empirically, however, the obvious successes in ocean fishery developments and harvest strategies, which employ the basic logical concepts revolving around species and size-specific physiological requirements, belie research results and commentaries about tuna fishery responses to the environment arising from poorly conceived environmental correlations or statistical tests that often examine perhaps only half the data required for meaningful analysis.

Surface and subsurface features, for example, thermocline properties, oxygen availability, local and regional gradients and dynamics, along with consequential prey responses, constitute the necessary information for understanding the behavioral dynamics that can lead to a successful tuna fishery. Even vessel origins, crew compositions and differences in fishing behavior (and esoterica such as a vessel's historical economic or maintenance information) can provide needed insights into catch efficiency (catch per unit effort – CPUE).

The following contrast that can be made in the available data from the western Indian Ocean fishery provides an interesting point of departure from theory and principles, to practice.

Comparison of the success of Spanish and French tuna seiners

Table 16.3 gives monthly catch statistics for the French and Spanish tuna seiners that operated in the western Indian Ocean from January 1985 to September 1986 (see Michaud, 1986), and the summary statistics for January 1988 to March 1989 (see the Tuna Bulletin of the Seychelles Fishing Authority, First Quarter 1989).

Note the dramatic differences in apparent effective fishing powers of the two fleets in recent years in comparison with the early data set and to one another. The dominance of the Spanish fleet has been characteristic of the eastern tropical Atlantic fishery for well over a decade. It also appears that it would be very difficult to use conventional CPUE-based statistics alone to sort out changes in catch trends relative to either changes in fleet fishing power, or dynamic climatic processes, or any other resource population variations independent of these. This series of problems remains the greatest challenge to fisheries management in this (and any other) tuna fishery, if not all fisheries.

There are, however, seeds for the study and differentiation of these three causes for concern in the year-to-year changes that are being documented in the Seychelles-based operations. In recognition of the importance of climate changes such as the location of the inter-tropical convergence zone (ITCZ), subsurface dynamics, and so forth, a full suite of environmental data is being collected, along with vessel operations and catch information, as part of a shipboard observer program that is specified within the licensing agreement for each vessel that wishes to fish the Seychelles exclusive economic zone (EEZ). This will provide a firm basis for

Table 16.3
Catch statistics for the Spanish and French seiner fleets
January 1985 to September 1986 compared with
January 1988 to March 1989

Month	Days fished		Total catch (mt)		Catch rate
1985	Spanish	French	Spanish	French	Spanish/French
Jan	383	611	5747	9083	15.00/14.87
Feb	349	421	3022	2090	8.66/4.96
Mar	467	547	5668	6117	12.14/11.18
Apr	321	542	3444	3179	10.73/5.86
May	273	549	2760	5169	10.11/9.41
Jun	237	479	2640	5637	11.14/11.77
Jul	156	353	173	5421	7.52/15.36
Aug	175	339	2058	4513	11.76/13.31
Sep	192	420	5809	9772	30.26/23.27
Oct	277	436	7802	10394	28.17/23.84
Nov	297	489	2446	7332	8.23/14.99
Dec	416	578	4375	4764	10.52/8.24
1986					
Jan	332	565	6435	8558	19.38/15.15
Feb	298	473	4007	8615	13.45/18.21
Mar	273	551	2542	7856	9.31/14.26
Apr	241	477	1650	7006	6.85/14.69
May	193	392	1726	4865	8.94/12.41
Jun	208	287	2704	3693	13.00/12.87
Jul	221	394	4128	6539	18.68/16.60
Aug	254	350	5433	4992	21.39/14.26
Sep	120	152	3156	2881	26.30/18.95
TOTALS	5683	9405	78724	128477	13.85/13.66
Jan 1988 to Mar 1989 Landings (Spanish/French):					
	5173	6274	147272	125906	28.47/20.07

the necessary studies of catch variation, and the attribution of the various contributions to this variability over time.

The evolution of a tuna fishery

The basic requirements for a successful fishery include shoreside facilities such as dependable fuel and provision supplies, access to mechanical and electronic parts and expertise (although most high seas tuna vessels employ highly qualified engineers and have shop facilities which provide for near autonomy at sea). If the fishery is operating far from processing facilities, as was the case in the Indian Ocean, an economic transshipping capability would also be needed.

Once the initial agreements about the prospection phase are worked out, and prospection proceeds beyond the initial search and test stages, the question arises as to when the prospection phase ends and commercial exploitation begins. For the western Indian Ocean tuna, this occurred more quickly than in most cases, because of the intense interest in the Law of the Sea accords that were being finalized in the early 1980s, a period when this fishery had begun to show signs of success.

This meant that the Seychellois would have to terminate the prospection phase, send foreign vessels out of the region, and negotiate from the position that they were the owners of an untapped resource, and that access to that resource should be beneficial to its owners. This scenario was played out between June and November 1982, during which period legal access agreements were negotiated and a mechanism for giving out access rights was implemented.

This was no trivial accomplishment, given the sorts of problems faced by small island nations having to deal with highly mobile, sophisticated fishing fleets, with large multinational, vertically integrated industries, and with the individual vessel's flag government. For example, by 1986 Seychelles had signed major fishing fee agreements with the European Economic Community (EEC) for 27 (French) vessels, with the Spanish government, with private Spanish firms, and with the Ivory Coast. Now that Spain is a member of EEC, another 20–30 vessels fall under EEC negotiations and considerable pressure has been brought to bear on the Seychellois negotiators, as the EEC conglomerate interests are better situated to wrangle political deals that tend to work against Seychellois interests.

To date, the relationships with early partners have been beneficial to the Seychelles in many regards and, given that the global economy is somewhat more stable than it was when these negotiations began, the future prospects are favorable.

With the initiation of such a major fishing activity, amidst historical political turmoil and ignorance about the resource base, it was imperative to develop an information system that would provide timely, accurate estimates not only of catches, but also of the whereabouts of fleet activities. The FAO, with funding from the Japanese Government, provided the technical assistance that was needed. More detailed information about the fleets and their catches are collected by ORSTOM (Office of Overseas Scientific and Technical Research) staff members based in Seychelles and Mauritius. As noted earlier, a full-scale Observer Program is being carried out, providing continuous at-sea monitoring of the fishery.

Recent economic analyses have shown that the benefits of the fishery are widespread within the Seychelles, and that less than one-fifth of these benefits derive from the license fees collected. Substantial benefits accrue from the payments for port fees, stevedores, fuel, food, and assorted other activities associated with the brief visits to port by each of the vessels.

Among the early concerns that faced the Seychellois were questions related to their own participation in this fishery: what functions might be most likely to help sustain the fishery over the long term, and what their level of participation might be. The costs of building and buying modern purse seine vessels have become prohibitive, particularly considering the up-front costs per job created (estimated to be about US$360,000 per job), all of which presumes that the global tuna markets remain at the least stable, and that expertise can be obtained quickly, and that technology transfer can proceed apace of the nation's needs.

The obvious benefits – such as those of building a transshipping facility as well as a cannery to provide a mechanism for producing value-added products, reducing shipping fees from refrigerated bulk round products to canned, readily marketable, dry shipped products – were recognized and carried out early. The cannery opened in 1987, and the transshipping facilities evolved from their start in 1983. Efficiency and rapid turn-around were also recognized as key ingredients for such capital intensive fishing activities as purse seining, where each day lost in port can be considered as a direct loss of more than US$15,000.

While direct participation in the fishery remains somewhat dubious, other than through apprenticeships as individual Seychellois venture into fishing activities, some pressures remain to provide a background tuna production capability. Such a move is necessary, given that the European-based vessels might, in some unforeseen situation, be forced to abandon the Indian Ocean fishing grounds for those closer to home.

Development of joint ventures throughout the Seychelles tuna fishery has helped to minimize both economic threats and problems of a lack of local expertise in many aspects of the fishing, processing, and global marketing operations.

The Seychelles government has set six major objectives: (1) to create the maximum number of work opportunities; (2) to maximize foreign exchange earnings; (3) to create optimal linkages with the other sectors of the economy; (4) to insure stable development of the industry; (5) to conserve marine resources in order to ensure the long-term viability of the industry; and (6) to establish Mahé as an important tuna center of the Indian Ocean.

A careful consideration of the available options by the Seychellois has provided a useful lesson in the economically and politically perilous tuna fishery investment ventures.

Other regional developments

There has been a series of important changes in the Indian Ocean tuna infrastructure and processing capabilities that has made the development of the Indian Ocean surface tuna fisheries viable. The first was the investment in and development of tuna processing facilities in Thailand. These facilities were established in response to the general collapse of the Hawaii- and California-based tuna processing capacities because of rising domestic labor costs and the recent decade's sharp increase in environmental concerns.

Thailand recorded zero canned tuna product exports in 1980; then 4,679 mt in 1981, and a subsequent doubling of output each year until 1986, when nearly 142,000 mt of products were shipped. It is estimated that Thailand shipped 170,000 mt of canned products in 1988, and that they would ship 180,000 mt in 1989 (T. Sakurai, IPTFMP, Colombo, Sri Lanka, personal communication). These exports were originally directed to the US market, but markets have since been expanded to include EEC countries, Canada and Japan.

Thailand has become a major fresh-frozen tuna importer, in spite of the continuous growth of its domestic tuna catches which range from 80–100,000 mt per year. At present it imports about 200,000 mt of frozen tuna per year. Price competition, trade protectionism, dependence on imported raw materials and varying market shares will continuously threaten this canning activity, as the global tuna markets in the future evolve and change.

Thailand is the single largest exporter of canned tuna in the world and now produces more canned tuna products than any other nation except the US, which has most of its canning operations in its island territories (Puerto Rico and Samoa) in order to capitalize on lower labor costs and less stringent facility and environmental standards.

The growing tuna processing capacities around the Indian Ocean provide an important link in the marketing chain, with value-added products being shipped (as opposed to products in the round, which are quite costly to transport due to requirements for refrigeration). However, proliferation of these facilities will lead to problems.

Recent investments in tuna processing facilities in the Maldives have aided that nation's economy and stabilized a situation that threatened the very existence of their age-old fisheries. Many of its problems arose as a result of foreign interests and interventions such as the mechanization of traditional *dhoni* fishing craft which, with the parallel development of a tourist industry, had severe immediate effects on the expectations and behavior of Maldivian boat owners and operators.

All of these problems affected the cultural traditions and social structure that had evolved in response to an unhealthy dependence on a single economic sector, that is, fishing or tourism. Progress all too often occurs at the expense of long-term traditional methods, and if there is one place in the world where a population had come to some semblance of balance with the natural stresses of the ocean climate, it was the Maldives – a collection of over 1,000 coral islands where the only resource and unit of barter, until recent decades, had been the skipjack tuna.

Early mechanization activities were initiated by the Japanese as part of a development effort that included a mobile tuna "collector vessel" marketing network that was eventually declared unworkable, but only after the majority of the sail-driven traditional Maldivian fleet had been converted to petrol-dependent propulsion systems. The fact that these islanders had conceded to change from a wind-driven mode to a petroleum-based propulsion system did not adequately prepare them for the consequences of "fickle" partners. They had, however, already weathered the recent collapse of their "Maldive fish" markets in Sri Lanka.

The collapse of the Japanese-sponsored collector boat service resulted in the loss of an important market (as well as the only hard currency source for petrol); as a result, the economic viability of that fishery simply collapsed. Many of the powered fishing vessels left the fishing fleet to engage in island-to-island ferry service, or to engage in other tourist-supported activities. A national crisis ensued which could only be resolved by somehow developing the processing capacity and market economy that would reconstitute the traditional fishery.

Maldivian tuna processing capacity was recently developed, along with appropriate technology transfer, such as the improved development and broader implementation of fish aggregation technology that was initiated by FAO Fishery Department staff in the early 1980s, and improved fish handling techniques. The Maldive tuna fishery has, thus, taken its place as a leader in island-based production, particularly based on the activities of local fleets.

Although rapid changes are now in progress, it is useful to mention the Indonesian tuna fishery developments, as a contrasting situation. In the extensive Indonesian archipelago and its surrounding EEZ, there exist many commercially valuable tunas and tuna-like species. The shoal eastern regions are limited because the habitat, though expansive, is not of sufficient volume to contain large abundances. The western regions and the contingent deep open ocean provide a remarkable potential, provided that Indonesian society could make the necessary changes to accommodate the needs of the world's tuna markets.

Indonesia lacks the high-seas fishing traditions and shore-based infrastructure to provide the appropriate platform for the developments that have occurred in the western Indian Ocean. For example, although the catch of a broad array of tunas and tuna-like species was reported at over 18,000 mt in 1979, with all the growth and development in the western Indian Ocean, Indonesia's catches totalled less than 58,000 mt in 1986, of which only about 22,000 mt were globally marketable tuna species.

To summarize, there is no "general" situation around the Indian Ocean basin. After several decades of failed tuna ventures, poor planning and investment, mostly attributable to misguidance related to technology transfer, or to the uncritical acceptance of assessment and conservation practices, the success of several of these Indian Ocean developmental efforts stands out as worthy of careful study *prior to initiating any other fishery development projects.* A major problem that has plagued developing regions has been difficult environmental settings, including extreme climates or bizarre oceanographic phenomena, such as the strong contrary wind and surface current structures found off the coast of East Africa.

For example, the author examined historical documents and scientists' field records available at the Zanzibar Marine Station. These notes and documents described strong seasonal current and wind patterns that would be quite difficult or damaging to seine nets, if deployed in the channel between Zanzibar and the mainland. Based on these reports and personal field visits, the Investment Bank did not develop a purse seine fishery based in or near the Zanzibar Channel. Collation of ocean thermal and wind field records from the MOODS file (and other data) showed that the current and wind speed dynamics were so extreme as to preclude safe, economic handling of expensive gear such as modern seines made of synthetic materials. After the author's visit to Zanzibar in 1982 with an FAO master fisherman (Captain Bobby Lee) and an experienced tuna vessel observer and aggregation-device deployment expert (Charles Peters), the consensus was that the only future for a more modern tuna fishery off Zanzibar was to the east, in the open ocean, and that would require extensive acculturation of local fishermen whose religious and social practices would not permit overnight fishing trips. In addition, Zanzibar did not have the port infrastructure to provide for even the most obvious on-demand needs such as fuel and food. Such issues, though extremely important to the development of the fishery, have rarely been considered. However, in this case the physical climate alone was sufficient reason to dissuade investment-oriented parties (Sharp, 1982b).

Indian Ocean climate histories and prognoses

The unique seasonality of the Indian Ocean and the subregional monsoonal patterns are affected by the differences between terrestrial and ocean temperatures. The double monsoon pattern reflects quite different seasonal hydrological regimes. The northeast monsoon which spans December through March is primarily driven by high sea surface temperatures and colder winter continental temperatures. This results in intense tropical storms and an area of high humidity ranging northward from the ITCZ, the position of which varies strongly from year to year (from 10° to 15°S latitude in its southernmost position).

The southwest monsoon, which dominates the ocean current system and climate off eastern Africa from June through September (but April through May from Thailand eastward), with peak seasonal rainfalls in June and July, is driven by terrestrial summer heating. The inter-monsoon periods, April–May and October–November, are relatively calm, with characteristic equatorial westerlies.

Ocean currents and oceanographic properties are functions of the seasonal wind changes. The northern Indian Ocean coastal areas are most affected, as reflected in its ocean dynamics. The Somalia Current at the western boundary of the Indian Ocean is very responsive to seasonal changes. The thermal structure, as measured by shifts in the thermocline depth, varies from 50 to over 200 m during the course of weeks to months.

Tropical tuna habitat in the Indian Ocean extends from the northern coastal boundaries to about 25° to 30°S latitude, with the greatest southerly extension occurring in the southern summer months. The seasonal thermocline varies dramatically from region to region, and in the the western region (west of 80°E), it is "shoalest" (most close to the sea surface) in the April–May inter-monsoon and deepest, and therefore less constraining to the fish during August–September, at the end of the southwest monsoon period.

Early prospection by the French vessel *Yves de Kerguellen* showed strong seasonal changes in the distributions and relative abundances of the two major tuna species, skipjack and yellowfin, as the initial fishing year progressed. As more vessels joined the fishery, a more opportunistic approach to fishing ensued and catch rates tended to even out, as evident from the statistics in Table 16.3.

There remains, however, considerable year-to-year catch variability. This is to be expected, particularly considering the historical perspectives that are provided by long-term climate summaries records such as those in the comprehensive ocean and atmosphere data set (COADS), which provides options for exploring historical observation information from the nineteenth to the twentieth century.

The dates cited in Fig. 16.2 correspond to major climatic changes and ecological shifts that are well recorded in the global fisheries literature. For example, the Pacific sardine off Japan and eastern Asia, and California, began to bloom in about the mid-1920s, peaking in the late 1930s. The 1940s period marked the beginning of the decline in the Pacific Basin sardine abundance, as well as the Gulf of Alaska Pacific halibut fisheries. The 1955–65 cool period marked the great harvests of the anchovy off the western coast of South America, while the recent epoch of Pacific Basin sardine and north Pacific halibut expansion and abundance started in the late 1960s. Basin-wide warming trends began in the late 1960s, and the sardine populations bloomed throughout the Pacific Ocean. Following the 1982–83 El Niño-Southern Oscillation (ENSO) event, ocean cooling, particularly off South America, has been accompanied by declining adult sardine abundance, primarily because of decreasing recruitment and increasing exploitation. Recruitment failure is inevitable, particularly if ocean changes continue, for example, leading to further cooling or intensified upwelling, as part of longer-term climatic cycles, or if warming and cooling cycles increase in frequency.

Similar but shorter time series – 1947 to present – for the western boundary of the Indian Ocean are shown in Fig. 16.3. These data were also extracted from the COADS, available through the Cooperative Institute for Research in Environmental Sciences (CIRES) at the University of Colorado at Boulder (USA).

Tropical tunas are very responsive to upper ocean thermal and oxygen gradients, as a function of their foraging behavior as well as for physiological reasons. It is important to consider the effects of any unusual or systematic climate-driven ocean changes on their fisheries.

Figure 16.4 provides the MOODS file climatology for monthly one degree square SST and 90 m temperature information needed to understand the seasonal dynamics of tuna fisheries around the globe. The months of March and September are shown in Fig. 16.4, as examples. The data are arrayed in a fashion that will allow

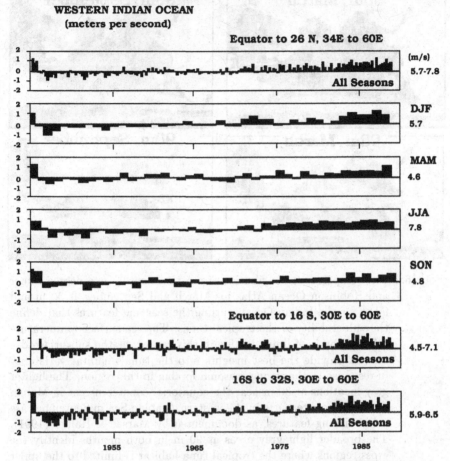

Fig. 16.3 The western coast of the Indian Ocean was divided into three climate–ocean regimes for the instrumental record period from 1947 to present, in which we have the most confidence. The four seasons are portrayed for the northernmost sector, and only the "All seasons" composites are shown for the other two.

casual browsing of the climatological mean one degree latitude–longitude, values of SST, and six levels, at 30 m depth intervals for the world ocean, as well as salinities, as devised.*

* The complete annual data are available from the NOAA Center for Ocean Analysis and Prediction, Monterey, California 93943-5005, to anyone with access to a MacIntosh II computer, and the spreadsheet software package, WingZ. The data were compiled by Margaret Robinson and Roger Bauer, of Compass Systems, Inc., San Diego, California. The MacIntosh software and formats for portraying these data were designed by G.D. Sharp and implemented by Mark Sutton, APEIRON, Inc., Dallas, Texas.

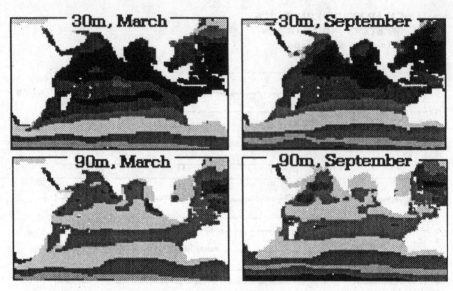

Fig. 16.4 The temperature distributions from the monthly one degree Bauer
and Robinson Ocean Atlas for March and September at 30 m and
90 m for the Indian Ocean outline the seasonal features that define
the vulnerability of the tropical tunas. The sea surface temperature
information is of limited value in this regard, as the subsurface dynamics provide the best insights into the tuna's habitat variability
in response to the strong seasonal forcing in this region. The lighter
gray locations at 30 m near the equatorial countercurrent in March,
and their counterparts in the western Arabian Sea in September are
prime fishing features, as documented by Marsac & Hallier (1991).
The broader light grey areas at 90 m in both months identify the
gross regions where the tropical tuna habitat is limited to the upper
ocean and, therefore, vulnerable to seine gear.

Resource population responses
to climate change

Issues involving climatic events and longer period climate–ocean
processes, especially those that induce much of the variability in
distribution and abundance changes in resource populations, are
quite complex.

Until recently, much of the conventional fishery assessment
community has been reluctant to embark on the necessary research and data integrations needed to promote reliable forecasting
paradigms. Since the mid-1970s, however, there have been several
determined efforts to expose these important fisheries problems.

Major advances that captured the interest of several fisheries modelers were made by Jones and Hall (1974) and Lasker (1975, 1980) to show that the major cause of fish population variability was varying reproductive success, suggesting that a closer look at what that meant in terms of critical environmental characteristics might be worthwhile.

These basic ideas are often credited to Johann Hjort (1914) under the aegis of his "critical period" early life history survival hypothesis. It is worth pointing out, however, that neither Lasker (1980) nor Hjort (1914) ignore sources of mortality either from predation or from effects of ocean transport processes. Both were interested (and rightly so) in defining the conditions necessary for survival for early life history stages, which in fact appear to be very unlikely co-occurrences in a constantly varying and very competitive biophysical milieu. They argued for the inclusion of biophysical structure at small time and space scales, scales at which few scientists even today are able to measure or discuss knowledgeably.

The observations that supported their hypotheses were almost serendipitous, in that microscale observations of ocean microplankton and chemical distributions were collected on cruises during which Lasker was carrying out early larval survival assays.

Owen (1981, 1989) shows that the microscale patchiness, as well as macroscale eddies and gyres, were of major importance to the general food particle distributions and densities and, in turn, fish population condition. Sharp (1981a, 1981b) and colleagues made the first focused attempt, since those of Sette (1943, 1960), to integrate the physics, ecological parameters, and first feeding larval fish data into a coherent, unified and verifiable research topic.

All this led to yet another effort to create the next level of relevant ocean information and perhaps to an informed consensus on how to begin to rethink the broader issues of aquatic population changes, whether climate–ocean driven, fishery-related or, most likely, a combination of the two (Bakun et al., 1982).

In response to the initial problem of stimulating awareness that fish populations vary in response to physical climatic processes, the FAO convened the Expert Consultation to Examine Changes in Distribution and Species Composition of Neritic Fish Resources, held in San José, Costa Rica in April 1983 (Sharp & Csirke, 1984). Of the 53 contributions, 20 were descriptive regional reviews (most of which implied that environmental fluctuations dominated the observed variability), 24 of the remainder were related to dominant climate–ocean variability and/or weather effects (as sources of resource population variability); five were technical–behavioral papers; and the remainder discussed either societal impacts or were quasi-theoretical.

During this Expert Consultation, it was noted that the El Niño (warm) event then in progress would provide the emphasis that no fishery was immune to climate variation. If the North Sea had been subjected to El Niño events, the development of fisheries science would most likely have evolved very differently.

Since the publication of the proceedings (Sharp & Csirke, 1984) and reports (Csirke & Sharp, 1984), there has been a remarkable convergence on issues surrounding the techniques for accounting for climate-driven ocean variability and subsequent fish population responses, particularly those associated with climate-related population recruitment issues. (See, for example, the proceedings of the International Symposium on the Long-Term Variability of Pelagic Fish Populations and Their Environment, held in Sendai, Japan, 14-17 November 1989.)

This is not to say, however, that the mostly academic, physical and biological oceanographic or the meteorological communities have decided, *en masse*, to contribute to the applied aquatic resource sciences (Wooster, 1987). It appears that nations with strong economic and cultural ties to fishery processes (e.g., Japan, Norway, Chile, Peru, Spain, and South Africa) have followed the advice of the 1983 FAO Consultations. They have begun major observation and research programs which integrate ecological processes with climate research studies in order to devise strategies that might carry their fisheries through extended periods of low abundance of their more valued fish resources.

Even in those fisheries for which it has been clear for decades that the populations behave or respond differently to different environmental changes, there has been a reluctance to accept the dominant role of environmental forcing in changes in fisheries catches as measured in the form of changes in size, apparent biomass, and/or catch rates. For example, in the eastern Pacific Ocean, where much of the basis for my approach was developed, the tropical tuna fisheries have waxed and waned for decades, with strong and weak year classes. However, conventional assessment methods (e.g., virtual population analysis and catch per unit effort) promote over-integration of the subregional population data, tending to smooth out the local reproductive and catch variabilities. Apparent "stability" results, as any local or subpopulation differences get lost in the integrations. While this may be a simple function of management philosophy, it is not necessarily good science. The difference is dependent more upon the separation of subregional data into coherent age and size classes, rather than distortions of the inherent relationships via gross integrations of poorly sampled catch data that are typically provided.

For example, Sharp and Francis (1976) showed that if the eastern Pacific Ocean fishery was divided into three geographic segments from north to south, each with distinctive size and age distributions, there would be three clearly independent patterns of response over time, in terms of numbers of recruits per semester (half year) by area; the intermediate area benefiting from sporadic large recruitments either from the northern or southern extremes, or vice versa, but the northern and southern areas would be completely independent.

Each of these three eastern Pacific subregions exhibits recruitment variations on the order of 5 to 10 times from year to year, as in many other pelagic fisheries. However, the annual catch rates for the fishery as a whole tended to decline almost monotonically from peaks in about 1972 until the peak of the warming periods that culminated with the 1982–83 ENSO (warm) event. The only exceptions were those years with cooler (shoaler) thermocline conditions, in which the tuna habitat was compressed toward the surface, increasing the visibility and vulnerability of the schools.

After the 1982–83 ENSO event, the ocean cooling period and shoaling of the tuna habitat within the region, beginning in late 1983 through 1986, in the eastern Pacific, a series of years occurred with very high catch rates, providing credence to my argument. The small ENSO warming of 1987–88 led to another catch rate decline, followed by a remarkable upward surge, as the cooling trend continued across the eastern tropical Pacific Ocean.

In fact, it is during these cooler periods, with shallow habitat and hence highest catch rates, that the maximum threat to the resource occurs, and not during the warming periods when the catch rates tend to decline. There is far more access to the resources, and less refuge for the fish. This goes entirely counter to the conventional logic of catch-rate based assessments, which is why it is imperative that ocean fisheries should not be managed under the same aegis as coastal pelagic fisheries. Climate is the dominant force in ocean fisheries, and its effects must be accounted for before anyone will be successful in managing these fisheries. The case of the Indian Ocean, in which intensive ocean monitoring has been carried out along with the fisheries sampling, provides the bases for this approach.

These types of dynamics and time scales are typical of tuna fishery responses, and are direct evidence that there needs to be far more attention paid to the environmental dynamics than those in management circles have advocated to date. The western Indian Ocean, a much more dynamic and energetic region than any other tropical region in the world, will certainly be affected by similar changes due to climate changes and interannual ocean dynamics. Such changes might well affect not only the fleet distribution, behavior, and catch composition, but also the processing capabilities of many of the processing plants.

The fact that the western Indian Ocean fishery attracted from the eastern Atlantic about 50 modern purse seine vessels (about half the total number operating in the eastern Atlantic) has been the economic salvation of the eastern tropical Atlantic tuna fishery, not only by decreasing competition, but also by allowing increased escapement that provides for a generally higher yield per recruit, and healthy catch rates that will sustain what was rapidly becoming another subsidy nightmare.

The slow, careful growth of the western Indian Ocean tuna fleet, controlled by access policies that favor nurturing not only of the resource but also the infrastructure, has made this region's tuna fishery the most likely candidate for truly rational management. Cynicism aside, however, greed, shortsightedness, and Western European market policies can quickly change this scene into typical North Sea management chaos, through overindulgence in conventional rituals of management that drive the competitive motivations and, hence, the resources into a typical economically inefficient spiral. Worse, of course, would be the over-building of processing capacity in order to sidestep the single-source issues that exist. All increased capacities will have diminished economic potentials, and lead to problems for all facilities during periods of low abundance.

While the marine mammal issues that plague the eastern Pacific Ocean tuna fisheries are of little consequence in the Indian Ocean, there are still many other problems that must be monitored so that this fishery will thrive. The majority of the problems will revolve around maintaining the strength of gear access legislation and care for the infrastructural health that has already made the western Indian Ocean regional situation a showcase among other examples around the world.

Observations, speculations and suggestions

Dr. Timothy Barnett of Scripps Institution of Oceanography (California) found that the recent decades' strong correlations between SST in the eastern Pacific Ocean and solar activity appeared to fall apart for earlier records from about the mid-1920s (Fig. 16.5). R.Y. Anderson (1990) posits that solar activity modulates ENSO frequency. The sunspot numbers (Fig. 16.6) are related with solar energy emissions, hence following the empirical engineering principle where oscillating systems are apparently modulated by solar input. In this case ENSO (warm) events occur less frequently, separated by extended tropical ocean cooling periods. If Anderson's interpretations are correct, Barnett's findings might even be expected, given that the annual sunspot numbers shot upward from about the 1920 period from a nearly constant, relatively lower value for the 1880–1920 period.

Long- and short-term trends, as well as forcing functions, complicate the issues surrounding recent emphases on anthropogenic changes. This, however, should not dull one's interests in separating out the latter from amongst the strong natural changes that are not only unmanageable, but also are of great magnitude. Since the upper ocean heat content is at least an order of magnitude greater than that of the atmosphere, the upper ocean mixing, and related heat transfers and transformations, have surely been underestimated by atmospheric modelers among the more critical climate mediating processes related to the earth's nonatmospheric energy contents and transfers on all time scales.

Similarly, fisheries variability has been documented in nearly every region of the world to vary on similar time scales, and in many cases can be shown to be changing in phase with climate signals, no matter what their source. Soutar and Isaacs (1969)

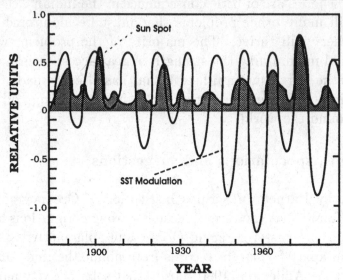

EASTERN PACIFIC SST MODULATION vs SUNSPOT NUMBER

Fig. 16.5 The sunspot frequency (gray pattern) for the period from the 1880s
to the late 1980s appears to be very well correlated with sea surface temperature (oscillating positive-negative anomaly record) in
the eastern Pacific Ocean during the period from the mid-1980s back
to about the mid-1920s. Prior to that period the records are far less
convincing (personal communication, Dr. T. Barnett, Scripps Institution of Oceanography).

documented 2,000 years of ecological changes of similar magnitude
in the sediment records from the Santa Barbara Basin off southern
California; similar studies are in progress around the world. Natural population states are demonstrably unstable, and they are not
always high in the absence of fishing. More frequently they are
low, with occasional periods of expansion, inevitably followed by
natural or human-induced collapse.

How to adapt to these processes and minimize human effects
demands that attention be paid to the entire global ecosystem, to
its major source of energy, and to the interactions from the high
stratosphere, through the marine layer and into the sea, if we are
ever to understand and adapt appropriately. Meanwhile, North
American fisheries science continues its struggle over "who was
right" and whether it is the Ricker or Beverton and Holt growth
model which applies in individual species assessments. In truth,
these models are inappropriate to address the problem of trying
to understand responses of entire regional ecosystems to climate-

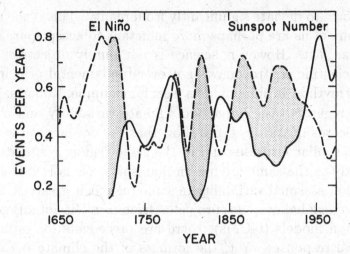

Fig. 16.6 Combining the estimates of El Niño frequency (dashed line) and
sunspot numbers (relative numbers portrayed by continuous line),
R.Y. Anderson (1990) showed that the records for the last 300 or
so years are strongly negatively correlated, from which he concludes
that solar energy may modulate ENSO frequencies.

driven ocean variabilities, with man pecking away at the biomass
along with an array of other competitors.

After the collapses of the California sardine fishery in the early
1950s and the Peruvian anchoveta fishery in the late 1960s and
early 1970s, it is clear that recovery from such conditions, if at all
possible, will require some combination of complete fishery closure
and a long waiting period for the next round of climatic changes
that favor a period of natural population growth. Many fishery
scientists now believe that single-species fisheries management and
associated assessments that ignore climate variation are outmoded
(Bax, 1991; Caddy & Sharp, 1986). Some within the marine ecological community have trouble dealing with the fact that climate
is the dominant force in long- and short-term fishery resource production in both oceanic and limnological contexts. Others, however, have accepted the relative importance of climate impacts on
these ecological changes. We remain relatively ignorant of specific
mechanisms that dominate local ecosystems at any place or point
in time, beyond the basic seasonal production patterns and their
perturbations. The last decade's collapse of support for ocean
monitoring and shipboard research has hindered understanding.

Where adequate population- and climate-related data have been
collected, the models, and their sustaining theories, have too often

been shown to deviate significantly from reality. This is the reason why many of us are perhaps more reluctant to accept pure model output as fact. However, science is not purely objective; much of the scientific controversy is not over fundamental principles of faith, or mythos, in spite of facts (see, for example, Roberts, 1989).

With regard to the impacts of climate variability and change, many records of various types show that the ocean, even within a basin at similar latitudes and in the same region, responds quite differently to the same forcing mechanisms such as ENSO events. Given that seasonal variability in some regions is greater than the annualized global averages projected from most (if not all) general circulation models (GCMs), there are precedents for estimating ecological responses. Yet, the formats of the climate–ocean scenarios produced by the present generations of GCMs are to date inappropriate for projecting with any degree of confidence the consequences of these scenarios.

The transitional nature of the ocean and the atmosphere is well embedded in the response mechanisms of all terrestrial and marine ecosystems; the paleoclimate records document this. What appears to be missing from the logic of the general circulation models is the importance of seasonality, regionality, and specified linkages to ecological systems that would permit reliable regional scenarios and their consequences to be developed.

For example, monsoon rains and floods account for sporadic occurrences of thousands of human deaths each year in many areas, particularly around the northern Indian Ocean. Changes in the frequencies, intensities, durations, and locations of these events will be of greater societal concern than "statistically smoothed" statements about global-averaged surface temperatures and soil moisture coefficients. Clearly, it is time to revise the approach being taken in order to reflect the most pressing requirements for defining impacts on society.

If global heating phenomena take place from the equatorial region, then we have some degree of experience and untold opportunities to obtain valuable insights from sediments and other proxy records with seasonal, and annual-to-centennial resolutions of these types of processes. If arguments can be proposed for another mode of change different from those that have been documented to have occurred in the recent glacial and inter-glacial periods, then it is important to get those mechanisms described as soon as possible. This requires carefully reasoned definitions of transitional processes, and their array of interactions.

It is clear that marine and terrestrial populations respond on short seasonal scales to all levels of perturbations. Biological indicators have been described that define nearly the entire spectrum of climatic regions, terrestrial as well as aquatic. Zoogeography is the basis for much of our current understanding of both near-term and paleoclimatic variability.

In the marine environment there are many clear-cut indicator species. Although our knowledge is greatest from harvested species, one does not have to depend only on commercially harvested species as status and trend indicators. This is so because some clearly identifiable species are already useful indicators of ocean boundary and vertical thermal partitioning processes (cf., Loeb & Rojas, 1987). For example, tuna fisheries around the world appear to benefit from increased recruitment in years following El Niño events (Suzuki, 1989). If this is because of the general warming of the upper ocean, and associated biophysical processes, then one might have to conclude that the ocean's higher trophic level production is in general actually enhanced during strong transitory heating periods.

In developing climate change (and consequences) scenarios, competing hypotheses need to be examined. For example, it could be that the apparent lack of new primary production during warm periods, shown from long-term studies, might actually be confounded because that new production may perhaps be quickly sequestered in the higher trophic levels (during the early life history stages when more direct advantage can be taken of local production blooms).

That so many species, both predators and prey, seem to "bloom" during the warming trends (e.g., sardines, Pacific halibut, tunas, etc.) belies the argument that warm periods are not favorable for biological productivity in the ocean. These species have extensive energetic demands, relative to their counterparts, and there is obviously some substantive mechanism that provides for blooms, colonizations, and broad spatial distributions of these populations on local to basin-wide scales. Cooling periods are tracked by other species such as those associated with coastal upwelling systems, that is, anchovies, or those in open ocean systems, such as the *Todarodes* squid.

This suggests that the warming of the upper ocean forces the interaction of water masses, both vertically and horizontally, and that the species that can forage over the broader opportunity fields have an advantage over those limited to, for example, coastal up-welling environments (e.g., anchovies). The opposite conditions benefit other species, as evidenced by the natural oscillations in dominance between these groups, for example, between sardine and anchovy or between tuna and squid.

Another group of species that would respond in synchrony with various scales of climate variations are those that depend upon freshwater flows in estuaries, such as shrimp (Garcia & LeReste, 1981), anadromous (Pearcy, 1984), and obviously also catadro-mous species. The Indian Ocean resource base is a mosaic of such estuarine and coastal species. As such, any seasonal changes in the hydrological cycle will affect each region to a greater or lesser extent (Pauly & Navaluna, 1984).

While there are many regions where tropical tunas appear to behave primarily as oceanic predators, the entire northern Indian Ocean coastal area is inhabited by *T. tonggol*, the long-tail tuna, which thrives on shrimp and many other estuarine species. In fact, the northern coast of Somalia is home to a tuna fishery which har-vests long-tail, yellowfin, and many other scombroids, within the nearshore (shelf) environment. From this one would infer that any changes that affect nearshore productivity would certainly af-fect these opportunistic predator species. Long-term monitoring of such areas is necessary, if one is to expect to project, as well as to compensate for, marine resource responses to climate-induced environmental change.

Open ocean tuna fisheries, on the other hand, are likely to be-come even more versatile. If the seasonal monsoonal gradients should shift, invoking greater surface winds, or greater current strengths than presently exist, there would be large areas of the western Indian Ocean that would become unfishable for purely geophysical reasons. The Somali Current provides an example; there are periods of the year (e.g., June and July) during which strong surface and subsurface currents produce shear forces which make it difficult, if not impossible, to deploy purse seine gear. However, the traditional *dhows*, *dhonis*, and outriggers that ply these coastal regions have evolved such that they employ break-away stays and rigging in order to minimize the adverse effects of extreme weather. They will likely persist, as have their crews' cultures, for millennia.

Without a doubt, the long-term trends with regard to climate change will affect local coastal environments and island communities within the Indian Ocean, yet each one is likely to respond in different ways to such changes in their regional environmental settings.

The Seychellois will have the advantage of the interests of the European Community, as "investors" in the western Indian Ocean fishery system, but it is not likely that the Maldivian culture will have become so changed over the next decades as to alter their traditional survival strategies. The Maldives' extended geographic position within the ocean ecosystem is such that they will be able to take advantage of nearly any latitudinal shifts in availability of their prey, and the geomorphology of their situation provides the unique focus for aggregations of tunas, and other species that will provide some level of needed support. However, a sea level rise may simply erase the Maldives from the surface of the ocean.

I have not emphasized, except for Thailand, the position of eastern Indian Ocean tuna in this chapter. However, it should be noted that, except for the southern bluefin tuna, there are few developed fisheries in the region, although there is a growing Indonesian-based fishery. Much of this fishery's growth will depend on the development of local technical expertise, and the evolution of collection and marketing systems. There is an opportunity within this region to harvest far more tuna, particularly skipjack and small yellowfin, than is presently taken.

Among other opportunities yet to be developed are subtropical fisheries for albacore tuna, particularly off southern Madagascar and the coast of South Africa. While these do not provide comparably dense schools, nor aggregate well enough for large fishing vessels to operate economically, smaller crafts with smaller crews, along with relevant operational oceanographic support via satellite, could provide a basis for significant catches in the future. It may be decades, however, before such a fishery evolves, given the harsh lifestyle that such a fishery would impose.

Climate change scenarios emanating from research institutions in industrialized countries in the Northern Hemisphere will not necessarily impress Indian Ocean fishermen who are already coping with some of the most dynamic systems in the world; these fishermen have come to expect change. One must wonder, however, about the stability of the infrastructure that has quickly developed in the region, and the possible effects of changes in climate and ocean on these fragile Indian Ocean ecological settings.

Conclusions

While many scientists have emphasized the importance of climate change on the production of fisheries, few have sought to utilize available knowledge in the construction of various possible alternative scenarios.

General circulation models have yet to include a coupled ocean–atmosphere interaction that can provide reliable regional projections about consequences one might expect to accompany a greenhouse gases-induced global warming.

Forecasting the ecological consequences based on climate change scenarios is not a particularly rigorous scientific activity within the contexts of what kinds of climate changes should be expected, that is, rates of change, or the causal links to the various potential responses by individuals, populations, economic sectors, and ecosystems.

Considering that the one characteristic of biogeochemical records is the relative strengths of the seasonal signals, it might be wise to search for empirically based solutions in order to improve the credibility and, more important, the reliability of climate change projections on which to base the development of ecological consequence models.

At the least, the climate modeling communities must focus more attention on the regional aspects of global warming, that is, on the local and seasonal weather patterns and their impacts. This is especially important for the coastal ocean and ocean transition zones, where the ecological responses will be strongest, and where the greatest impacts on marine ecosystems and, hence, on fisheries, will occur.

International meetings have recently focused on long- and short-term variations of marine populations in regional contexts. One fact persists. The species complexes within these dynamic regions respond to oceanic and atmospheric forcing on a much broader scale and more rapidly than our climate observing systems or our understanding of their information can discern.

Analogous problems and approaches have been the basis of many arguments within the ecological and economic modeling communities for decades (Clark, 1989). The historical resolution of these arguments has almost always favored the empiricist modelers rather than the mathematical modelers. Both groups use tools that are relatively blunted through "necessary" but often unrealistic simplifications, as well as by simplistic methods of parameterizing mostly unstable variables.

There is already a considerable wealth of knowledge about effects of changing such parameters as light levels (cloud cover), temperature (greenhouse or long wave re-emissions), or even rainfall, soil moisture, or surface water flow patterns (hydrological cycle), particularly on local terrestrial ecosystems. However, our knowledge of the remaining 70 percent of the earth's surface (the oceans), and the thermally active upper ocean in particular, is rudimentary at best.

We must face the problem of monitoring the ocean, and its contents, with an eye toward building data-driven empirical models, from which future scenarios might be devised. Meanwhile, we must rebuild the *in situ* ocean and near-shore observation system needed to better understand the effects of climate change on biogeochemical processes and their climatic interactions. The importance of aquatic organisms, as indicators of status and trends in periods of potentially rapid change, cannot be overstated. It is important to convince technologists, physicists, and bureaucrats, among others, that these well-adapted populations are far more responsive, and informative, over appropriate time and space scales than newly hatched technologies and satellite sensors which tend to degrade at rates faster than the regional climate can change.

Acknowledgment

I continue to appreciate that there have been many before me who have contributed to my knowledge and subsequent applications of these basic principles, from Kishinouye (1923), Uda and Nakamura (1974), Hela and Laevastu (1962), Green (1967), Magnuson (1973), Saito (1973, 1975), Saito and Sasaki 1974, Hanamoto (1974, 1975), and so on, to the lifeworks of Fred Fry and his followers. Each has tried to teach and apply the underlying logic which I learned somewhat independently, from many years of work aboard many types of tuna vessels all around the world, from empirical study, happenstance, and examination of many records and even anecdotal histories of the dynamics of many tuna fisheries.

References

Anderson, R.Y. (1990). Solar-cycle modulation of ENSO: A mechanism for Pacific and global change. In *Proceedings of the Sixth Annual Pacific Climate (PACLIM) Workshop*, ed. J.L. Betancourt & A.M. MacKay, pp. 77–82.

Bakun, A., Beyer, J., Pauly, D., Pope, J.G. & Sharp, G.D. (1982). Ocean sciences in support of living marine resources: A report. *Canadian Journal of Fisheries and Aquatic Sciences*, **39**, 1059–70.

Bax, N. (1991). What can models tell us of interactions between the environment and fisheries? Paper presented at *Environmental Influences on Marine Biological Resources: Models and Methods*, Murmansk, USSR, 1–8 June 1989. (In press.)

Caddy, J.F. & Sharp, G.D. (1986). An ecological framework for fishery investigations. FAO Fisheries Technical Paper No. 283.

Carey, F.G. & Olson, R.J. (1982). Sonic tracking experiments with tunas. *International Commission for the Conservation of Atlantic Tunas, Collected Volume of Scientific Papers*, **17**, 444–57.

Clark, M.E. (1989). Letter in *Science*, **246**, 10.

Cort, J.L. (1983). *Informe de la campaña de prospección de tunidos effectuada por caneros españoles en Seychelles (julio 1981–marzo 1982)*. Madrid: Instituto Español Oceanográfico.

Csirke, J. & Sharp, G.D. (Eds.) (1984). *Reports of the Expert Consultation to Examine the Changes in Abundance and Species Composition of Neritic Fish Resources*. Seminar, 18–29 April 1983 in San José, Costa Rica. FAO Fisheries Report Series No. 291.

Garcia, S. & LeReste, L. (1981). Life cycles, dynamics, exploitation and management of coastal *penaeid* shrimp stocks. FAO Fisheries Technical Paper No. 203.

Green, R.E. (1967). Relationship of the thermocline to success of purse seining for tuna. Transactions of the US Fish and Wildlife Service. *Fishery Bulletin*, **96**, 126–30.

Hanamoto, E. (1974). Fishery oceanography of bigeye tuna, I: Depth of capture by tuna long-line gear in the eastern tropical Pacific Ocean. *La Mer* (Bulletin of the French–Japanese Oceanographic Society), **12**, 10–8.

Hanamoto, E. (1975). Fishery oceanography of bigeye tuna, II: Thermocline and dissolved oxygen content in relation to tuna long-line fishing grounds in the eastern tropical Pacific Ocean. *La Mer* (Bulletin of the French–Japanese Oceanographic Society), **13**, 58–71.

Hastenrath, S. & Lamb, P.J. (1979). *Climate Atlas of the Indian Ocean, Part I: Surface Climate and Atmospheric Circulation*. Madison: University of Wisconsin Press.

Hela, I. & T. Laevastu (1962). *Fisheries Hydrography.* London: Fishing News Ltd.

Hjort, J. (1914). The fluctuations in the great fisheries of northern Europe viewed in the light of biological research. *Rapports et Proès-Verbaux des Réunions. Conseil International pour l'Exploration de la Mer*, **20**, 1–228.

IPTDMP (Indo–Pacific Tuna Development and Management Programme) (1988). *Collective Volumes of Working Documents*, Vol. 3. Presented at the seminar on Expert Consultation on Stock Assessment of Tunas in the Indian Ocean, 22–27 June 1988, Mauritius. Colombo, Sri Lanka: IPTDMP.

Jones, R. & Hall, W.B. (1974). Some observations on the population dynamics of the larval stag in the common gadoids. In *Early Life History of Fish*, ed. J.H.S. Blaxter, pp. 87–102. Berlin: Springer-Verlag.

Kishinouye, K. (1923). Contributions to the comparative study of the so-called scombroid fishes. *Journal of the College of Agriculture, Tokyo Imperial University*, **8**, 293–475.

Lasker, R. (1975). Field criteria for survival of anchovy larvae: The relation between inshore chlorophyll maximum layers and successful first feeding. *Fishery Bulletin*, **75**, 453–62.

Lasker, R. (1980). Factors contributing to variable recruitment of the northern anchovy, *Engraulis mordax*, in the California current: Contrasting years 1975 through 1978. *Rapports et Procès-Verbaux des Réunions. Conseil International pour l'Exploration de la Mer*, **178**, 375–88.

Lee, R.E.K.D. (1982). Live bait pole and line for tuna. Report prepared for the technology transfer in Zanzibar Fisheries Project. Report FI:URT/81/T01 (UNO/URT/001/STD). Rome: UN FAO.

Loeb, V. & Rojas, O. (1987). Interannual variation of ichthyoplankton composition and abundance off northern Chile, 1964–83. *Fishery Bulletin*, **86**, 1–24.

Magnuson, J.J. (1973). Comparative study of adaptations for continuous swimming of scombroid fishes. *Fishery Bulletin*, **71**, 337–56.

Marcille, J. & De Reviers, X. (1981). *Secteurs favorables à la pêche au thon a la senne dans l'océan Indien.* Paris: COFREPECHE.

Marcille, J. & Uktolseja, J.C.B. (1984). Tuna fishing in Sumatra, Indonesia. Paper presented to the Ad Hoc Workshop on the Stock Assessment of Tuna in the Indo–Pacific Region, 20–22 August 1984, Jakarta, Indonesia. Colombo, Sri Lanka: IPTDMP.

Marsac, F. & Hallier, J.P. (1991). The recent drop in the yellowfin catches by the western Indian Ocean purse seine fishery: overfishing or oceanographic changes? In *Proceedings of the Expert Consultation on the Stock Assessment of Tunas in the Indian Ocean, Bangkok, Thailand, 2–6 July 1990.* Colombo, Sri Lanka: Indian Ocean Commission.

Messieh, S.N. (1983). Fisheries in the UAE: Present status and future outlook. Technical Report FI:DB/RAB/80/015/4. Rome: UN FAO/UNDP.

Michaud, P. (1986). Seychelles' response to rapid development in industrial fishing. *Proceedings of the Ninth Meeting of the Indian Ocean Fisheries Commission, Committee on the Management of Indian Ocean Tuna, December 1986.* Colombo, Sri Lanka: IPTDMP.

Owen, R.W. (1981). Patterning of flow and organisms in the larval anchovy environment. In *Report and Documentation of the Workshop on the Effects of Environmental Variation on the Survival of Larval Pelagic Fishes*, ed. G.D. Sharp, 167–200. IOC Workshop Report Series No. 28. Paris: UNESCO.

Owen, R.W. (1989). Microscale and finescale variations of small plankton in coastal and pelagic environments. *Journal of Marine Research*, **47**, 197–240.

Pauly, D. & Navaluna, N.A. (1984). Monsoon-induced seasonality in the recruitment of Philippine fishes. In *Proceedings of the Expert Consultation to Examine the Changes in Abundance and Species Composition of Neritic Fish Resources*, ed. G.D. Sharp & J. Csirke, 823–35. FAO Fisheries Report Series No. 291.

Pearcy, W.G. (ed.) (1984). *The Influence of Ocean Conditions on the Production of Salmonids in the North Pacific: A Workshop*, 8–10 November 1983, Newport, Oregon. Corvallis: Oregon State University Sea Grant College Program.

Roberts, L. (1989). Global warming: Blaming the sun. *Science*, **246**, 992–3.

Saito, S. (1973). Studies of the fishing of albacore, *Thunnus alalunga*, (Bonnaterre) by experimental deep-sea tuna long-line. *Memoirs, Faculty of Fisheries, Hokkaido University*, **21**, 107–82.

Saito, S. (1975). On the depth of capture of bigeye tuna by further improved vertical long-line in the tropical Pacific. *Bulletin of the Japanese Society of Scientific Fisheries*, **41**, 831–41.

Saito, S. & Sasaki, S. (1974). Swimming depth of large size albacore in the south Pacific Ocean, II: Vertical distribution of albacore catch by an improved vertical long-line. *Bulletin of the Japanese Society of Scientific Fisheries*, **40**, 643–9.

Sette, O. (1943). Biology of the Pacific mackerel (*Scomber scombrus*) of North America, Part II: Migrations and habits. US Fish & Wildlife Service. *Fishery Bulletin*, **51**, 251–8.

Sette, O. (1960). The long-term historical record of meteorological, oceanographic and biological data. CalCOFI Report No. VII. Monterey: California Cooperative Oceanic Fishery Investigations.

Sharp, G.D. (1976). Vulnerability of tunas as a function of environmental profiles. In *Maguro Gyogyo Kyogikay Gijiroku, Suisano-Enyo Suisan Kenkyusho* (Proceedings of Tuna Fishery Research Conference) Shimizu, Japan: Fisheries Agency, Far Seas Fisheries Research Laboratory.

Sharp, G.D. (1978). Behavior and physiological properties of tunas and their effects on vulnerability to fishing gear. In *The Physiological Ecology of Tunas*, ed. G.D. Sharp & A.E. Dizon, pp. 397–449. San Francisco: Academic Press.

Sharp, G.D. (1979). Areas of potentially successful exploitation of tunas in the Indian Ocean with emphasis on surface methods. Technical Report No. IOFC/DEV/79/47. Rome: UN FAO.

Sharp, G.D. (1981a). Report of the Workshop on Effects of Environmental Variation on the Survival of Larval Pelagic Fishes. In *Report and Documentation of the Workshop on the Effects of Environmental Variation on the Survival of Larval Pelagic Fishes*, ed. G.D. Sharp, pp. 1–47. IOC Workshop Report Series No. 28. Paris: UNESCO.

Sharp, G.D. (1981b). Colonization: Modes of opportunism in the ocean. In *Report and Documentation of the Workshop on the Effects of Environmental Variation on the Survival of Larval Pelagic Fishes*, ed. G.D. Sharp, pp. 125–48. IOC Workshop Report Series No. 28. Paris: UNESCO.

Sharp, G.D. (1982a). Foreword to the Symposium (November 1981, Tenerife, Canary Islands) of the International Commission for the Conservation of Atlantic Tunas, *Collected Volume of Scientific Papers*, Vol. XVII (SCRS-1981), **2**, 431–8.

Sharp, G.D. (1982b). Developing tuna fisheries and oceanographic considerations in the Zanzibar–Pembra area. Report to Project URT/81/T01 concluding contract UNP/URT/001-1(STD)/FI with Compass Systems, Incorporated. Rome: UN FAO.

Sharp, G.D. & Csirke, J. (Eds.) (1984). *Proceedings of the Expert Consultation to Examine the Changes in Abundance and Species Composition of Neritic Fish Resources*, 18–29 April 1983, San José, Costa Rica. FAO Fisheries Report Series No. 291, Vols. 2–3.

Sharp, G.D., Csirke, J. & Garcia, S. (1984). Modelling fisheries: What was the question? In *Proceedings of the Expert Consultation to Examine Changes in Abundance and Species Composition of Neritic Fish Resources*, eds. G.D. Sharp & J. Csirke, 1177–224. *FAO Fisheries Report Series*, 291.

Sharp, G.D. & Dizon, A.E. (Eds.) (1978). *The Physiological Ecology of Tunas*. San Francisco: Academic Press.

Sharp, G.D. & Francis, R.C. (1976). An energetics model for the exploited yellowfin tuna population in the eastern Pacific Ocean. *Fishery Bulletin*, **75**, 447–50.

Shomura, R.S., Menasveta, D., Suda, A. & Talbot, F. (1967). The present status of fisheries and assessment of potential resources of the Indian Ocean and adjacent seas. Report of the IPFC Group of Experts on the Indian Ocean. FAO Fisheries Report No. 54. Rome: UN FAO.

Silas, E.G. & Pillai, P.P. (1982). Resources of tunas and related species and their fisheries in the Indian Ocean. CMFRI Bulletin No. 32. Cochin, India: Central Marine Fisheries Research Institute.

Sivasubramanian, K. (1965). Exploitation of tunas in Ceylon's coastal waters. *Bulletin of the Fisheries Research Station (Ceylon)*, **18**, 59–73.

Sivasubramanian, K. (1970). Surface and subsurface fisheries for young and immature yellowfin tuna (*T. Affinis Cantor*) off the southwest region of Ceylon. *Bulletin of the Fisheries Research Station (Ceylon)*, **21**, 113–22.

Sivasubramanian, K. (1972). Shipjack tuna (*K. pelamis L.*) resources in the areas around Ceylon. *Bulletin of the Fisheries Research Station (Ceylon)*, **23**, 19–28.

Sivasubramanian, K. (1981). Large pelagics in the Gulf and Gulf of Oman. In *Pelagic Fish Resources of the Gulf and the Gulf of Oman*, 122–44. Technical Report FI:DP/RAB/71/278/11. Rome: UN FAO/UNDP.

Soutar, A. & Isaacs, J.D. (1969). History of populations inferred from fish scales in anaerobic sediments off California. CalCOFI Report No. XIII, 63–70. Monterey: California Cooperative Oceanic Fishery Investigations.

Stequert, B. & Marsac, F. (1986). *La pêche de surface des thonides tropicaux dans l'océan Indien. FAO Document Technique sur les Pêches* No. 282.

Suda, A. (1973). Recent status of resources of tuna exploited by long-line fishery in the Indian Ocean. *Bulletin of the Far Seas Fisheries Research Laboratory (Shimizu)*, **1**, 99–114.

Suzuki, Z. (1989). Symposium on tuna fishery and El Niño. *Japanese Society of Fisheries Oceanography*, **53**, 56–87.

Uda, M. & Nakamura, Y. (1974). Hydrography in relation to tuna fisheries in the Indian Ocean. Special Publication of *Marine Biological Association of India*, 276–92.

UN FAO/UNDP (1970). Fishery survey – Somalia: the tuna fishery of the Gulf of Aden. Report prepared for Government of Somalia by UN FAO. Technical Report FI:SF/SOM/3. Rome: UN FAO/UNDP.

UN FAO/UNDP (1973). The People's Democratic Republic of Yemen: fishery resources of Aden and some adjacent areas. Report prepared for the Fishery Development of the Gulf of Aden Project. Technical Report FI:SF/DP9/12, PDY/64/501/7. Rome: UN FAO/UNDP.

UN FAO/UNDP (1981). Pelagic fish resources of the Gulf and the Gulf of Oman. Technical Report FI:DP/RAB/71/278/11. Rome: UN FAO/UNDP.

UN FAO/UNDP (1982). Pole and line tuna fishing in southern Thailand: project findings and recommendations. Report prepared for the Government of Thailand. Terminal Report FI:DP/THA/77/008. Rome: UN FAO/UNDP.

UN FAO/UNDP (1985). Bay of Bengal programme – tuna fishery in the EEZs of India, Maldives and Sri Lanka: Colombo, Sri Lanka, marine fishery resources management in the Bay of Bengal. Technical Report RAS/81/051, BOBP/WP/31. Rome: UN FAO/UNDP.

Wooster, W.S. (1987). Immiscible investigators: Oceanographers, meteorologists, and fishery scientists. *BioScience*, **37**, 18–21.

17

Climate variability, climate change, and fisheries: a summary

MICHAEL H. GLANTZ

Environmental and Societal Impacts Group
National Center for Atmospheric Research
Boulder, CO 80307, USA

and

LUCY E. FEINGOLD

College of Marine Studies
University of Delaware
Lewes, DE 19958, USA

Because we are concerned about what might be the local and regional effects of and responses to a global climate change, there is a need at the outset to identify how well societies have in the past dealt with local climate-related environmental changes, regardless of cause. Such an identification will provide researchers as well as policymakers with societal baseline data. To know how best to prepare for future changes, we must know how well we cope with present-day changes, identifying societal strengths and weaknesses. To accomplish this, the "forecasting by analogy" approach is used, not to forecast future states of the atmosphere or the ocean, but to forecast how well prepared societies are to cope with changes in abundance or availability of living marine resources, regardless of the causes of those regional changes.

Fish populations are influenced by many elements of their natural environments during all phases of their life cycle. Subtle changes in key environmental variables such as temperature, salinity, wind speed and direction, ocean currents, strength of upwelling, as well as predators, can sharply alter their abundance, distribution and availability. Human activities can also affect the sustainability of these populations through, for example, the application of a variety of management schemes or new technologies, each of which could have a different (either beneficial or adverse) impact on the state of the fishery.

Interactions within the marine environment are acknowledged to be extremely complex. A global warming of the atmosphere or, more broadly, a climate change adds to that complexity. The relationship between climate change and fisheries will not be easy to define and most likely will have to depend, at least for the near future, on generalizations derived from case-by-case assessments of past and present experience.

This climate and fisheries study is based on an assessment of 15 case studies around the globe. Most of the changes in abundance in these fisheries were induced by a combination of natural and anthropogenic causes (i.e., temperature changes and overfishing, respectively). Some changes in the availability of living marine resources were induced by political and legal action. The studies are as follows: *Alaska king crab* (Warren Wooster), *California sardine* (Ed Ueber and Alec MacCall), *Pacific Northwest salmon* (Kathleen Miller and Dave Fluharty), *Gulf of Mexico shrimp* (Richard Condrey and Deborah Fuller), *Atlantic menhaden* (Lucy Feingold), *Maine lobster* (James Acheson), *Mexican oyster* (James McGoodwin), *Great Lakes sea lamprey* (Henry Regier and John Goodier), *North Sea herring* (Roger Bailey and John Steele), *Atlanto-Scandian herring* (Andrei Krovnin and Sergei Rodionov), *Anglo-Icelandic Cod Wars* (Michael Glantz), *Polish distant-water fisheries* (Zdzislaw Russek), *Far Eastern sardine* (Tsuyoshi Kawasaki), *Peruvian–Chilean fisheries* (César Caviedes and Timothy Fik), and *Indian Ocean tuna* (Gary Sharp).

We have used the forecasting by analogy approach because the present generation of general circulation models (GCMs) of the atmosphere does not have, among other factors, the spatial resolution required for climate-related impact assessments at the regional and local levels. In addition, present-day GCMs, while showing relative agreement on temperature changes that might accompany a continued loading of the atmosphere with radiatively active trace (greenhouse) gases, show much less agreement on changes in regional precipitation and soil moisture. Furthermore, the linkage of a realistic oceanic component to GCMs has yet to be accomplished. While these scientific problems are being addressed, we have turned to forecasting by analogy.

There is a need to produce information that will be of value regardless of the magnitude (or direction) of those changes. In this regard, forecasting by analogy might be viewed as providing a win/win approach (as opposed to win/lose) to researchers as well as to policymakers. It underscores the value of improving our understanding of how societies respond to environmental stress. It provides decision-makers with baseline information about how well societies have responded to the consequences of environmental change even in the absence of a warming of the atmosphere. Whether the atmosphere warms, cools, or stays as it has been for the past several decades, it is important to improve our understanding of the interactions between human activities and climate variability.

An assessment of the general lessons provided by each case study author(s) suggests that the lessons tend to fall into one of the following categories: the economic and societal setting at the time of the change, institutions, technology, information, conflicting interests, management, and societal responses. Several of the lessons, it should be pointed out, could have been placed in different categories depending on the perceptions of the observer. Although we do not identify each lesson in the case studies by these categories, the categories were useful in identifying general areas of concern. These categories of lessons can be briefly described as follows:

1. *Setting:* Lessons in this category relate to the socioeconomic and political setting at the time of the environmental change. The degree of dependence of a community or a country on the commercial exploitation of a particular fish population could be a major determinant in the resolution of a conflict over that fish population. Lessons here are also related to the importance of the environmental setting in which the societal responses to environmental changes have occurred.

2. *Institutions:* Lessons relating to institutions focused on the role of regional scientific and political organizations in dealing with the implications of environmental changes and their preparedness to cope with these resource-related changes.

3. *Technology:* These lessons generally related to the development of new technologies and to technology transfer. Newly developed technological innovations incorporated into a fishery should perhaps be treated as if they were additional predators of the exploited population. This transfer also includes the application of existing technologies to fisheries that have, heretofore, been exploited by less efficient fishing technologies.

4. *Information:* Lessons drawn from these case studies suggest a need for improvements in biological information as well as better environmental monitoring (not just of sea surface temperature) in order to understand responses of exploited fish populations. It was also suggested that there was a need for improved socioeconomic information about specific fisheries in order to better understand the pressures that are placed on commercially exploited fish stocks. One cannot readily forecast the future status of fish populations based on catch information alone.

5. *Conflicting interests:* Several conflicts were identified in each of these case studies: scientists vs. political leaders, fisheries managers vs. fisheries biologists, politicians vs. politicians, one group of harvesters vs. another, and so forth. These conflicts often restrict timely and effective responses (even if these were known) to environment-related marine resource crises.

6. *Management:* A major concern was voiced about ecological perspectives. Should the fishery be managed from an holistic perspective or from a much narrower perspective, such as focusing primarily on biological information or short-term economic realities?

7. *Societal responses:* Lessons in this category involve responses taken in the face of environmental as well as legal change. Responses have usually included a shift in the fishing effort to other living marine resources within and outside of traditional fishing grounds, as well as infrastructural changes to take advantages of new fishing opportunities. In addition, societies will be responding to the effects of environmental variability or change long after the change itself has disappeared.

The general lessons drawn from these case studies tend to focus on issues that have traditionally been of concern to the fisheries community, and are not necessarily unique to climate change. Thus, many traditional concerns to the fisheries community will continue to require attention in the face of uncertain environmental change. At the least, this study underscores the need to seek resolutions to many of these chronic, lingering issues. Collectively, the case studies underscore the importance of assessing the impacts on different economic sectors of changes in the availability or abundance of living marine resources. They also provide a richness of information on possible responses to the consequences of climate-related environmental changes that could occur at the local and regional levels in the fishing sector.

The general findings are followed by a section in which each chapter and some of its key findings have been succinctly summarized.

General findings

- It is important to take societal changes into account when considering societal responses to the consequences of climate variability and extreme climatic events. Societies are constantly changing and will continue to do so, regardless of whether the global climate changes.

- Problems related to the management of fish populations that fall within two or more political jurisdictions suggest the need for the establishment of effective interstate or multinational regional organizations to manage shared resources.

- One approach to assessing the pressures on a particular fish population is to consider those pressures as predators. These might include environmental variability on several time scales, fishing, pollution, new technological inputs (gear, vessels, training), market demands, and so forth, each one claiming a share of the existing population. This suggests that with the addition of a new technological input, management of the resource must compensate for changes. Adaptability of the fishing industry is imperative.

- Climate change will likely have an uneven impact on the marine environment. As a result, some commercially exploited populations will decrease and collapse in response to environmental changes, while others might expand and prosper in response to those same changes.

- International fisheries managers must be prepared to make objective decisions based on scientific information at hand and not on political expediency. To wait for researchers to significantly reduce scientific uncertainty could jeopardize the long-term viability of a fishery.

- In order to improve the management of living marine resources, there is a strong need for an improvement in the scientific understanding of the interactions between societies, fish populations, and environmental changes.

- The importance of the local (and sectoral) nature of the societal impacts of a global warming, and responses to those impacts, underscores the need to educate relevant decision-makers in those localities (and sectors) about the potential effects of a climate change on their normal way of "doing business."

- Scientific uncertainty that surrounds the dynamics and exploitation of fish populations creates the setting for disagreement on policies related to their management. When management decisions are made, it is better to err on the side of the living marine resources than on the side of the commercial interests exploiting them.

- When national leaders perceive a threat to their country's economic viability, long-standing agreements and habits may no longer be tolerated. As a result, the economic, social and political settings at the time of a climate-related environmental change can determine the actions taken by decision-makers involved in fisheries management.

- Excessive, unrestricted competitive fishing effort destroys a fish population's breeding potential as well as the economic viability of the commercially exploited fish population.

- Proactive (as opposed to reactive) adaptation and flexibility in the exploitation of marine resources are keys to ecological and economic viability.

- The impacts on society of changes in abundance or availability of a living marine resource often last much longer than the perturbations causing those changes.

- *Ad hoc* responses are generally favored over longer term planned responses. As a result, there is a tendency to "muddle through," often with the hope that the favored but declining exploited fish population would return to higher levels.

- When a commercially exploited fish population collapses, there is a shift in resources (vessels, processing capacity, capital) to the exploitation of other fish populations. Such a shift generally adds pressures to those populations already subjected to heavy exploitation.

- In most of these case studies a catalyst prompted action by individuals, organizations and governments (often too late to achieve their desired goals). Usually, that catalyst was a perception of a change in abundance or availability of a valued living marine resource.

Specific case study findings

Alaska king crab fishery (Wooster)

This study focuses on the red king crab (*Paralithodes camtschatica*) stock in the eastern Bering Sea and the Gulf of Alaska. King crab abundance has fluctuated tenfold in the past 25 years. From the late 1960s king crab was the second most valuable marine resource in Alaska, surpassing all salmon stocks combined. In 1981 stock abundance fell sharply and has recovered only very slightly since then. Many reasons have been offered for the collapse: overfishing, predation, disease, and environmental change. The collapse of the fishery was devastating because the fleet was too large, boats carried big loans, and fishermen were unable to pay their debts. The eventual solution to this dilemma for the industry was to transfer effort and investment to other resources. In this study, management responses to the changes in abundance of red king crab are evaluated. Given concern about climate change and speculation about potential impacts on living marine resources, the Alaska king crab case study might provide some lessons to fisheries managers about problems that could arise with a global warming.

Lessons

- The fate of any specific fishery under changed environmental conditions is difficult to predict, but large changes in abundance and distribution of many fishery resources are likely to continue.

- The traditional response of the fishing industry to collapse is to diversify, to target other stocks, and to develop new fisheries. Success requires that transition costs be kept tolerable.

- Present methods for fishery management in the US are clearly ineffective in matching catching capacity to fluctuating resource potentials.

- An underlying concern is whether there will continue to be alternative fishery resources available as the climate changes.

California sardine fishery (Ueber and MacCall)

The California sardine fishery began in the last decades of the 1800s, peaked in the 1930s, and began to collapse after World War II. It is a classic case of the rise and fall of a fishery dependent on a pelagic species, of overcapitalization of an industry, and of too many fishing boats using new technologies to harvest a fragile, if not dwindling, resource. Its collapse spawned the rapid development of similar fisheries in Peru, Chile, and South Africa, each of which then underwent essentially the same kind of growth and decline as the California sardine fishery. This fishery can be used as an analogy of potential changes that might accompany the regional impacts of a global warming of the atmosphere and could provide lessons for proactive as well as reactive responses to changes in abundance of a pelagic industrial fishery.

Lessons

- Development-oriented government agencies can contribute toward delayed and ineffective fisheries management responses to changing environmental conditions.

- A substitute fishery will develop more rapidly than a newly developed independent fishery, because existing capital, labor, technology, and markets can readily be transferred to the substitute fishery.

- The instant availability of technology and expertise eliminates the "learning curve" and a rapid transfer of expertise, technology and processing capacity will exacerbate inherent instability in the fishery.

- The political process of establishing management institutions and the scientific process of developing predictive fishery models are much slower than the processes associated with industrial development of substitute fisheries.

- Internationally, governments and their fishery management agencies should be prepared to adopt politically difficult (and industry-resisted) management policies of deliberately constrained fishery development and should avoid politically popular but economically destabilizing subsidies.

Pacific Northwest salmon fishery (Miller and Fluharty)

The 1982–83 El Niño–Southern Oscillation (ENSO) event in the eastern and central equatorial Pacific affected sea surface temperatures and upwelling in the higher latitudes of the northeast Pacific. This is believed to have contributed to poor salmon harvests along the coast of North America, from northern California to Washington in 1983 and 1984. It has also been associated with improved salmon landings in Alaska. The purpose of this study is to evaluate the extent to which this ENSO event was responsible for poor runs of coho and chinook salmon and for socioeconomic distress experienced by commercial fishermen along the coasts of states in the Pacific Northwest. The complex interactions between climate, biological processes and socioeconomic impacts described in this study enable us to gain insights into the types of societal effects that might be anticipated with climate change associated with a global warming. This study also focuses on how this complex fishery system is affected by and responds to climate and other perturbations. The fishery system as defined here encompasses the life history and oceanographic setting of salmon, scientific research and monitoring, harvesting, processing, marketing, consumption, and governmental management of the commercial and sports fisheries.

Lessons

- Climate change will have uneven impacts. While individuals in one region may lose as a result of climate change, there may be winners elsewhere. It is easy to misconstrue the net societal effects of climatic events if one focuses only on the adversely affected regions.

- The societal impacts of changes in salmon abundance or availability in any given region are intertwined with the effects of changes in market conditions, in salmon abundance elsewhere and in regulatory programs.

- Since climate change may increase the frequency with which resource managers confront unusual climatic conditions, this case suggests that it may be valuable to devote increased attention to anticipating the effects of currently unusual climate-related conditions and to plan possible response strategies.

- Climate and other sources of variability affect the structure and management of the salmon harvesting industry. A variety of adaptations to interannual variability already exists in the fishery. These could provide insights into the possible responses to regional impacts of a global climate change.

Gulf of Mexico shrimp fishery (Condrey and Fuller)

The US Gulf of Mexico shrimp fishery is one of the most diverse and valuable in the nation. It is primarily dependent on the harvest of three closely related, estuarine-dependent species: brown shrimp (*Penaeus aztecus*), white shrimp (*P. setiferus*) and pink shrimp (*P. duroraum*). The present-day fishery is a classic example of an open access fishery which has been allowed and, in some cases, encouraged to expand well beyond the point of maximum net economic return. The US Gulf shrimp fishery had its origins in the seventeenth and eighteenth centuries. It centers around the early New World colonization of New Orleans (Louisiana) and Biloxi (Mississippi) and the settling of the surrounding cypress swamps, grassy marshes, and barrier islands. Nothing in the 300-year history of the fishery, until the mid-1970s, prepared the shrimpers for anything less than a larger cumulative harvest. Today, regional management is beginning to deal with the finite, fragile nature of these resources and their habitats. If global warming were to result (as suggested) in a severe, prolonged Great Plains drought, it may result in at least a partial drying of Louisiana's estuaries and a concommitant widespread loss of marshes and decline in shrimp yields.

Lessons

- The critical threat to this fishery – the ultimate dependence of the resources on a threatened and deteriorating environment – has been recognized for a long time. The consequences are either being felt now as with pink shrimp, or may be felt in the near future for brown and white shrimp. The industry is unprepared for a reduction in shrimp yield.

- The recent scientific consensus on the critical link between shrimp production and habitat could come in time to allow managers and industry to deal with an impending crisis in a constructive manner.

- A scientific consensus must be sought across a broad array of backgrounds and experiences and provide forums for constructive scientific debate.

- It is important to know the history, culture, and practices of those you are dealing with. For example, before the signing of the turtle exclusion device (TED) mediation report, little work had been conducted towards introducing the device into Louisiana, despite the fact that it has the largest number of shrimpers.

Atlantic menhaden fishery (Feingold)

Although uncommon as a food fish, for at least the last century menhaden have supported the largest (by weight) and most important industrial US commercial fishery along the Atlantic coast. There are two separate US menhaden fisheries, the Atlantic Coast fishery (*Brevoortia tyrannus*) and the Gulf of Mexico fishery (*B. patronus*). By 1963 Gulf menhaden landings exceeded those from the Atlantic, in terms of total catch. This case study focuses on Atlantic menhaden. Society has direct effects on menhaden's success or failure; from the simplest effects of demand for by-products, location of factories, and overfishing to more complex effects of change and degradation of habitat. In times of high product demand, menhaden are overfished. The presence of northern factories located along the coast, necessary because of proximity to fishing grounds, was overpowered by urban development, changes in land use from commercial to recreational, and the desire to end odor problems caused by the factories. Finally, the destruction of environments (particularly marsh and estuarine habitats) upon which the menhaden is dependent, as well as potential adverse changes in the ocean environment that might accompany a climate change, lead to changes in the fishery.

Lessons

- An industry must be able to adapt to changes in availability and abundance of its resource, and to create new products and uses for the resource to meet changes in market conditions in order to ensure viability of the industry.

- The use of newly available technologies, in the absence of compensating management restrictions, places a tremendous burden on the exploited living marine resource.

- Land use changes can affect the overall efficiency of fishing industries which often require specific geographic locations for their operations.

- An industry dependent on highly migratory resources with an extensive geographical range needs to be managed on a total range scale to ensure persistence of the resource.

Maine lobster fishery (Acheson)

The American lobster (*Homarus americanus*), is a hard-shelled crustacean that lives on the ocean bottom. It is found off the Atlantic coast of North America from Newfoundland to the Carolinas. Maine is the center of action for this fishery, producing more lobsters than any other state. Lobster fishing has been the most important fishing industry in Maine for more than a century. It underwent a major decline in the 1920s and 1930s but has been remarkably stable since 1947. In the 1980s lobster was the most important fishery in Maine by any economic measure. For example, in 1981 Maine fishermen landed 22 million pounds (10 million kg) of lobster, accounting for almost 44 percent of Maine's total value of all fish landed. Review of the lobster industry during its decline earlier this century could well provide insights into the kinds of environmental changes and societal responses to them that might be associated with a global warming.

Lessons

- The kinds of changes observed in the 1920s and 1930s, and the societal responses to them, are similar to those which could be observed if climatic shifts occurred in ways that damaged the lobster stock or lowered the catch. However, changes in ocean temperatures alone probably did not play a key role in the decline in the industry.

- Although lobstering, along with other major industries in the central coastal area of Maine, was depressed during the 1920s and 1930s, it still may have been a major regional employer.

- People adapted to the lack of economic opportunities by intensifying their subsistence activities or migrating from the area.

- The cause of the lobster decline in the 1920s and 1930s is not known, although overexploitation and economic factors are assumed to have played a primary role in reducing catches.

Mexican oyster fishery (McGoodwin)

Human exploitation of marine resources along Mexico's Pacific coast began about 8,000 years ago and large aboriginal populations inhabited South Sinaloa's coastal region for more than a millenium before European colonization. Archaeological studies as well as contemporary records show that the region is subjected to severe catastrophic flooding. Today, the coastal plain in the south of the modern-day Mexican state of Sinaloa is the site of Mexico's most productive inshore marine fisheries, producing shrimp, fish and, until recently, oysters and similar mollusks. The economy of the region is considerably less flexible today than it was just a few decades ago. There has been a steady shift throughout the twentieth century from an economic reliance on diverse, individualistic subsistence and small-scale commercial activities, to a predominant reliance on wage labor. At the turn of the century practically the entire coastal population had independent and diverse means of subsistence. Today, more than 75 percent is wage-dependent. These wage-dependent people rely to a great extent on income earned from working on corporate-run farms, plantations and cattle-raising enterprises, organizations which are far less able to cope with sudden ecological change, regardless of cause, than were the former subsistence-oriented fishermen and farmers.

Lessons

- Local fishermen should retain cultural attributes which stress diversified and flexible approaches to fishing, as well as diversity in their local economies (i.e., occupational pluralism).

- The setting in which these climate-related events occur determines the level of impacts as well as the effectiveness of coping strategies. Socioeconomic settings must be periodically reassessed to update coping mechanisms developed for changes in the environment.

- If South Sinaloa's rural-coastal population continues to increase while the region's economic strength and socioeconomic and political flexibility continue to decline, the impacts of future catastrophic floods in this region will be even more severe than they have been in the recent past.

- Most local people and their governments seem unconcerned about the possibility that such an event could adversely impact their lives. However, their interest could be heightened if the frequency of such catastrophic floods were to increase.

Great Lakes sea lamprey (Regier and Goodier)

As a possible analogue of some unpredictable phenomena that might accompany a global warming of the atmosphere, this case study focuses on the impact of sea lamprey (*Petromyzon marinus*) on fisheries in the Upper Great Lakes (Huron, Michigan, and Superior) between the late 1930s and the early 1960s. Prey populations in the Great Lakes include lake trout, lake whitefish, suckers and introduced salmon. Sea lamprey compete with fishermen for all the species other than suckers. In the 25-year period under review there have been several natural and social stresses on fish

populations in the Great Lakes in addition to the sea lamprey –
primarily, overfishing and pollution. The fortunes of most Great
Lakes fisheries ebbed in the early 1960s. In the Upper Lakes the
direct as well as indirect cause for the collapse of various salmonid
stocks was the sea lamprey. In the shallower areas pollution and
eutrophication also played a role. The harm done by the sea lamprey has only partially been reversed during the 50 years following
their initial irruption. It appears that, as of 1989, the sea lamprey
is under partial control in most parts of the Great Lakes.

Lessons

- The objective of controlling sea lamprey provided an ap-
  proriate focus for the establishment of joint regional interjurisdictional action with respect to Great Lakes fisheries.

- No reduction of fishing pressure was made in order to compensate for the added pressures that the sea lamprey placed
  on the lake trout and lake whitefish fisheries.

- Some of the larger fisheries consolidated their vertically integrated operations, diversified their fishing practices, and diversified their business interests to include non-fishing enterprises in order to survive.

- With the loss of a preferred fish population, there is a shift
  to other stocks, which are often already under great fishing
  pressure.

- The consequences of the sea lamprey irruption contributed
  to the disintegration of small human settlements along the
  shores.

- An ecosystem can undergo serious restructuring with the intrusion of a parasitic species. Climate-related environmental
  changes may also lead to ecosystem restructuring.

North Sea herring fishery (Bailey and Steele)

The North Sea herring is one of the world's most important
marine fish resource. It has supported major fisheries in many
countries of northwest Europe for hundreds of years. Yet, in 1977
the directed fisheries were closed following a collapse of the stocks

to a small fraction of their earlier levels. If climatic change were to result in more persistent changes in the ecosystem, then changes in herring and other stocks are likely to persist over much longer periods. In particular, this study addresses the relevance of environmental changes to the North Sea herring collapse and whether action could have been taken to prevent (or mitigate) the stock's collapse. The role of perceptions of representatives of management bodies and of the fishing industry are important concerns and an evaluation is given of what might be expected in similar instances in the future. While the scenario presented here does not directly address this eventuality, it is relevant to situations in which anomalous events recur with a greater frequency.

Lessons

- As a fishery collapses in one area, the fleet will move to other areas adding additional pressures to the existing commercially exploited fish stocks.

- General consensus on concepts of conservation and of environmental change does not necessarily lead to agreement on particular policies, as long as there is a high level of uncertainty.

- If the worst effects of a stock collapse are to be mitigated in the future, fisheries managers have to react to strong inference as opposed to scientific proof. This is likely to be contentious because there is more than one way to respond to uncertainty.

Atlanto-Scandian herring fishery (Krovnin and Rodionov)

The history of the Atlanto-Scandian herring may be considered to be an appropriate analogue in an attempt to better understand the interactions between climate variability and change, fish populations, and human activities. The warming in the first half of the twentieth century may provide a first approximation of ecological and societal responses to changing environmental conditions. During warm epochs, the appearance of strong year classes is more frequent than during cold periods, but there are exceptions. This suggests that the expected global warming will possibly be favorable for the development of the Atlanto-Scandian herring fishery.

Lessons

- Technological development in a fishery is at such a high level that the survival of a schooling species such as herring is threatened, even at high levels of abundance. Coupling these developments with climatic conditions unfavorable for recruitment, the threat of collapse is magnified.

- Climatic changes may lead to aberrations of traditional migration routes of fish. This suggests that there will be a need to strengthen regional organizations and improve regional cooperation mechanisms. In this regard the joint Soviet–Norwegian efforts in support of rehabilitating herring stocks may provide a useful model for other regions facing similar resource management situations.

- There is a need to accept efficient and timely fishery regulation measures. Symptoms of problems in the fishery appeared in the 1960s but their importance was underestimated. As a result, regulatory measures were often too little, too late.

- Uncertainties in our scientific knowledge of Atlanto-Scandian herring led to erroneous stock assessments (too optimistic) which fostered a lack of concern about the future state of the stock.

- It is necessary to take an ecosystems approach to the analysis of population dynamics, which must be taken in turn as a key component for scientifically based fishery resource management.

Anglo-Icelandic Cod Wars (Glantz)

On four occasions since World War II Iceland unilaterally extended its fishing jurisdiction, putting that nation in direct conflict with other European countries such as the UK, West Germany, and Belgium. Only the UK took a militant stance against these extensions. The ensuing conflicts have been popularly referred to as the Anglo-Icelandic Cod Wars. The analogy of the Cod Wars to the potential impacts of a global warming on marine resources lies in societal responses to changes in abundance or availability of critically important resources. Lessons can be drawn from this analogy for possible responses to the regional impacts of a global warming.

Lessons

- The economic, political and social settings at the time of a change in resource availability will be crucial to the outcome of a conflict situation. Depending, for example, on the state of the economy or the health of the fishing sector, different (even diametrically opposed) responses could be justified.

- One cannot rely on traditional international law in times of environmental change. For example, the International Court of Justice was not useful to these disputes' outcomes and, in fact, may have exacerbated conflicts in the short term.

- Countries will resort to "tie-in" tactics, bringing in other issues in order to strengthen their chance for success. Iceland, for example, brought NATO into the Cod Wars conflict and exacerbated Cold War fears by trading with the USSR, Poland, and East Germany.

- When a nation's economy is perceived by its leaders to be threatened, long-standing agreements and "habits" may no longer be tolerated.

- Regional arrangements should maintain flexibility in the face of changes in environmental conditions and must not become too bureaucratized to act. Regional political or military organizations are not effective at dealing with environment- or natural resources-related conflicts.

- Resolution of ecological problems in one location could generate new problems for fish stocks and fishing fleets in other areas.

Polish distant-water fisheries (Russek)

In the late 1970s and early 1980s the Polish distant-water fleet was confronted by what was essentially a lock-out from many of the fishing grounds it had been exploiting for decades, as a result of widespread international support for the establishment of national exclusive economic zones (EEZs). This relatively abrupt change in the availability of living marine resources to Polish vessels can be viewed as analogous to the impacts of a global climate change on local fisheries, as species shift in abundance and locations and as habitats change. Thus, the establishment of the EEZs, its impacts on Polish fishing activities, and Poland's responses to those impacts can provide useful insights into societal responses to the impacts of a global warming.

Lessons

- Access to the appropriate level of technology required for catching and processing different species in remote areas needs to be available.

- In the event of a shift in resources associated with a global warming, fishery managers should consider exploiting fauna, such as krill, at lower trophic levels.

- One should not expect to find homogeneous views about, and actions in, coping with the impacts of a global warming on fishing industries and communities.

- In order to alleviate regional differences and to make social responses more effective: (1) International cooperation should be strengthened through relevant institutional arrangements. Perhaps the existing 29 international fisheries bodies could establish a coordinating committee. (2) Knowledge about the possible consequences of a global warming must be widely disseminated to fishery administrators and managers.

Far Eastern sardine fishery (Kawasaki)

The Far Eastern sardine fishery extends back to the early part of the seventeenth century. In the earliest days of the fishery, fortunes of the fishing villages rose and fell with the increase and decrease in abundance of sardine populations. In the 1940s sardine catches started to decline sharply, ebbing in 1965 at 9,000 mt. Beginning in 1970, the sardine catch began to increase once again. Substantial catches have been obtained off the coasts of South Korea since 1976 and the USSR since 1978. If sardine landings decline again, Japanese society, as well as the fisheries, will be seriously affected.

Lessons

- The Japanese fishing industry has become dependent on large-quantity fisheries and is, therefore, vulnerable to both environmental and societal changes.

- An improved understanding of the interaction between fish populations and global and regional environmental factors needs to be fostered to protect the fisheries (and society) from dislocations caused by a sudden decline in sardine landings. This will enable scientists to make better forecasts of biological productivity on which fishing communities have become dependent.

- Ports and industries dependent on sardine landings should develop measures to anticipate and cope with problems that might accompany sharp changes in sardine availability or abundance. For example, such changes in abundance of sardines could adversely affect the availability of low-value feed, generating economic problems in the production of *hamachi*, an important protein source to the Japanese, which is only economically accessible because of the abundance, low cost, and high volume catches of sardine.

Peruvian–Chilean fisheries (Caviedes and Fik)

Major ENSO events in the equatorial Pacific Ocean alter the structure of marine ecosystems, noticeably depressing long-term yield and catch levels. Paleoecological evidence supports the view that ENSO events have affected the coastal upwelling regions off the coasts of Peru and Chile for millenia. However, only since the devastating ENSO event of 1972-73 has scientific and policymaking attention become highly focused on this phenomenon. Heightened research interests have resulted in a greater understanding of ENSO and its environmental and societal impacts. It is clear that ENSO has a major impact on regional fisheries, specifically, anchovy in Peru and sardine in Chile and that forecasts of ENSO events might enable fisheries managers to better mitigate the impacts of these potentially devastating events.

Lessons

- The combined interactions of overfishing, economic pressures, and environmental variability can lead to the demise of an otherwise productive fishery.

- Diversification with respect to targeted stocks builds resilience into the fishing industry (as well as exploited stocks) in the event of fluctuations in environmental conditions.

- Concern about overfishing and stock conservation practices in a regional context demands concerted action by Peru and Chile. These two countries have been antagonistic as opposed to cooperative neighbors. In the light of the regional shifts in living marine resources that accompany decadal climate variability, this study suggests that a climate change might best be dealt with through improved regional cooperation.

- ENSO frequency or intensity are superimposed upon these longer-term environmental changes, limiting the utility of conventional equilibrium management concepts.

- The search for a single cause of a collapse is often futile as well as misleading.

Indian Ocean tuna fishery (Sharp)

The development of the western Indian Ocean tuna fishery, particularly around the Seychelles Plateau, is recent and unique. For a variety of reasons this fishery is thriving, while similar developments in other oceans over recent decades have either been marginally successful or have failed. During the last few decades, there have been several national efforts to develop near-shore fisheries for tuna. These have been successful only in locating resources but have not been very effective in transferring the technology needed in order to extend local fishing grounds into the open ocean. Most recently, tuna development projects began in the Maldives and the Seychelles, both of which have been successful but for different reasons. After several decades of failed tuna ventures, poor planning and investments, and overly optimistic or misleading resource assessments, the success of several of these Indian Ocean development efforts merit careful study before initiating any other fishery development efforts.

Lessons

- Fish populations rise and collapse with climate changes. The concept of stabilizing them through management is only an optimistic artifact of expectations resulting from two decades of somewhat stable climate (1947–67), when fisheries science was becoming a quantitative exercise.

- It is usually the case that while one array of populations is in decline, another array will be in transition to greater abundance. Adaptability of the fishery is imperative.

- The message from the fisheries science community, which has remained valid for decades, is that too much competitive fishing effort destroys population breeding potential as well as the economics of fisheries.

- We know a lot more about the relationships between the ocean as habitat and the responses of various species, than many researchers use in their assessments of existing and potential ocean resources.

- The challenge is to recognize the precursors of systemic change, mount the appropriate changes in behavior in preparation, and to shift behavior once these processes have occurred. Pre-adaptation and versatility are the keys to ecological and economic survival.

INDEX